Characterization of Probability Distributions on Locally Compact Abelian Groups

Mathematical
Surveys
and
Monographs

Volume 273

Characterization of Probability Distributions on Locally Compact Abelian Groups

Gennadiy Feldman

AMS AMERICAN MATHEMATICAL SOCIETY
Providence, Rhode Island

2020 *Mathematics Subject Classification.* Primary 60B15, 62E10, 43A25, 43A35; Secondary 39B52.

For additional information and updates on this book, visit
www.ams.org/bookpages/surv-273

Library of Congress Cataloging-in-Publication Data

Names: Fel'dman, G. M. (Gennadiĭ Mikhaĭlovich), author.
Title: Characterization of probability distributions on locally compact Abelian groups / Gennadiy Feldman.
Description: Providence, Rhode Island : American Mathematical Society, 2023. | Series: Mathematical surveys and monographs, 0076-5376 ; volume 273 | Includes bibliographical references and index.
Identifiers: LCCN 2022059184 (print) | LCCN 2022059185 (ebook) | ISBN 9781470472955 (paperback) | ISBN 9781470473266 (ebook)
Subjects: LCSH: Locally compact Abelian groups. | Distribution (Probability theory) | AMS: Probability theory and stochastic processes – Probability theory on algebraic and topological structures – Probability measures on groups or semigroups, Fourier transforms, factorization. | Statistics – Distribution theory – Characterization and structure theory. | Abstract harmonic analysis – Abstract harmonic analysis – Fourier and Fourier-Stieltjes transforms on locally compact and other abelian groups. | Abstract harmonic analysis – Abstract harmonic analysis – Positive definite functions on groups, semigroups, etc. | Difference and functional equations – Functional equations and inequalities – Equations for functions with more general domains and/or ranges.
Classification: LCC QA387 .F445 2023 (print) | LCC QA387 (ebook) | DDC 519.2/4–dc23/eng20230328
LC record available at https://lccn.loc.gov/2022059184
LC ebook record available at https://lccn.loc.gov/2022059185

∞ The paper used in this book is acid-free and falls within the guidelines
established to ensure permanence and durability.
Visit the AMS home page at `https://www.ams.org/`

10 9 8 7 6 5 4 3 2 1 28 27 26 25 24 23

Contents

Preface

It is well known that if two independent identically distributed random variables are Gaussian, then their sum and difference are also independent. It turns out that only Gaussian random variables have such property. This statement, known as the Kac–Bernstein theorem, is a typical example of a so-called characterization theorem. Characterization theorems in mathematical statistics are statements in which the description of possible distributions of random variables follows from properties of some functions of these random variables.

The first characterization theorems in mathematical statistics are associated with the names of such classics of mathematics of the 20th century as G. Pólya, M. Kac, S.N. Bernstein, Yu.V. Linnik. The proofs of these theorems required the development of powerful methods of complex analysis. By now, the corresponding theory on the real line has basically been constructed. The problem of extending the classical characterization theorems to various algebraic structures has been actively studied in recent decades. Characterization theorems were studied on Lie groups, quantum groups, symmetric spaces, and Banach spaces. Many papers were especially devoted to the study of characterization theorems on locally compact Abelian groups.

The purpose of the book is to present the results obtained in this direction in the last 15 years. We have also included some important results obtained earlier. On the whole, the book gives a comprehensive and self-contained overview of the current state of the theory of characterization problems on locally compact Abelian groups.

When proving characterization theorems in mathematical statistics on groups, two related problems arise. The first problem is the same as in the classical case of the real line: the description of possible distributions of random variables through some properties of functions of these random variables. As a rule, this problem is solved either in a certain class of groups or for a particular group. From this point of view, the real line is just a very important but a particular locally compact Abelian group. The second problem is to find out when the distributions arising in this description, depending on the properties of a group, are such known distributions as either Gaussian distributions or idempotent distributions or their convolutions.

We shall briefly describe the main contents of the book.

Chapter I is of an auxiliary nature. In order to make the presentation independent, we present in Chapter I well-known facts related to abstract harmonic analysis and results on probability distributions on topological Abelian groups. A separate section is devoted to Gaussian distributions on locally compact Abelian groups.

In Chapter II we study the distributions of independent random variables ξ_1 and ξ_2 with values in a locally compact Abelian group X and having independent sum and difference. If, in addition, ξ_1 and ξ_2 are identically distributed with distribution μ, then μ is called a Gaussian distribution in the sense of Bernstein. First, for various classes of locally compact Abelian groups we solve the main problems on Gaussian distributions in the sense of Bernstein. For groups X whose connected component of the zero contains a finite number of elements of order 2 we prove a decomposition theorem for such distributions: every Gaussian distribution in the sense of Bernstein is a convolution of a Gaussian distribution, an idempotent distribution, and a signed measure supported in a subgroup generated by elements of order 2. Based on this result, we give a complete description of locally compact Abelian groups on which every Gaussian distribution in the sense of Bernstein is represented as a convolution of a Gaussian distribution and an idempotent distribution. These are groups whose connected component of the zero contains no more than one element of order 2. We describe supports of Gaussian distributions in the sense of Bernstein and also prove a zero-one law for them. Next, we study the distributions of independent random variables with independent sum and difference. In so doing, we do not assume that the random variables are identically distributed.

The Kac–Bernstein theorem is a special case of a more general statement known as the Skitovich–Darmois theorem. In this theorem the Gaussian distribution on the real line is characterized by the independence of two linear forms L_1 and L_2 of independent random variables ξ_j with distributions μ_j. In Chapter III we study analogues of the Skitovich–Darmois theorem for various classes of locally compact Abelian groups X. In so doing, coefficients of L_1 and L_2 are topological automorphisms of X. First, we assume that the characteristic functions of ξ_j do not vanish. We prove that if a group X contains no subgroups topologically isomorphic to the circle group $\mathbb{T}$, then the independence of L_1 and L_2 implies that all distributions μ_j are Gaussian. We also prove that for an arbitrary locally compact Abelian group X the description of distributions which are characterized by the independence of L_1 and L_2 is reduced to the case when independent random variables take values in a group of the form $\mathbb{R}^n \times \mathbb{T}^m$, where $\mathbb{T}$ is the circle group. For two independent random variables we prove that if a group X has no subgroups topologically isomorphic to the group $\mathbb{T}^2$, then the independence of L_1 and L_2 implies that μ_j are either Gaussian distributions or convolutions of Gaussian distributions and signed measures supported in a subgroup of X generated by an element of X of order 2. In the classes of discrete and compact totally disconnected Abelian groups we prove the series of theorems where shifts of idempotent distributions are characterized. In so doing, the characteristic functions of independent random variables can vanish. To conclude this chapter, we prove that in contrast to the case of two linear forms of two independent random variables, if we consider three linear forms of three independent random variables with values in the cylinder $\mathbb{R} \times \mathbb{T}$, then only Gaussian distributions are characterised by the independence of these linear forms.

Chapter IV is devoted to analogues of the well-known Heyde theorem, where the Gaussian distribution on the real line is characterized by the symmetry of the conditional distribution of one linear form of independent random variables given another. We study Heyde's theorem for various classes of locally compact Abelian groups X. Coefficients of linear forms, as in the Skitovich–Darmois theorem, are topological automorphisms of X. Particular attention is paid to the case of two independent random variables. If we consider the linear forms $L_1 = \xi_1 + \xi_2$ and $L_2 = \xi_1 + \alpha\xi_2$, where α is a topological automorphism of X, then the kernel

$\mathrm{Ker}(I + \alpha)$ plays an important role in the description of distributions μ_j. In addition, it is important whether a group X contains an element of order 2. This is the fundamental difference between the group situation and the case of the real line. First, we assume that the characteristic functions of independent random variables do not vanish and X contains no elements of order 2. We prove a series of theorems characterizing the Gaussian distribution. Particular attention is paid to Heyde's theorem on $\boldsymbol{a}$-adic solenoids. Then we study Heyde's theorem for discrete Abelian groups and compact totally disconnected Abelian groups. In this case the characteristic functions of independent random variables can vanish, but X still contains no elements of order 2. For these classes of groups we prove some theorems, where shifts of idempotent distributions are characterized. In the final section of the chapter we study Heyde's theorem for some groups containing an element of order 2.

Chapter V is devoted to characterization theorems for probability distributions in the case when independent random variables ξ_1 and ξ_2 take values in the additive group of the field of p-adic numbers Ω_p. We study analogues of the Skitovich–Darmois and Heyde theorems for the group Ω_p. We also prove for the field Ω_p, a theorem, where shifts of idempotent distributions are characterized through the independence of sum and difference squared of ξ_1 and ξ_2. We also prove a similar theorem for discrete fields.

According to the well-known Rao theorem, if we have two linear forms L_1 and L_2 of three independent random variables ξ_j with nonvanishing characteristic functions, then under certain conditions on the coefficients of the forms, the distribution of the random vector (L_1, L_2) determines the distributions of ξ_j up to a shift. C.R. Rao also proved that if four independent random variables ξ_j are considered, then the distribution of the random vector (L_1, L_2) determines the distributions of ξ_j up to a convolution with a Gaussian distribution. In Chapter VI we prove for an arbitrary locally compact Abelian group X an analogue of the first Rao theorem and for $\boldsymbol{a}$-adic solenoids an analogue of the second Rao theorem. In so doing, coefficients of the linear forms are continuous endomorphisms of X. We also prove for $\boldsymbol{a}$-adic solenoids an analogue of the well-known generalized Pólya theorem, where the Gaussian distribution on the real line is characterized by the property of equidistribution of a monomial and a linear form of independent identically distributed random variables.

Comments are given at the end of each chapter.

Characterization problems studied in the book are reduced to solving various functional equations. These equations are considered either on the character group of the original group or on the group itself. In most cases, solutions are sought in the class of characteristic functions (Fourier transforms) of the corresponding distributions.

We hope that the book will be useful to everyone who is interested in abstract harmonic analysis, probability distributions on groups or functional equations on groups.

Gennadiy Feldman

CHAPTER I

Preliminaries

1. Results of abstract harmonic analysis

We present in this section the summary of the basic results of abstract harmonic analysis which we use in the book. The monograph by E. Hewitt and K. A. Ross [**63**] is the main source for us. Unless the contrary is stated, we consider in the book only locally compact Abelian groups.

Basic definitions and notation. Let X be a locally compact Abelian group. Unless otherwise stated, we use the additive notation for the operation in X and denote by "0" the neutral element of X. Let $x \in X$. The element x is said to be *compact* if the smallest closed subgroup of X containing x is compact. Denote by b_X the set of all compact elements of X and denote by c_X the connected component of the zero of X. If every element of X has finite order, then X is called a *torsion group*. If every element of X except the zero has infinite order, then X is called a *torsion-free group*. Let $f : \mathbb{R} \to X$ be a continuous homomorphism. The image $f(\mathbb{R})$ is called a *one-parameter subgroup* of X.

A finite subset $\{x_1, \dots, x_n\}$ of a group X is said to be *independent* if it does not contain 0 and if for any integers $k_1, \dots, k_n$ the equality $k_1x_1 + \dots + k_nx_n = 0$ implies that $k_1x_1 = \dots = k_nx_n = 0$. An infinite subset A of X is said to be *independent* if every finite subset of A is independent. The cardinality of a maximal independent subset of the group X containing only elements of infinite order is called the *torsion-free rank* of X. Denote by $r_0(X)$ the torsion-free rank of X. Denote the dimension of a connected group X by $\dim(X)$.

Let $\{X_\iota : \iota \in \mathcal{I}\}$ be a nonvoid family of groups. Denote by $\mathop{\mathbf{P}}\limits_{\iota \in \mathcal{I}} X_\iota$ the *direct product of the groups* X_ι, i.e., the group coinciding with the Cartesian product of the sets X_ι and with the coordinate-wise operation. Let $\mathop{\mathbf{P}^*}\limits_{\iota \in \mathcal{I}} X_\iota$ be the subset of all $(x_\iota) \in \mathop{\mathbf{P}}\limits_{\iota \in \mathcal{I}} X_\iota$ such that $x_\iota = 0$ for all but a finite set of indices ι. Then $\mathop{\mathbf{P}^*}\limits_{\iota \in \mathcal{I}} X_\iota$ is a subgroup of $\mathop{\mathbf{P}}\limits_{\iota \in \mathcal{I}} X_\iota$. It is called the *weak direct product of the groups* X_ι. If $X_\iota = X$ for all $\iota \in \mathcal{I}$, then we denote by $X^{\mathfrak{n}}$ the direct product of the groups X_ι, and we denote by $X^{\mathfrak{n}*}$ the weak direct product of the groups X_ι, where $\mathfrak{n}$ is the cardinal number of the set $\mathcal{I}$. Let $\{K_\iota : \iota \in \mathcal{I}\}$ be a nonvoid family of compact groups. We always consider the group $\mathop{\mathbf{P}}\limits_{\iota \in \mathcal{I}} K_\iota$ in the product topology. By the Tychonoff theorem the group $\mathop{\mathbf{P}}\limits_{\iota \in \mathcal{I}} K_\iota$ is also compact. If $\{D_\iota : \iota \in \mathcal{I}\}$ is a nonvoid family of discrete groups, then we always consider the group $\mathop{\mathbf{P}^*}\limits_{\iota \in \mathcal{I}} D_\iota$ in the discrete topology.

Let n be an integer. Denote by f_n the mapping of X into X defined by $f_nx = nx$ for all $x \in X$. Put $X_{(n)} = \operatorname{Ker} f_n$ and $X^{(n)} = f_n(X)$. A group X is said to be a

Corwin group if $X^{(2)} = X$. A group X is said to be *divisible*, if $X^{(n)} = X$ for all natural n.

If G is a subgroup of X, then we denote either by $x+G$ or by $[x]$ elements of the factor-group X/G. We also the use notation $x + G$ for cosets G in X. Let A be a subset of X. The set of all elements of the form $x = l_1x_1 + \cdots + l_kx_k$, where $x_j \in A$ and l_j are integers, is said to be the *subgroup of X generated by A*. Let A and B be subsets of X. Denote by $A+B$ the set $A+B = \{x \in X : x = a+b,\ a \in A,\ b \in B\}$. If A is a subset of X, then denote by $\overline{A}$ the closure of A. For a finite set A denote by $|A|$ the number of elements of A. We use the symbol $\aleph_0$ to denote the least infinite cardinal number. Denote by $\mathcal{P}$ the set of prime numbers. We denote the set of complex numbers by $\mathbb{C}$. A topological isomorphism of groups, i.e. an algebraic isomorphism which is a homeomorphism, we denote by "$\cong$". Denote by $\langle \cdot, \cdot \rangle$ the scalar product in the space $\mathbb{R}^n$.

Examples of locally compact Abelian groups. We give examples of the most important locally compact Abelian groups. We need these groups in proving the characterization theorems.

(a) $\mathbb{R}$: the additive group of real numbers with the natural topology of the real line.

(b) $\mathbb{T}$: the circle group (the one-dimensional torus), i.e. $\mathbb{T} = \{z \in \mathbb{C} : |z| = 1\}$ with multiplication as operation and with the usual topology.

(c) $\mathbb{Z}$: the additive group of integers (the infinite cyclic group) with the discrete topology.

(d) $\mathbb{Z}(n)$: the additive group of the integers modulo n (the finite cyclic group). The elements of the group $\mathbb{Z}(n)$ we denote by $\{0, 1, \dots, n-1\}$. Observe also that the group $\mathbb{Z}(n)$ is isomorphic to the multiplicative group of nth roots of unity. We retain for this group the notation $\mathbb{Z}(n)$ and denote by $\exp\{2\pi ik/n\}$, $k = 0, 1, \dots, n-1$, its elements.

(e) $\mathbb{Z}(p^\infty)$: the quasicyclic group (the group of type p^∞). Let p be a prime number. Consider a set of rational numbers of the form $\{k/p^n : k = 0, 1, \dots, p^n - 1,\ n = 0, 1, \dots\}$ and denote this set by $\mathbb{Z}(p^\infty)$. If we define the operation in $\mathbb{Z}(p^\infty)$ as addition modulo 1, then $\mathbb{Z}(p^\infty)$ is transformed into an Abelian group which we consider in the discrete topology. Obviously, this group is isomorphic to the multiplicative group of p^nth roots of unity, where n goes through the nonnegative integers, considered in the discrete topology. This group we also denote by $\mathbb{Z}(p^\infty)$, and its elements by $\exp\{2\pi ik/p^n\}$, $k = 0, 1, \dots, p^n - 1$, $n = 0, 1, \dots$.

(f) $\Delta_{\boldsymbol{a}}$: the group of $\boldsymbol{a}$-adic integers. Let $\boldsymbol{a} = (a_0, a_1, \dots, a_n, \dots)$ be an arbitrary infinite sequence of integers, where each of a_n is greater than 1. We define the group of $\boldsymbol{a}$-adic integers $\Delta_{\boldsymbol{a}}$. As a set $\Delta_{\boldsymbol{a}}$ coincides with the Cartesian product $\mathop{\mathbf{P}}_{n=0}^{\infty} \{0, 1, \dots, a_n - 1\}$. Consider $x = (x_0, x_1, \dots, x_n, \dots)$, $y = (y_0, y_1, \dots, y_n, \dots) \in \Delta_{\boldsymbol{a}}$ and define the sum $z = x + y$ as follows. Let $x_0 + y_0 = t_0a_0 + z_0$, where $z_0 \in \{0, 1, \dots, a_0 - 1\}$, $t_0 \in \{0, 1\}$. Assume that the numbers $z_0, z_1, \dots, z_k$ and $t_0, t_1, \dots, t_k$ have already been determined. Let us then set $x_{k+1} + y_{k+1} + t_k = t_{k+1}a_{k+1} + z_{k+1}$, where $z_{k+1} \in \{0, 1, \dots, a_{k+1} - 1\}$, $t_{k+1} \in \{0, 1\}$. This defines by induction a sequence $z = (z_0, z_1, \dots, z_n, \dots)$. The set $\Delta_{\boldsymbol{a}}$ with the addition defined above is an Abelian group, whose neutral element is the sequence $(0, 0, 0, \dots,)$. Consider $\Delta_{\boldsymbol{a}}$ in the product topology. The obtained group is called the ***a**-adic integers*. The group $\Delta_{\boldsymbol{a}}$ is compact and totally disconnected [**63**, (10.2)].

If all of the integers a_n are equal to some fixed prime integer p, then we write Δ_p instead of $\Delta_{\boldsymbol{a}}$ and call this object the group of *p-adic integers*. Each element $x = (x_0, x_1, \dots, x_n, \dots) \in \Delta_p$ is thought of as a formal power series $\sum\limits_{n=0}^{\infty} x_n p^n$. The addition of formal power series defining in the usual way corresponds to the addition of the corresponding sequences. We can define a multiplication in Δ_p in the natural way as the multiplication of formal power series. Then Δ_p is transformed into a commutative ring [**63**, (10.10)].

(g) $\Sigma_{\boldsymbol{a}}$: the $\boldsymbol{a}$-adic solenoid. Let $\boldsymbol{a} = (a_0, a_1, \dots, a_n, \dots)$ be an arbitrary infinite sequence of integers, where each a_n is greater than 1. Consider the group $\mathbb{R} \times \Delta_{\boldsymbol{a}}$. Let B be the subgroup of $\mathbb{R} \times \Delta_{\boldsymbol{a}}$ of the form $B = \{(n, nu)\}_{n=-\infty}^{\infty}$, where $u = (1, 0, 0, \dots) \in \Delta_p$. The factor-group $\Sigma_{\boldsymbol{a}} = (\mathbb{R} \times \Delta_{\boldsymbol{a}})/B$ is called the ***a**-adic solenoid* [**63**, (10.12)]. In the particular case when $\Delta_{\boldsymbol{a}} = \Delta_p$, the factor-group $(\mathbb{R} \times \Delta_p)/B$ is called the *p-adic solenoid* and denoted by Σ_p. The group $\Sigma_{\boldsymbol{a}}$ is compact, connected [**63**, (10.13)], and has dimension 1 ([**63**, (24.28)]).

(h) $\mathbb{Q}$: the additive group of rational numbers with the discrete topology.

Character group. A *character* of a locally compact Abelian group X is a continuous homomorphism of X into the circle group $\mathbb{T}$. We denote by X^* the set of all characters of the group X. Put $Y = X^*$. The value of a character $y \in Y$ at an element $x \in X$ is denoted by (x, y). Then Y is an Abelian group with respect to the pointwise multiplication. For every compact set F of X and every $\varepsilon > 0$, let $P(F, \varepsilon)$ be the set $\{y \in Y : |(x, y) - 1| < \varepsilon \text{ for all } x \in F\}$. With all sets $P(F, \varepsilon)$ taken as an open basis at the zero in Y, Y is a locally compact Abelian group [**63**, (23.15)]. The group Y is called the *character group* of X.

Let τ be the mapping $\tau : X \to Y^*$ defined by $(y, \tau x) = (x, y)$ for all $x \in X$, $y \in Y$.

Pontryagin duality theorem. The following theorem is central to abstract harmonic analysis.

THEOREM 1.1 (Pontryagin duality theorem [**63**, (24.8)]). *The mapping τ is a topological isomorphism of the group X onto Y^*.*

According to Pontryagin's duality theorem every topological or algebraic property of a locally compact Abelian group can be described in terms of topological or algebraic properties of its character group. The following theorems hold.

THEOREM 1.2 ([**63**, (23.17), (24.14)]). *A locally compact Abelian group is compact if and only if its character group is discrete. A compact Abelian group X is second countable if and only if its character group is countable.*

THEOREM 1.3 ([**63**, (24.25)]). *A compact Abelian group is connected if and only if its character group is torsion-free.*

THEOREM 1.4 ([**63**, (24.25), (24.28)]). *The dimension of a compact connected Abelian group is equal to the torsion-free rank of its character group.*

THEOREM 1.5 ([**63**, (24.26)]). *A compact Abelian group is totally disconnected if and only if its character group is a torsion group.*

We can compute the character group of a direct product of groups knowing the character groups of factors. The following theorems hold.

THEOREM 1.6 ([**63**, (23.18)]). *Let X_j, $j = 1, 2, \ldots, n$, be locally compact Abelian groups and let $Y_j = X_j^*$. Then*

$$(X_1 \times \cdots \times X_n)^* \cong Y_1 \times \cdots \times Y_n.$$

THEOREM 1.7 ([**63**, (23.21), (23.22)]). *Let $\{K_\iota : \iota \in \mathcal{I}\}$ be a nonvoid family of compact Abelian groups and let $K_\iota^* = D_\iota$. Then*

$$\Big(\mathop{\mathbf{P}}_{\iota\in\mathcal{I}} K_\iota\Big)^* \cong \mathop{\mathbf{P}^*}_{\iota\in\mathcal{I}} D_\iota.$$

Let $\{D_\iota : \iota \in \mathcal{I}\}$ be a nonvoid family of discrete Abelian groups and let $D_\iota^* = K_\iota$. Then

$$\Big(\mathop{\mathbf{P}^*}_{\iota\in\mathcal{I}} D_\iota\Big)^* \cong \mathop{\mathbf{P}}_{\iota\in\mathcal{I}} K_\iota.$$

Let X be a locally compact Abelian group with character group Y and let B be a nonvoid subset of X. Put

$$A(Y, B) = \{y \in Y : (x, y) = 1 \text{ for all } x \in B\}.$$

The set $A(Y, B)$ is called the *annihilator* of B in Y. The following theorems are true.

THEOREM 1.8 ([**63**, (24.10)]). *Let G be a closed subgroup of the group X. Then $G = A(X, A(Y, G))$.*

THEOREM 1.9 ([**63**, (24.11), (23.25)]). *Let G be a closed subgroup of the group X. The character group of the group G is topologically isomorphic to the factor-group $Y/A(Y, G)$. Every continuous character of G has the form $x \to (x, y)$ for some $y \in Y$. Two characters y_1 and y_2 of Y define the same character of the subgroup G if and only if $y_1 - y_2 \in A(Y, G)$. The character group of the factor-group X/G is topologically isomorphic to the group $A(Y, G)$.*

THEOREM 1.10 ([**63**, (24.17)]). *The sets b_X and c_X are closed subgroups of X. Moreover, $b_Y = A(Y, c_X)$ and $c_X = A(X, b_Y)$.*

THEOREM 1.11 ([**63**, (23.29)]). *Let K be a compact subgroup of the group X. Then the annihilator $A(Y, K)$ is an open subgroup of Y. If H is an open subgroup of Y, then $A(X, H)$ is a compact subgroup of X.*

THEOREM 1.12 ([**63**, (24.22)]). *For all natural n the equalities*

$$A(Y, X^{(n)}) = Y_{(n)}, \quad A\left(Y, X_{(n)}\right) = \overline{Y^{(n)}}$$

hold.

THEOREM 1.13 ([**63**, (24.25)]). *If X is a connected group, then X is a divisible group, i.e. $X^{(n)} = X$ for all natural n.*

Character groups of some locally compact Abelian groups. We consider examples of character groups of some locally compact Abelian groups.

(a) $\mathbb{R}^* \cong \mathbb{R}$, $(x, y) = \exp\{ixy\}$, where $x \in \mathbb{R}$, $y \in \mathbb{R}$;
(b) $\mathbb{Z}^* \cong \mathbb{T}$, $(n, z) = z^n$, where $n \in \mathbb{Z}$, $z \in \mathbb{T}$;
(c) $\mathbb{T}^* \cong \mathbb{Z}$, $(z, n) = z^n$, where $z \in \mathbb{T}$, $n \in \mathbb{Z}$;

(d) $\mathbb{Z}(n)^* \cong \mathbb{Z}(n)$, $(k,l) = \exp\{2\pi ikl/n\}$, where $k \in \mathbb{Z}(n)$, $l \in \mathbb{Z}(n)$;

(e) $\Delta_p^* \cong \mathbb{Z}(p^\infty)$,

$$(x,y) = \exp\left\{\left(x_0 + x_1 p + \cdots + x_{n-1}p^{n-1}\right)\frac{2\pi i}{p^n}\right\},$$

where

$$x = (x_0, x_1, \dots, x_n, \dots) \in \Delta_p, \quad y = \exp\left\{\frac{2\pi il}{p^n}\right\} \in \mathbb{Z}(p^\infty)$$

[**63**, (25.2)];

(f) Consider the group $\Sigma_{\boldsymbol{a}} = (\mathbb{R} \times \Delta_{\boldsymbol{a}})/B$, where $\boldsymbol{a} = (a_0, a_1, \dots, a_n, \dots)$. Then $\Sigma_{\boldsymbol{a}}^* \cong H_{\boldsymbol{a}}$, where

$$H_{\boldsymbol{a}} = \left\{\frac{m}{a_0 a_1 \cdots a_n} : n = 0, 1, \dots;\ m \in \mathbb{Z}\right\}$$

is a subgroup of the group $\mathbb{Q}$, and

$$(x,y) = \exp\left\{\left(t - (x_0 + a_0 x_1 + \cdots + a_0 a_1 \cdots a_{n-1} x_n)\right)\frac{2\pi im}{a_0 a_1 \cdots a_n}\right\},$$

where $x = (t; x_0, x_1, \dots, x_n, \dots) + B \in \Sigma_{\boldsymbol{a}}$, $(t; x_0, x_1, \dots x_n, \dots) \in \mathbb{R} \times \Delta_{\boldsymbol{a}}$, $y = \dfrac{m}{a_0 a_1 \cdots a_n} \in H_{\boldsymbol{a}}$ [**63**, (25.3)].

Observe also that if $\boldsymbol{a} = (2, 3, 4, \dots)$, then $\Sigma_{\boldsymbol{a}}^* \cong \mathbb{Q}$.

Structure theorems. We need the following structure theorems for some classes of locally compact Abelian groups.

THEOREM 1.14 ([**63**, (24.30)]). *A locally compact Abelian group X is topologically isomorphic to a group of the form $\mathbb{R}^m \times G$, where $m \geq 0$ and G is a locally compact Abelian group containing a compact open subgroup. Furthermore, m is uniquely determined.*

THEOREM 1.15 ([**63**, (9.14)]). *A locally compact connected Abelian group X is topologically isomorphic to a group of the form $\mathbb{R}^m \times K$, where $m \geq 0$ and K is a compact connected Abelian group.*

THEOREM 1.16 ([**63**, (25.8)]). *A compact torsion-free Abelian group is topologically isomorphic to a group of the form*

$$(\Sigma_{\boldsymbol{a}})^{\mathfrak{n}} \times \mathop{\mathbf{P}}_{p \in \mathcal{P}} \Delta_p^{\mathfrak{n}_p},$$

where $\boldsymbol{a} = (2, 3, 4, \dots)$ and $\mathfrak{n}$, $\mathfrak{n}_p$ are cardinal numbers.

Small subgroups in locally compact Abelian groups. The following theorems on the existence of small subgroups in locally compact Abelian groups are also valid.

THEOREM 1.17 ([**63**, (7.7)]). *Let X be a locally compact totally disconnected Abelian group. Then every neighborhood of the zero in X contains a compact open subgroup.*

THEOREM 1.18 ([**63**, (24.7)]). *Let X be a compact Abelian group. Then every neighborhood of the zero in X contains a closed subgroup G such that the factor-group X/G is topologically isomorphic to a group of the form $\mathbb{T}^m \times F$, where $m \geq 0$ and F is a finite Abelian group.*

Adjoint homomorphisms. We need the following statement.

THEOREM 1.19 ([**63**, (24.38)]). *Let X_1 and X_2 be locally compact Abelian groups with character groups Y_1 and Y_2 respectively. Let $p : X_1 \to X_2$ be a continuous homomorphism. For each $y_2 \in Y_2$ define the mapping $\widetilde{p} : Y_2 \to Y_1$ by the equality $(px_1, y_2) = (x_1, \widetilde{p}y_2)$ for all $x_1 \in X_1$, $y_2 \in Y_2$. Then $\widetilde{p}$ is a continuous homomorphism. Moreover,* $\operatorname{Ker}\widetilde{p} = A(Y_2, p(X_1))$.

The homomorphism $\widetilde{p}$ is called the *adjoint* of p. We formulate some properties of adjoint homomorphisms as a theorem.

THEOREM 1.20 ([**63**, (24.41)]). *Let X_1 and X_2 be locally compact Abelian groups with character groups Y_1 and Y_2 respectively. Let $p : X_1 \to X_2$ be a continuous homomorphism. The following statements are valid.*

(a) *The homomorphism p satisfies $p = \widetilde{\widetilde{p}}$.*
(b) *The homomorphism $\widetilde{p}$ is a monomorphism if and only if the subgroup $p(X_1)$ is dense in X_2.*
(c) *The homomorphism $\widetilde{p}$ is a topological isomorphism from the group Y_2 into Y_1 if and only if p is a topological isomorphism from the group X_1 into X_2.*
(d) *Let X be a locally compact Abelian group with character group Y. If $f_n x = nx$ for all $x \in X$, then $\widetilde{f_n} y = ny$ for all $y \in Y$, i.e. $\widetilde{f_n} = f_n$ for all integers n.*

Topological automorphism groups of some locally compact Abelian groups. Let X be a locally compact Abelian group with character group Y. Denote by $\mathrm{Aut}(X)$ the group of all *topological automorphisms* of the group X, i.e. the group of all mappings of X into X that are simultaneously algebraic automorphisms and homeomorphisms. Denote by I the identical automorphism of a group. It is possible to define a topology in the group $\mathrm{Aut}(X)$ in such a manner that $\mathrm{Aut}(X)$ becomes a topological group [**63**, (26.1)]. In studying characterization problems, we do not need to consider the group $\mathrm{Aut}(X)$ as a topological group. Therefore, when we say that a group $\mathrm{Aut}(X)$ is isomorphic to some group, we keep in mind an algebraic rather than topological isomorphism, in spite of the fact that in the cases under consideration a topological isomorphism also holds. Let $\alpha \in \mathrm{Aut}(X)$. The mapping $\alpha \to \widetilde{\alpha}$ is an anti-isomorphism of the groups $\mathrm{Aut}(X)$ and $\mathrm{Aut}(Y)$ [**63**, (26.9)]. Assume K is a closed subgroup of the group X such that $\alpha(K) = K$, i.e. the restriction of α to K is a topological automorphism of the group K. Denote by either α_K or $\alpha|_K$ this restriction. A subgroup G of the group X is said to be *characteristic* if G is invariant under each automorphism $\alpha \in \mathrm{Aut}(X)$. The subgroups c_X, b_X, $X_{(n)}$, and $X^{(n)}$ are characteristic.

We consider examples of automorphism groups of some locally compact Abelian groups.

(a) Every topological automorphism α of the group $\mathbb{R}$ is of the form $\alpha t = ct$, for all $t \in \mathbb{R}$, where $c \in \mathbb{R}$, $c \neq 0$. It follows from this that the group $\mathrm{Aut}(\mathbb{R})$ is isomorphic to the multiplicative group of all nonzero real numbers, i.e. $\mathrm{Aut}(\mathbb{R})$ is isomorphic to the group $\mathbb{R} \times \mathbb{Z}(2)$.
(b) The group $\mathbb{Z}$ admits just two automorphisms $\pm I$. This implies that the group $\mathrm{Aut}(\mathbb{Z})$ is isomorphic to the group $\mathbb{Z}(2)$.

(c) Let p be an arbitrary prime number. Denote by $\Delta_p^\times$ the multiplicative group of all invertible elements of the ring Δ_p. It is easy to see that $\Delta_p^\times = \{c = (c_0, c_1, \dots, c_n, \dots) \in \Delta_p : c_0 \neq 0\}$. The group $\mathrm{Aut}(\Delta_p)$ is isomorphic to $\Delta_p^\times$. For the proof we note that if $c \in \Delta_p^\times$ we can define the automorphism $\alpha_c \in \mathrm{Aut}(\Delta_p)$ by the formula $\alpha_c x = cx$ for all $x \in \Delta_p$. Let $\alpha \in \mathrm{Aut}(\Delta_p)$ be an arbitrary automorphism. Denote by u the element $(1, 0, 0, \dots) \in \Delta_p$. Since the subgroup generated by u is dense in Δ_p, the same is also true for the element αu. Hence $\alpha = c$, where $c \in \Delta_p^\times$. It follows from this that $\alpha = \alpha_c$ [**63**, (26.18e)].

(d) To find the group $\mathrm{Aut}(\Sigma_{\boldsymbol{a}})$, we note that $\Sigma_{\boldsymbol{a}}^* \cong H_{\boldsymbol{a}}$, where $H_{\boldsymbol{a}}$ is a subgroup of $\mathbb{Q}$. It is easy to see that either the group $H_{\boldsymbol{a}}$ admits just two automorphisms $\pm I$ and hence the group $\mathrm{Aut}(H_{\boldsymbol{a}})$ is isomorphic to the group $\mathbb{Z}(2)$ or every automorphism α of the group $H_{\boldsymbol{a}}$ is of the form $\alpha = f_m f_n^{-1}$, where m, n are some mutually prime numbers. Furthermore, $f_m, f_n \in \mathrm{Aut}(H_{\boldsymbol{a}})$ and at least one of the automorphisms f_m, f_n is different from $\pm I$. If the number of primes p such that $f_p \in \mathrm{Aut}(H_{\boldsymbol{a}})$ is finite and is equal to l, it is easily seen that the group $\mathrm{Aut}(H_{\boldsymbol{a}})$ is isomorphic to $\mathbb{Z}^l \times \mathbb{Z}(2)$. If this number is infinite, then the group $\mathrm{Aut}(H_{\boldsymbol{a}})$ is isomorphic to $\mathbb{Z}^{\aleph_0 *} \times \mathbb{Z}(2)$. Inasmuch as the groups $\mathrm{Aut}(\Sigma_{\boldsymbol{a}})$ and $\mathrm{Aut}(H_{\boldsymbol{a}})$ are isomorphic, we found the group $\mathrm{Aut}(\Sigma_{\boldsymbol{a}})$. Moreover, since every automorphism $\alpha \in \mathrm{Aut}(H_{\boldsymbol{a}})$ is of the form $\alpha = f_m f_n^{-1}$, every automorphism $\widetilde{\alpha} \in \mathrm{Aut}(\Sigma_{\boldsymbol{a}})$ is of the same form.

Let us now formulate some theorems on locally compact Abelian groups, which we need in what follows.

THEOREM 1.21. *Let G be a closed subgroup of the group $\mathbb{R}^m$. Then G is topologically isomorphic to a group $\mathbb{R}^p \times \mathbb{Z}^q$, where $p + q \leq m$.*

THEOREM 1.22 ([**63**, (25.31)]). *Let X be a locally compact Abelian group. If X contains a subgroup G such that G is topologically isomorphic to $\mathbb{T}^{\mathfrak{n}}$, where $\mathfrak{n}$ is a cardinal number, then G is a topological direct factor of X.*

THEOREM 1.23 ([**63**, (5.29)]). *Let X_1 be a locally compact σ-compact group. Let f be a continuous epimorphism of X_1 onto a locally countably compact T_0 group X_2. Then f is an open mapping.*

Let X be an Abelian group. A group X is said to be *reduced* if X contains no nonzero divisible subgroups. Let p be a prime number. A group X is called a *p-group* if the order of every element of X is a power of p. By a *free Abelian* group we mean a weak direct product of infinite cyclic groups, i.e. a countable free Abelian group is isomorphic to either $\mathbb{Z}^n$ or $\mathbb{Z}^{\aleph_0 *}$.

Decomposition theorems. The theorems listed below are valid for arbitrary Abelian groups.

THEOREM 1.24 ([**58**, Theorem 8.4]). *Let X be an Abelian torsion group. For each prime number p let X_p be the subgroup of X consisting of all elements of X whose order is a power of p. Then*

$$X = \mathop{\mathbf{P}^*}_{p \in \mathcal{P}} X_p.$$

The subgroups X_p are called the *p-components* of X. If X is not a torsion group, the p-component of its torsion part is called the p-component of X.

THEOREM 1.25 ([**58**, Theorem 21.3]). *An Abelian group X is the direct product of a divisible group D and a reduced group N,*

$$X = D \times N.$$

Here D is a uniquely determined subgroup of X, while N is unique up to isomorphism.

THEOREM 1.26 ([**58**, Corollary 19.3]). *A countable Abelian group X can be decomposed into the direct product*

$$X = N \times M,$$

where M is free and N has no free factor-groups. The subgroup N is uniquely determined.

THEOREM 1.27 ([**58**, Theorem 15.5]). *Assume that an Abelian group X is generated by elements $x_1, \dots, x_n$. Then*

$$X = X_1 \times \cdots \times X_m,$$

where each X_j is isomorphic to either $\mathbb{Z}$ or $\mathbb{Z}(n_j)$.

THEOREM 1.28 ([**58**, Theorem 23.1]). *A divisible Abelian group X is isomorphic to a week direct product*

$$\mathbb{Q}^{\mathfrak{n}_o *} \times \mathop{\mathbf{P}^*}_{p \in \mathcal{P}} \mathbb{Z}(p^\infty)^{\mathfrak{n}_p *},$$

where $\mathfrak{n}_o$ and $\mathfrak{n}_p$ are cardinal numbers.

$\boldsymbol{a}$-adic solenoid as a subgroup of the infinite dimensional torus $\mathbb{T}^{\aleph_0}$. An $\boldsymbol{a}$-adic solenoid $\Sigma_{\boldsymbol{a}}$ is an important example of a compact Abelian group. We will be dealing with $\boldsymbol{a}$-adic solenoids throughout the book. It is well-known that every second countable compact Abelian group X is topologically isomorphic to a compact subgroup of the infinite-dimensional torus $\mathbb{T}^{\aleph_0}$. Although we will not use this, we describe this subgroup for an $\boldsymbol{a}$-adic solenoid. For the sake of simplicity, we consider the case when $\boldsymbol{a} = (p, p, p, \dots)$, where p is a prime number, i.e., the case of a p-adic solenoid Σ_p. The corresponding reasoning can be carried out for an arbitrary $\boldsymbol{a} = (a_0, a_1, \dots, a_n, \dots)$.

Let $u = (1, 0, 0, \dots) \in \Delta_p$, $B = \{(n, nu)\}_{n=-\infty}^{\infty} \subset \mathbb{R} \times \Delta_p$, and let $\Sigma_p = (\mathbb{R} \times \Delta_p)/B$ be the corresponding p-adic solenoid. Consider the mapping

$$\tau : \mathbb{R} \times \Delta_p \to \mathbb{T}^{\aleph_0}$$

defined by

$$\tau(t, x) = (z_1, z_2, \dots, z_n, \dots), \quad t \in \mathbb{R}, \ x = (x_0, x_1, \dots, x_n, \dots) \in \Delta_p,$$

$$z_n = \exp\left\{\frac{2\pi i}{p^{n-1}}\left(t - \sum_{k=0}^{n-1} x_k p^k\right)\right\}.$$

Then τ is a continuous homomorphism, $\operatorname{Ker} \tau = B$,

$$G = \tau(\mathbb{R} \times \Delta_p) = \{z = (z_1, z_2, \dots, z_n, \dots) \in \mathbb{T}^{\aleph_0} : z_k^p = z_{k-1}, \ k \geq 2\}$$

is a closed subgroup of $\mathbb{T}^{\aleph_0}$, and $G \cong \Sigma_p$. The consideration of the p-adic solenoid Σ_p as the subgroup G allows us to verify the following:

(a) if $h = \dfrac{m}{p^n}$ is a character of the group Σ_p, then

$$(z, h) = z_{n+1}^m, \ z = (z_1, z_2, \dots, z_n, \dots) \in G;$$

(b) the automorphism f_p operates on G by the formula

$$f_p z = (z_1^p, z_2^p, \dots, z_n^p, \dots), \ z = (z_1, z_2, \dots, z_n, \dots) \in G;$$

(c)

$$\tau(\mathbb{R}) = \{(e^{2\pi i t}, e^{2\pi i t/p}, e^{2\pi i t/p^2}, \dots), \ t \in \mathbb{R}\}$$

is the dense one-parameter subgroup of G.

Polynomials. Unless otherwise stated, the word "function" means a mapping taking either real or complex values. Let Y be an Abelian group, let $f(y)$ be a function on Y, and let h be an element of Y. Denote by Δ_h the *finite difference operator*

$$\Delta_h f(y) = f(y+h) - f(y), \quad y \in Y.$$

A function $f(y)$ on Y is called a *polynomial* if

$$\Delta_h^{n+1} f(y) = 0$$

for some n and for all $y, h \in Y$. The minimal n for which this equality holds is called the *degree* of the polynomial $f(y)$.

A function $g(y_1, \dots, y_k)$ defined on Y^k is said to be *k-additive* if the equality

$$g(y_1, \dots, u_l + v_l, \dots, y_k) = g(y_1, \dots, u_l, \dots, y_k) + g(y_1, \dots, v_l, \dots, y_k)$$

holds for all $l = 1, 2, \dots, k$. By definition, 0-additive functions are constants.

THEOREM 1.29 ([**21**], see also [**36**, Theorem 5.5]). *Let Y be an Abelian group and let $f(y)$ be a polynomial on Y of degree n. Then there exist symmetric k-additive functions $g_k(y_1, \dots, y_k)$, $k = 0, 1, \dots, n$, such that*

$$f(y) = \sum_{k=0}^{n} g_k^*(y), \quad y \in Y, \tag{1.1}$$

where $g_k^*(y) = g_k(y, \dots, y)$.

PROPOSITION 1.30. *Let Y be a locally compact Abelian group and let $f(y)$ be a continuous polynomial on Y. Then $f(y+h) = f(y)$ for all $y \in Y$, $h \in b_Y$. In particular, $f(y) = const$ for all $y \in b_Y$.*

PROOF. First we note that if H is a compact Abelian group and $l(h)$ is a continuous additive function on H, then $l(h) = 0$ for all $h \in H$. Indeed, in this case $l(H)$ is a compact subgroup, but $\{0\}$ is the only compact subgroup of the groups $\mathbb{R}$ and $\mathbb{C}$. It follows from this that if a function $A(y_1, y_2, \dots, y_n)$ defined on Y^n is n-additive, then $A(y_1, \dots, y_n) = 0$ once $y_k \in b_Y$ for some k.

Let $f(y)$ be a continuous polynomial on the group Y. By Theorem 1.29, representation (1.1) is valid. It follows from (1.1) that it suffices to prove that $g_k^*(y+h) = g_k^*(y)$ for all $y \in Y$, $h \in b_Y$, $k = 1, 2, \dots, n$. We have

$$g_k^*(y+h) - g_k^*(y) = g_k(y+h, \dots, y+h) - g_k(y, \dots, y)$$
$$= \sum_{l=1}^{k} C_k^l g_k(\underbrace{h, \dots, h}_{l}, \underbrace{y, \dots, y}_{k-l}) = 0$$

in view of what has been said above. □

2. Probability distributions on topological Abelian groups

In this section, we summarize known results on probability distributions on topological Abelian groups, mainly on locally compact Abelian groups. We assume that all groups considered in this section are second countable, in spite of the fact that a part of the statements below is true without this restriction.

Let X be a topological Abelian group. Let $\mathfrak{B}(X)$ be the Borel σ-algebra on X, i.e., the smallest σ-algebra of subsets of X which contains all open subsets of X. By a *measure* on the group X we understand a nonnegative countably additive set function defined on $\mathfrak{B}(X)$. The difference of two measures is called a *signed measure*. By a *probability distribution* or *distribution* on X we mean a probability measure on X, i.e., a measure μ such that $\mu(X) = 1$. Throughout the book, by distributions we assume only probability distributions. Denote by $\mathrm{M}^1(X)$ the set of all distributions on the group X. Denote by $\sigma(\mu)$ the *support* of a distribution μ, i.e., the smallest closed subset $A \subset X$ such that $\mu(A) = \mu(X)$. Let $\mu, \nu \in \mathrm{M}^1(X)$. We define their *convolution* $\mu * \nu \in \mathrm{M}^1(X)$ by the formula

$$(\mu * \nu)(B) = \int_X \mu(B - x)d\nu(x), \quad B \in \mathfrak{B}(X)$$

(we note that $\mu(B - x)$ is a Borel measurable function in x for every $B \in \mathfrak{B}(X)$ due to the assumption that X is a second countable group). Observe that the convolution is commutative, i.e., $\mu * \nu = \nu * \mu$ for all $\mu, \nu \in \mathrm{M}^1(X)$.

The set $\mathrm{M}^1(X)$ is a semigroup with respect to the convolution. Let $\mu \in \mathrm{M}^1(X)$. Denote by μ^{*n} the convolution of n distributions each of which is equal to μ. If $\mu = \mu_1 * \mu_2$, where $\mu_j \in \mathrm{M}^1(X)$, then the distributions μ_j are called *factors* of the distribution μ. For $\mu \in \mathrm{M}^1(X)$ define the distribution $\bar{\mu} \in \mathrm{M}^1(X)$ by the formula $\bar{\mu}(B) = \mu(-B)$ for all $B \in \mathfrak{B}(X)$.

We say that a *distribution μ is concentrated on a set* $A \in \mathfrak{B}(X)$ if $\mu(B) = 0$ for all $B \in \mathfrak{B}(X)$ such that $B \cap A = \emptyset$. In general the set A need not be closed, and A is not uniquely determined. Let $x \in X$. Denote by E_x the *degenerate distribution* concentrated at the element $x \in X$. The convolution $\mu * E_x$ of a distribution μ with a degenerate distribution E_x is called a *shift of the distribution* μ.

Let $(\Omega, \mathfrak{A}, P)$ be a probability space, where Ω is a set, $\mathfrak{A}$ is a σ-algebra of subsets of Ω, and P is a probability measure defined on $\mathfrak{A}$. By a *random variable* on $(\Omega, \mathfrak{A}, P)$ with values in X we understand a mapping $\xi(\omega)$ defined on Ω with values in X and such that $\xi^{-1}(B) \in \mathfrak{A}$ for any Borel subset $B \subset X$. The random variable ξ defines a distribution μ_ξ on $\mathfrak{B}(X)$ by the formula

$$\mu_\xi(B) = P\{\omega \in \Omega : \quad \xi(\omega) \in B, \ B \in \mathfrak{B}(X)\}.$$

Two random variables ξ and η with values in X are called *independent* if the equality

$$P\{\omega : \xi(\omega) \in A, \ \eta(\omega) \in B\} = P\{\omega : \xi(\omega) \in A\}P\{\omega : \eta(\omega) \in B\}$$

is fulfilled for all $A, B \in \mathfrak{B}(X)$. If random variables ξ and η are independent, then

$$\mu_{\xi+\eta} = \mu_\xi * \mu_\eta.$$

The independence of n random variables is defined in a similar way.

PROPOSITION 2.1. *Let X be a topological Abelian group and let G be a Borel subgroup of X. Assume $\mu \in \mathrm{M}^1(G)$ and $\mu = \mu_1 * \mu_2$, where $\mu_j \in \mathrm{M}^1(X)$. Then the* *distributions* μ_j *can be replaced by their shifts* λ_j *in such a manner that* $\mu = \lambda_1 * \lambda_2$ *and* $\lambda_j \in \mathrm{M}^1(G)$, $j = 1, 2$.

PROOF. We deduce from the equality

$$1 = \mu(G) = \int_X \mu_1(G - x) d\mu_2(x)$$

that $\mu_2\{x \in X : \mu_1(G - x) = 1\} = 1$. Hence $\mu_1(G - x_1) = 1$ for some $x_1 \in X$, i.e., the distribution μ_1 is concentrated on the set $G - x_1$. This implies that $\lambda_1 = \mu_1 * E_{x_1} \in \mathrm{M}^1(G)$. Put $\lambda_2 = \mu_2 * E_{-x_1}$. Then $\mu = \lambda_1 * \lambda_2 = \lambda_2 * \lambda_1$ and

$$1 = \mu(G) = \int_X \lambda_2(G - x) d\lambda_1(x) = \int_G \lambda_2(G - x) d\lambda_1(x).$$

It follows from this that $\lambda_1\{x \in G : \lambda_2(G - x) = 1\} = 1$. Therefore $\lambda_2(G - x_2) = 1$ for some $x_2 \in G$. Since G is a group, we have $\lambda_2(G) = 1$, i.e., $\lambda_2 \in \mathrm{M}^1(G)$. □

THEOREM 2.2 (Suslin theorem [**71**, §39, IV]). *Let* X_1 *be a separable complete metric space, let* X_2 *be a metric space, and let* $p : X_1 \to X_2$ *be a continuous one-to-one mapping. If* B *is a Borel set in* X_1*, then* $p(B)$ *is a Borel set in* X_2.

PROPOSITION 2.3. *Let* X_1 *and* X_2 *be topological Abelian groups. Let* $p : X_1 \to X_2$ *be an algebraic isomorphism with the property that the images and preimages of Borel sets under the mapping* p *are Borel. Then* p *generates an isomorphism of the semigroups* $\mathrm{M}^1(X_1)$ *and* $\mathrm{M}^1(X_2)$ *(we keep the notation* p *for this isomorphism) by the formula*

$$p(\mu)(B) = \mu(p^{-1}(B)), \tag{2.1}$$

where $\mu \in \mathrm{M}^1(X_1)$, $B \in \mathfrak{B}(X_2)$.

The proof of Proposition 2.3 is some standard verifications. The statement below follows directly from Suslin's theorem and Proposition 2.3.

COROLLARY 2.4. *Let* X_1 *be a separable complete topological Abelian metric group, let* X_2 *be a topological Abelian metric group, and let* $p : X_1 \to X_2$ *be a continuous monomorphism. Then, by formula* (2.1), p *generates an isomorphism of the semigroups* $\mathrm{M}^1(X_1)$ *and* $\mathrm{M}^1(p(X_1))$.

Let X be a locally compact Abelian group with character group Y. For every $\mu \in \mathrm{M}^1(X)$ the *characteristic function (Fourier transform)* of μ is defined by the formula

$$\hat{\mu}(y) = \int_X (x, y) d\mu(x) \quad y \in Y.$$

A function $f(y)$ on the group Y is called *characteristic* if $f(y) = \hat{\mu}(y)$ for a distribution $\mu \in \mathrm{M}^1(X)$. We note that if ξ is a random variable with values in X and distribution μ, then the characteristic function of μ is the expectation

$$\hat{\mu}(y) = \mathbf{E}[(\xi, y)].$$

The *characteristic function of a random variable* is the characteristic function of its distribution. If ξ and η are random variables with values in X, then ξ and η are independent if and only if the equality

$$\mathbf{E}[(\xi, u)(\eta, v)] = \mathbf{E}[(\xi, u)]\mathbf{E}[(\eta, v)]$$

holds for all $u, v \in Y$.

It is convenient for us to formulate some well-known properties of the characteristic functions in the form of a theorem.

THEOREM 2.5. *Let X be a locally compact Abelian group with character group Y. Let $\mu, \nu \in \mathrm{M}^1(X)$. The following statements are valid:*

(i) $|\hat\mu(y)| \le \hat\mu(0) = 1$;
(ii) *μ is uniquely defined by the function $\hat\mu(y)$;*
(iii) $(\widehat{\mu * \nu})(y) = \hat\mu(y)\hat\nu(y)$, $\mu, \nu \in \mathrm{M}^1(X)$;
(iv) $\hat{\bar\mu}(y) = \overline{\hat\mu(y)} = \hat\mu(-y)$;
(v) *if H is a closed subgroup of the group Y and $|\hat\mu(y)| = 1$ for all $y \in H$, then there exists an element $x \in X$ such that $\hat\mu(y) = (x, y)$ for all $y \in H$;*
(vi) *$\hat\mu(y)$ is a uniformly continuous function;*
(vii) *the inequality*

$$|\hat\mu(u) - \hat\mu(v)|^2 \le 2(1 - \mathrm{Re}\ \hat\mu(u - v))$$

holds for all $u, v \in Y$.

We note that the convolution of two signed measures and the characteristic function of a signed measure are defined in the same way as for probability distributions. Furthermore, properties (ii)–(iv) and (vi) of the characteristic functions of probability distributions are also fulfilled for signed measures.

Let Y be an arbitrary Abelian group. A function $f(y)$ defined on Y is said to be *positive definite* if for any natural n, any complex numbers $z_1, \dots, z_n$, and any elements $y_1, \dots, y_n \in Y$ the inequality

$$\sum_{i,j=1}^{n} f(y_i - y_j) z_i \bar z_j \ge 0$$

holds.

THEOREM 2.6 (Bochner theorem [**64**, (33.3)]). *Let X be a locally compact Abelian group with character group Y. A function $f(y)$ on the group Y is the characteristic function of a distribution $\mu \in \mathrm{M}^1(X)$, i.e., $f(y) = \hat\mu(y)$, if and only if the following conditions hold:*

(i) *$f(y)$ is continuous;*
(ii) *$f(y)$ is positive definite;*
(iii) $f(0) = 1$.

Thanks to Bochner's theorem we can make no distinction between characteristic functions and normalized continuous positive definite functions.

PROPOSITION 2.7. *Let X_1 and X_2 be locally compact Abelian groups with character groups Y_1 and Y_2 respectively. Let $p : X_1 \to X_2$ be a continuous homomorphism generating the mapping $p : \mathrm{M}^1(X_1) \to \mathrm{M}^1(X_2)$ by formula (2.1). If $\mu \in \mathrm{M}^1(X_1)$, then the characteristic function of the distribution $p(\mu)$ is of the form*

$$\widehat{p(\mu)}(y_2) = \hat\mu(\widetilde{p} y_2), \quad y_2 \in Y_2,$$

where $\widetilde{p} : Y_2 \to Y_1$ is the homomorphism adjoint of p.

COROLLARY 2.8. *Let X be a locally compact Abelian group with character group Y. Let G be a closed subgroup of X and let $H = A(Y, G)$. Let $p : X \to X/G$*

be the natural homomorphism and let $\mu \in \mathrm{M}^1(X)$. *Then the restriction of the characteristic function* $\hat{\mu}(y)$ *to* H *is the characteristic function of the distribution* $p(\mu) \in \mathrm{M}^1(X/G)$.

The following statement is often used in the construction of positive definite functions.

PROPOSITION 2.9 ([**64**, (32.43)]). *Let* Y *be an Abelian group and let* H *be a subgroup of* Y. *Let* $f_0(y)$ *be a positive definite function on* H *and let* $f(y)$ *be the function on the group* Y *of the form*

$$f(y) = \begin{cases} f_0(y) & \text{if } \ y \in H, \\ 0 & \text{if } \ y \notin H. \end{cases}$$

Then $f(y)$ *is a positive definite function.*

PROPOSITION 2.10. *Let* X *be a locally compact Abelian group with character group* Y. *Let* $\mu \in \mathrm{M}^1(X)$. *Then the set* $E = \{y \in Y : \hat{\mu}(y) = 1\}$ *is a closed subgroup of* Y, *the characteristic function* $\hat{\mu}(y)$ *is* E*-invariant, i.e.,* $\hat{\mu}(y+h) = \hat{\mu}(y)$ *for all* $y \in Y$, $h \in E$. *Moreover,* $\sigma(\mu) \subset A(X, E)$.

PROOF. The inequality

$$1 - \mathrm{Re}\ (x, y_1 + y_2) \leq 2[(1 - \mathrm{Re}\ (x, y_1)) + (1 - \mathrm{Re}\ (x, y_1))]$$

holds for all $x \in X$, $y_1, y_2 \in Y$. This implies the inequality

$$1 - \mathrm{Re}\ \hat{\mu}(y_1 + y_2) \leq 2[(1 - \mathrm{Re}\ \hat{\mu}(y_1)) + (1 - \mathrm{Re}\ \hat{\mu}(y_1))]. \tag{2.2}$$

It results from (2.2) that E is a subgroup. Obviously, E is closed. Statement (vii) of Theorem 2.5 implies that if $y_1 - y_2 \in E$, then $\hat{\mu}(y_1) = \hat{\mu}(y_2)$, i.e., the function $\hat{\mu}(y)$ is E-invariant. Put $H = Y/E$. By Theorem 1.9, we have $H^* \cong A(X, E)$. The function $\hat{\mu}(y)$ can be considered as a function $f([y])$ on H. The function $f([y])$ is continuous, positive definite, and $f([0]) = 1$. It follows from Bochner's theorem that $f([y]) = \hat{\lambda}([y])$, where $\lambda \in \mathrm{M}^1(A(X, E))$. We have

$$\hat{\mu}(y) = \int\limits_X (x, y) d\mu(x) = f([y]) = \int\limits_{A(X,E)} (x, [y]) d\lambda(x) = \int\limits_X (x, y) d\lambda(x) = \hat{\lambda}(y).$$

This implies that $\mu = \lambda$. □

COROLLARY 2.11. *Let* X *be a locally compact Abelian group with character group* Y. *Let* $\mu \in \mathrm{M}^1(X)$. *Then the set* $B = \{y \in Y : |\hat{\mu}(y)| = 1\}$ *is a closed subgroup of* Y.

PROOF. Put $\nu = \mu * \bar{\mu}$. Then $\hat{\nu}(y) = |\hat{\mu}(y)|^2$ for all $y \in Y$ and hence $\{y \in Y : |\hat{\mu}(y)| = 1\} = \{y \in Y : \hat{\nu}(y) = 1\}$. □

Let X be a locally compact Abelian group. There exists a measure m_X on X with the following properties:

(a) $m_X(B + x) = m_X(B)$ for all $B \in \mathfrak{B}(X)$ and $x \in X$;
(b) $m_X(-B) = m_X(B)$ for all $B \in \mathfrak{B}(X)$.

The measure m_X is called a *Haar measure.* If X is a compact group, then $m_X(X) < \infty$ [**63**, Chapter 4]. We agree to assume that in this case $m_X \in \mathrm{M}^1(X)$.

Denote by Y the character group of the group X. Let K be a compact subgroup of X. We observe that the characteristic function of the distribution m_K is of the form

$$\widehat{m}_K(y) = \begin{cases} 1 & \text{if} \quad y \in A(Y, K), \\ 0 & \text{if} \quad y \notin A(Y, K). \end{cases} \tag{2.3}$$

A distribution $\mu \in \mathrm{M}^1(X)$ is said to be *idempotent* if $\mu^{*2} = \mu$. Denote by $I(X)$ the set of all shifts of idempotent distributions on the group X.

PROPOSITION 2.12. *Let X be a locally compact Abelian group. A distribution $\mu \in \mathrm{M}^1(X)$ is idempotent if and only if $\mu = m_K$, where K is a compact subgroup of X.*

PROOF. Denote by Y the character group of the group X. It follows from $\mu^{*2} = \mu$ that $\hat{\mu}^2(y) = \hat{\mu}(y)$ for all $y \in Y$. This implies that either $\hat{\mu}(y) = 0$ or $\hat{\mu}(y) = 1$. Put $E = \{y \in Y : \hat{\mu}(y) = 1\}$. In view of Proposition 2.10, E is an open-closed subgroup. By Theorem 1.11, $K = A(X, E)$ is a compact subgroup. By Theorem 1.8, $E = A(Y, K)$. It follows from (2.3) that $\hat{\mu}(y) = \widehat{m}_K(y)$ for all $y \in Y$. Hence $\mu = m_K$.

In view of (2.3), the converse statement is obvious. □

Let G be a locally compact Abelian group. Consider the group $\mathbb{R}^m \times G$. Denote by (t, g), where $t \in \mathbb{R}^m$, $g \in G$, elements of the group $\mathbb{R}^m \times G$ and by (s, h), where $s \in \mathbb{R}^m$, $h \in H$, elements of the group $(\mathbb{R}^m \times G)^* \cong \mathbb{R}^m \times H$, where $H = G^*$.

PROPOSITION 2.13. *Let $\mu \in \mathrm{M}^1(\mathbb{R}^m \times G)$ and let $\hat{\mu}(s, 0)$ be an entire function in s. Then $\hat{\mu}(s, h)$ is an entire function in s for each fixed $h \in H$, the representation*

$$\hat{\mu}(s, h) = \int\limits_X e^{i\langle t, s\rangle}(g, h) d\mu(t, g) \tag{2.4}$$

is valid for all $s \in \mathbb{C}^m$, $h \in H$, and the inequality

$$\max_{s \in \mathbb{C}^m,\ |s| \le r} |\hat{\mu}(s, h)| \le \max_{s \in \mathbb{C}^m,\ |s| \le r} |\hat{\mu}(s, 0)|, \quad h \in H, \tag{2.5}$$

holds. Moreover, if $m = 1$, then the function $\hat{\mu}(-iy + x, h)/\hat{\mu}(-iy, 0)$, where $x, y \in \mathbb{R}$, for any fixed y is a characteristic function in variable $(x, h) \in \mathbb{R} \times H$.

PROOF. It is obvious that

$$\hat{\mu}(s, 0) = \int\limits_{\mathbb{R}^m} e^{i\langle t, s\rangle} d\mu_1(t), \quad s \in \mathbb{R}^m, \tag{2.6}$$

where μ_1 is a distribution on $\mathbb{R}^m$ such that $\mu_1(E) = \mu(E \times G)$ for every Borel set E in $\mathbb{R}^m$. Since $\hat{\mu}(s, 0)$ is an entire function in s, the integral in the right side of (2.6) converges absolutely and uniformly on every compact set in $\mathbb{C}^m$ [**73**, Theorem 6.1.3]. Consider the integral

$$I(s, h) = \int\limits_X e^{i\langle t, s\rangle}(g, h) d\mu(t, g), \quad s \in \mathbb{C}^m,\ h \in H.$$

It follows from what has been said that the integral $I(s,h)$ converges absolutely and uniformly on every compact set in $\mathbb{C}^m$ and admits the estimate

$$|I(s,h)| \leq \int\limits_{\mathbb{R}^m} \exp\left\{ -\sum_{j=1}^{m} t_j \ \mathrm{Im}\ s_j \right\} d\mu_1(t). \tag{2.7}$$

This implies that $I(s,h)$ is an entire function in every s_j for each fixed h. This implies that $I(s,h)$ is an entire function in s for each fixed h. Inequality (2.5) follows from (2.7).

Assume that $m = 1$. The fact that (2.4) is true for all $s \in \mathbb{C}$, $h \in H$, implies that the function $\hat{\mu}(-iy + x, h)/\hat{\mu}(-iy, 0)$, where $x, y \in \mathbb{R}$, for any fixed y is a characteristic function in variable $(x,h) \in \mathbb{R} \times H$. □

Let X be a topological Abelian group. Let ξ and η be random variables taking values in X. By the *conditional distribution* of the random variable η given ξ we understand a function $P_{\eta|\xi}(x,B)$ defined on $X \times \mathfrak{B}(X)$ such that

(i) for every fixed $x \in X$ the function $P_{\eta|\xi}(x,B)$ is a distribution on X;
(ii) for each fixed $B \in \mathfrak{B}(X)$ and each $A \in \mathfrak{B}(X)$ the equality

$$P\{\omega \in \Omega : \xi(\omega) \in A,\ \eta(\omega) \in B\} = \int\limits_A P_{\eta|\xi}(x,B) d\mu_\xi(x)$$

holds.

We deduce from the definition of the conditional distribution of the random variable η given ξ that the distribution $P_{\eta|\xi}(x,B)$ is symmetric, i.e.,

$$P_{\eta|\xi}(x,B) = P_{\eta|\xi}(x,-B),$$

if and only if the equality

$$P\{\omega \in \Omega : \xi(\omega) \in A,\ \eta(\omega) \in B\} = P\{\omega \in \Omega : \xi(\omega) \in A,\ \eta(\omega) \in -B\}$$

holds for all $A, B \in \mathfrak{B}(X)$ [**3**, Chapter XIV, §5].

3. Gaussian distributions on locally compact Abelian groups

Let X be a second countable locally compact Abelian group. The Gaussian distributions, along with the idempotent distributions, are very important in the study of characterization problems in mathematical statistics on the group X. In this section we define the Gaussian distribution on the group X and study its properties.

Definition 3.1. Let X be a locally compact Abelian group with character group Y. A distribution γ on X is called *Gaussian*, if its characteristic function can be represented in the form

$$\hat{\gamma}(y) = (x,y)\exp\{-\varphi(y)\}, \quad y \in Y, \tag{3.1}$$

where $x \in X$ and $\varphi(y)$ is a continuous nonnegative function on the group Y satisfying the equation

$$\varphi(u+v) + \varphi(u-v) = 2[\varphi(u) + \varphi(v)], \quad u, v \in Y. \tag{3.2}$$

Denote by $\Gamma(X)$ the set of Gaussian distributions on the group X. A Gaussian distribution $\gamma \in \Gamma(X)$ is called *symmetric* if $x = 0$ in (3.1). Denote by $\Gamma^s(X)$ the set of symmetric Gaussian distributions on the group X.

Definition 3.1 of the Gaussian distribution for the groups $\mathbb{R}^k$ and $\mathbb{T}^k$ coincides with the classical one. It is clear that a continuous nonnegative function $\varphi(s)$ on $\mathbb{R}^k$ satisfying equation (3.2) is of the form

$$\varphi(s) = \langle As, s\rangle, \quad s \in \mathbb{R}^k,$$

where A is a symmetric positive semidefinite matrix. It follows from this that Definition 3.1 of the Gaussian distribution for the group $\mathbb{R}^k$ coincides with the classical one. In particular, if $\mu \in \Gamma(\mathbb{R})$, then the characteristic function $\hat{\mu}(s)$ can be written in the form

$$\hat{\mu}(s) = \exp\left\{ias - \frac{\sigma^2 s^2}{2}\right\}, \quad s \in \mathbb{R}, \tag{3.3}$$

where $a \in \mathbb{R}$, $\sigma \geq 0$. If in (3.3) $\sigma > 0$, then μ has a density $f(t)$ with respect to the Lebesgue measure

$$f(t) = \frac{d\mu}{dt} = \frac{1}{\sqrt{2\pi}\sigma} \exp\left\{-\frac{(t-a)^2}{2\sigma^2}\right\} \quad t \in \mathbb{R}.$$

For the group $\mathbb{T}^k$, in accordance with the classical definition, γ is a Gaussian distribution on $\mathbb{T}^k$ if γ is an image of a Gaussian distribution μ on $\mathbb{R}^k$ under the natural homomorphism $p : \mathbb{R}^k \to \mathbb{T}^k \cong \mathbb{R}^k/(2\pi\mathbb{Z})^k$. If the characteristic function of a Gaussian distribution μ is of the form

$$\hat{\mu}(s) = \exp\{i\langle t, s\rangle - \langle As, s\rangle\}, \quad s \in \mathbb{R}^k,$$

where $t \in \mathbb{R}^k$ and A is a symmetric positive semidefinite matrix, then by Proposition 2.7, we have

$$\hat{\gamma}(n) = \widehat{p(\mu)}(n) = \exp\{i\langle t, n\rangle - \langle An, n\rangle\}, \quad n \in \mathbb{Z}^k.$$

We see that γ is a Gaussian distribution in accordance with Definition 3.1. It is easy to see that the converse statement is also true (compare with Proposition 3.7).

It should be noted that the Gaussian distributions on the group $\mathbb{R}^n$ belong to the class of stable distributions ([**62**, §1.3]).

REMARK 3.2. Let Y be an Abelian group. A function $\varphi(y)$ on the group Y satisfying equation (3.2) is also called a quadratic form on Y. If we know a quadratic form $\varphi(y)$, then we can construct a 2-additive function $\psi(u, v)$ by the formula

$$\psi(u, v) = \frac{1}{2}[\varphi(u+v) - \varphi(u) - \varphi(v)], \quad u, v \in Y. \tag{3.4}$$

The function $\psi(u, v)$ satisfies the conditions:

(i) $\psi(u, v) = \psi(v, u)$;
(ii) $\psi(u + v, w) = \psi(u, w) + \psi(v, w)$.

It is easy to see that if a function $\psi(u, v)$ satisfies conditions (i) and (ii), then the function

$$\varphi(y) = \psi(y, y) \tag{3.5}$$

satisfies equation (3.2). Thus, there is a one-to-one correspondence between functions $\varphi(y)$ satisfying equation (3.2) and 2-additive functions $\psi(u, v)$ satisfying (i) and (ii).

REMARK 3.3. Let Y be a locally compact Abelian group with character group X. By Pontryagin's duality theorem, Y is topologically isomorphic to the character group of X. Let $\varphi(y)$ be a continuous nonnegative function on the group Y satisfying equation (3.2) and let $f(y) = \exp\{-\varphi(y)\}$. Then $f(y)$ is a characteristic function. Indeed, let $y_1, \dots, y_k$ be fixed elements of Y. Consider the function $\varphi(n_1y_1+\dots+n_ky_k)$ as a function in integer-valued variables n_j. Define the function $\psi(y_1, y_2)$ by formula (3.4). It results from (3.5) that

$$\varphi(n_1y_1 + \dots + n_ky_k) = \langle An, n\rangle, \quad n = (n_1, \dots, n_k) \in \mathbb{Z}^k,$$
$$A = (\alpha_{ij})_{i,j=1}^k, \quad \alpha_{ij} = \psi(y_i, y_j).$$

Inasmuch as $\varphi(y) \geq 0$, the matrix A is positive semidefinite. Hence the function

$$\exp\{-\varphi(n_1y_1 + \dots + n_ky_k)\}$$

considered as a function in integer-valued variables n_j is positive definite on the group $\mathbb{Z}^k$. It follows from what has been said that $f(y)$ is a positive definite function. By Bochner's theorem, $f(y)$ is a characteristic function and hence the characteristic function of a Gaussian distribution on the group X.

When proving characterization theorems on groups, we need the following assertion.

LEMMA 3.4. *Let H be an open subgroup of a locally compact Abelian group Y and let $\varphi_0(h)$ be a continuous nonnegative function on H satisfying equation* (3.2). *Then there exists a continuous nonnegative function $\varphi(y)$ on Y satisfying equation* (3.2) *and such that its restriction to H coincides with $\varphi_0(h)$.*

PROOF. Let $y_1 \notin H$. Standard reasoning shows that it suffices to extend the function $\varphi_0(h)$ retaining its properties from the subgroup H to the open subgroup

$$H_1 = \{y \in Y : y = ny_1 + h, \ n \in \mathbb{Z}, \ h \in H\}.$$

Two cases are possible:

1. $ny_1 \cap H = \emptyset$ for all $n \in \mathbb{Z}$, $n \neq 0$. Set $\varphi(ny_1 + h) = \varphi_0(h)$ for all $n \in \mathbb{Z}$, $h \in H$.

2. $ny_1 \in H$ for some $n \in \mathbb{Z}$, $n \neq 0$. Let n_0 be the minimal natural number such that $n_0y_1 \in H$. Then $n_0y \in H$ for all $y \in H_1$. Set $\varphi(y) = \varphi_0(n_0y)/n_0^2$ for all $y \in H_1$.

The function $\varphi(y)$ defined above is continuous, nonnegative, and in case 1 satisfies equation (3.2). In case 2 we have

$$\begin{aligned}\varphi(u+v) + \varphi(u-v) &= \varphi_0(n_0(u+v))/n_0^2 + \varphi_0(n_0(u-v))/n_0^2 \\ &= 2\left(\varphi_0(n_0u)/n_0^2 + \varphi_0(n_0v)/n_0^2\right) = 2[\varphi(u) + \varphi(v)], \quad u, v \in H_1.\end{aligned}$$

Thus, the function $\varphi(y)$ has the desired properties. □

Let us prove now a statement describing supports of Gaussian distributions.

PROPOSITION 3.5. *Let X be a locally compact Abelian group. If $\gamma \in \Gamma(X)$, then $\sigma(\gamma)$ is a coset of a closed connected subgroup of the group X.*

PROOF. In view of Definition 3.1, we can assume without loss of generality that $\gamma \in \Gamma^s(X)$. Denote by Y the character group of the group X. Let $\varphi(y)$ be a continuous function on the group Y satisfying equation (3.2). It follows from (3.2) that $\varphi(y)$ is a continuous polynomial. By Proposition 1.30, $\varphi(y+h)=\varphi(y)$ for all $y\in Y$, $h\in b_Y$. In particular, we have $\varphi(y)=0$ for all $y\in b_Y$. Let

$$\hat{\gamma}(y)=\exp\{-\varphi(y)\},\quad y\in Y.$$

Consider the set $E=\{y\in Y:\hat{\gamma}(y)=1\}$ and put $G=A(X,E)$. It results from Proposition 2.10 that $\sigma(\gamma)\subset G$. Hence we can consider γ as a Gaussian distribution on the group G. Set $H=G^*$. The characteristic function $\hat{\gamma}(y)$, $y\in H$, has the property

$$\{h\in H:\hat{\gamma}(h)=1\}=\{0\}.$$

Taking into account that $\varphi(h)=0$ for all $h\in b_H$ and hence $\hat{\gamma}(h)=1$ for all $h\in b_H$, we conclude that $b_H=\{0\}$. By Theorem 1.10, this implies that $c_G=G$, i.e., G is a connected group.

The proposition will be proved if we verify that $\gamma(U)>0$ for any open subset $U\subset G$. Suppose $\gamma(U)=0$ for some open subset $U\subset G$. Choose an open subset U_0 in U and a neighborhood of the zero V in G such that $U_0+V\subset U$. By Theorem 1.15, $G=L\times K$, where $L\cong\mathbb{R}^m$, $m\geq 0$, and K is a compact connected Abelian group. Apply Theorem 1.18 and choose in $K\cap V$ a closed subgroup S such that $K/S\cong\mathbb{T}^k\times F$, where F is a finite Abelian group. Since the group K is connected, $F=\{0\}$. It follows from this that the factor-group G/S is topologically isomorphic to the group $\mathbb{R}^m\times\mathbb{T}^k$. Denote by τ this topological isomorphism and set $p=\tau\pi$, where π is the natural homomorphism $\pi:G\to G/S$. It is obvious that $p(\gamma)\in\Gamma(\mathbb{R}^m\times\mathbb{T}^k)$ and

$$\{y\in\mathbb{R}^m\times\mathbb{Z}^k:\widehat{p(\gamma)}(y)=1\}=\{0\}.$$

Moreover,

$$p(\gamma)(p(U_0))=\gamma\{p^{-1}(p(U_0))\}=\gamma\{U_0+S\}\leq\gamma\{U_0+V\}\leq\gamma\{U\}=0.$$

This is impossible, because $p(U_0)$ is an open set. The obtained contradiction proves that $\gamma(U)>0$. □

The following statement follows directly from the proof of Proposition 3.5.

COROLLARY 3.6. *Let X be a locally compact Abelian group with character group Y. Let $\gamma\in\Gamma^s(X)$. Assume that $\{y\in Y:\hat{\gamma}(y)=1\}=\{0\}$. This implies that the group X is connected and $\sigma(\gamma)=X$.*

Thus, studying Gaussian distributions on a locally compact Abelian group X we can restrict ourselves to the case when X is a locally compact connected Abelian group. We prove now that an arbitrary symmetric Gaussian distribution on a locally compact connected Abelian group X is a continuous homomorphic image of a Gaussian distribution in a linear space. This space is either finite-dimensional or infinite-dimensional depending on the dimension of the group X. First we consider the case when the group X has finite dimension.

PROPOSITION 3.7. *Let X be a locally compact connected Abelian group of finite dimension l. Then there exists a continuous homomorphism $p:\mathbb{R}^l\to X$ with the property: for any $\gamma\in\Gamma^s(X)$ there is $\mu\in\Gamma^s(\mathbb{R}^l)$ such that $\gamma=p(\mu)$.*

PROOF. By Theorem 1.15, the group X is topologically isomorphic to a group of the form $\mathbb{R}^m\times K$, where $m\geq 0$ and K is a compact connected Abelian group. First assume $X=K$. Put $D=K^*$. We conclude from Theorems 1.2 and 1.3 that D is a discrete torsion-free Abelian group. By Theorem 1.4, $r_0(D)=l$. Choose in

D a maximal independent set of elements $\{d_1, \dots, d_l\}$. Then for every $d \in D$ there exist integers $k, k_1, \dots, k_l$ such that

$$kd = k_1 d_1 + \cdots + k_l d_l. \tag{3.6}$$

The independence of the set $\{d_1, \dots, d_l\}$ implies that the rational numbers $\{k_j/k\}$ are uniquely determined by d. Inasmuch as D is a torsion-free group, the mapping $\pi : D \to \mathbb{R}^l$, defined by the formula

$$\pi d = (k_1/k, \dots, k_l/k), \tag{3.7}$$

is a monomorphism.

Let a function $\varphi(d)$ on the group D satisfy equation (3.2). Taking into account Remark 3.2, we define the function $\psi(y_1, y_2)$ by formula (3.4). It results from (3.5) and (3.6) that

$$k^2\varphi(d) = k^2\psi(d,d) = \psi(kd, kd)$$
$$= \psi\left(k_1 d_1 + \cdots + k_l d_l, k_1 d_1 + \cdots + k_l d_l\right) = \sum_{i,j=1}^{l} \alpha_{ij} k_i k_j,$$

where $\alpha_{ij} = \psi(d_i, d_j)$, $1 \le i \le l$, $1 \le j \le l$. Hence

$$\varphi(d) = \sum_{i,j=1}^{l} \alpha_{ij}(k_i/k)(k_j/k) = \langle A\pi d, \pi d\rangle, \quad d \in D, \tag{3.8}$$

where $A = (\alpha_{ij})_{i,j=1}^{l}$. If $\varphi(d) \ge 0$ for all $d \in D$, then the symmetric matrix A is positive semidefinite, i.e., the quadratic form $\langle As, s\rangle$ is nonnegative for all $s \in \mathbb{R}^l$.

Let $\gamma \in \Gamma^s(K)$. Assume that the function $\varphi(d)$ in (3.1) corresponds to the characteristic function $\hat{\gamma}(d)$. We deduce from (3.8) that

$$\hat{\gamma}(d) = \exp\{-\langle A\pi d, \pi d\rangle\}, \quad d \in D. \tag{3.9}$$

Let μ be the Gaussian distribution on the group $\mathbb{R}^l$ with the characteristic function

$$\hat{\mu}(s) = \exp\{-\langle As, s\rangle\}, \quad s \in \mathbb{R}^l. \tag{3.10}$$

Put $p = \widetilde{\pi}$. It follows from Proposition 2.7 and (3.10) that

$$\widehat{p(\mu)}(d) = \exp\{-\langle A\pi d, \pi d\rangle\}, \quad d \in D. \tag{3.11}$$

We conclude from (3.9) and (3.11) that $\gamma = p(\mu)$.

Assume now that the group X is noncompact. Then $X \cong \mathbb{R}^m \times K$, where $m \ge 1$ and K is a compact connected Abelian group of finite dimension n, $m + n = l$. Put $D = K^*$. Denote by Y the character group of the group X. Then $Y \cong \mathbb{R}^m \times D$. In order not to complicate the notation, suppose $Y = \mathbb{R}^m \times D$. Denote by $y = (s, d)$, where $s = (s_1, \dots, s_m) \in \mathbb{R}^m$, $d \in D$, elements of the group Y. Let $e_1 = (1, 0, \dots, 0), \dots, e_m = (0, \dots, 0, 1)$ be the natural basis in $\mathbb{R}^m$ and let $\{d_1, \dots, d_n\}$ be a maximal independent set of elements in D. Take $y = (s, d) \in Y$. If $kd = k_1 d_1 + \cdots + k_l d_n$, define the mapping $Y \to \mathbb{R}^l$ by the formula

$$\pi y = \pi(s, d) = (s_1, \dots, s_m, k_1/k, \dots, k_n/k), \quad (s, d) \in Y. \tag{3.12}$$

Then π is a continuous monomorphism. Reasoning as in the case when $X = K$, we make sure that

$$\varphi(y) = \langle A\pi y, \pi y\rangle, \quad y \in Y, \tag{3.13}$$

where $A = (\alpha_{ij})_{i,j=1}^{l}$ is a symmetric positive semidefinite matrix, $\alpha_{ij} = \psi(e_i, e_j)$, $1 \le i \le m$, $1 \le j \le m$, $\alpha_{i,j+m} = \psi(e_i, d_j)$, $1 \le i \le m$, $1 \le j \le n$, $\alpha_{i+m,j+m} = \psi(d_i, d_j)$, $1 \le i \le n$, $1 \le j \le n$. The proof can be completed as in the case $X = K$. □

COROLLARY 3.8. *Let X be a locally compact connected Abelian group of finite dimension l. Then every symmetric Gaussian distribution γ on X is a continuous monomorphic image of a symmetric Gaussian distribution on a group $\mathbb{R}^m \times \mathbb{T}^n$.*

PROOF. Let $\mu \in \Gamma^s(\mathbb{R}^l)$ and let $p : \mathbb{R}^l \to X$ be as in Proposition 3.7. We observe that $L = \sigma(\mu)$ is a subspace in $\mathbb{R}^l$. Denote by q the restriction of p to L. Put $E = \operatorname{Ker} q$. Then by Theorem 1.21, $E \cong \mathbb{R}^a \times \mathbb{Z}^n$. Denote by a the natural homomorphism $a : L \to L/E$. It is obvious that $L/E \cong \mathbb{R}^m \times \mathbb{T}^n$. In order not to complicate the notation, suppose $L/E = \mathbb{R}^m \times \mathbb{T}^n$. Then $a(\mu) \in \Gamma^s(\mathbb{R}^m \times \mathbb{T}^n)$. Set $b = qa^{-1}$. It is obvious that b is a continuous monomorphism from $\mathbb{R}^m \times \mathbb{T}^n$ to X and $\gamma = b(a(\mu))$. □

EXAMPLE 3.9. We illustrate Proposition 3.7 with the following example. Consider an $\boldsymbol{a}$-adic solenoid $\Sigma_{\boldsymbol{a}}$ with character group $H_{\boldsymbol{a}}$, where

$$H_{\boldsymbol{a}} = \left\{ \frac{m}{a_0 a_1 \cdots a_n} : n = 0, 1, \ldots;\ m \in \mathbb{Z} \right\}$$

is a subgroup of $\mathbb{Q}$. The monomorphism $\pi : H_{\boldsymbol{a}} \to \mathbb{R}$ is the identity embedding $\pi(r) = r$. It follows from (3.9) that the characteristic function of a Gaussian distribution γ on an $\boldsymbol{a}$-adic solenoid $\Sigma_{\boldsymbol{a}}$ is of the form

$$\hat{\gamma}(r) = (g, r) \exp\{-\sigma r^2\}, \quad r \in H_{\boldsymbol{a}}, \tag{3.14}$$

where $g \in \Sigma_{\boldsymbol{a}}$, $\sigma \ge 0$. Obviously, we get the same result by solving the equation (3.2) on the group $H_{\boldsymbol{a}}$.

It should be noted that the homomorphism p in this case is a monomorphism. Moreover, it is easy to prove that the Gaussian distribution γ is singular with respect to $m_{\Sigma_{\boldsymbol{a}}}$.

Consider now the case when a locally compact connected Abelian group X has infinite dimension. In what follows we need the concept of the Gaussian distribution in the space $\mathbb{R}^{\aleph_0}$.

Denote by $\mathbb{R}^{\aleph_0}$ the space of all sequences of real numbers in the product topology. The convergence of elements

$$t^{(k)} = (t_1^{(k)}, \ldots, t_n^{(k)}, \ldots) \to t = (t_1, \ldots, t_n, \ldots)$$

in this topology is the coordinate-wise convergence. The space $\mathbb{R}^{\aleph_0}$ can be regarded as the projective limit of a directed set of spaces $\mathbb{R}^n$. We observe that the topological group $\mathbb{R}^{\aleph_0}$ is not locally compact. It is possible to introduce a metric in the group $\mathbb{R}^{\aleph_0}$ inducing the product topology by the formula

$$d(u, v) = \sum_{n=1}^{\infty} \frac{1}{2^n} \frac{|u_n - v_n|}{1 + |u_n - v_n|}, \quad u = (u_1, \ldots, u_n, \ldots), v = (v_1, \ldots, v_n, \ldots) \in \mathbb{R}^{\aleph_0}.$$

With respect to this metric $\mathbb{R}^{\aleph_0}$ is a complete separable group.

Denote by $\mathbb{R}^{\aleph_0 *}$ the space of all finitary sequences of real numbers with the topology of the strictly inductive limit of spaces $\mathbb{R}^n$. The convergence of elements

$$s^{(k)} = (s_1^{(k)}, \ldots, s_n^{(k)}, 0, \ldots) \to s = (s_1, \ldots, s_n, 0, \ldots)$$

in this topology means that all $s^{(k)}$ belong to some subspace $\mathbb{R}^n$ and converge there. The topological group $\mathbb{R}^{\aleph_0 *}$ is not locally compact either. Let $t = (t_1, \dots, t_n, \dots) \in \mathbb{R}^{\aleph_0}$ and $s = (s_1, \dots, s_n, 0, \dots) \in \mathbb{R}^{\aleph_0 *}$. Set

$$\langle t, s\rangle = \sum_{j=1}^{\infty} t_j s_j.$$

Fix $s \in \mathbb{R}^{\aleph_0 *}$ and consider the function

$$(t, s) = \exp\{i\langle t, s\rangle\} \tag{3.15}$$

on the group $\mathbb{R}^{\aleph_0}$. This function is a continuous homomorphism of the group $\mathbb{R}^{\aleph_0}$ into the circle group $\mathbb{T}$, i.e., it is a character of the group $\mathbb{R}^{\aleph_0}$. It is easily seen that every character of the group $\mathbb{R}^{\aleph_0}$ is of the form (3.15). Denote by $\mathfrak{B}(\mathbb{R}^{\aleph_0})$ the Borel σ-algebra in $\mathbb{R}^{\aleph_0}$, i.e., the smallest σ-algebra generated by the sets of the form $U_J^n \times \mathbb{R}^{\aleph_0 \setminus J}$, where U_J^n is an open set in $\mathbb{R}_J^n$, $J = (j_1, \dots, j_n)$, $\mathbb{R}_J^n = \{t \in \mathbb{R}^{\aleph_0} : t_j = 0$ for $j \notin J\}$, and $\mathbb{R}^{\aleph_0 \setminus J} = \{t \in \mathbb{R}^{\aleph_0} : t_j = 0$ for all $j \in J\}$. Taking into account the definitions of the topology in $\mathbb{R}^{\aleph_0}$ and the σ-algebra $\mathfrak{B}(\mathbb{R}^{\aleph_0})$, it is easy to see that every continuous function on the group $\mathbb{R}^{\aleph_0}$ is $\mathfrak{B}(\mathbb{R}^{\aleph_0})$-measurable. Let μ be a distribution on $\mathbb{R}^{\aleph_0}$. We define the characteristic function of μ by the formula

$$\hat{\mu}(s) = \int\limits_{\mathbb{R}^{\aleph_0}} (t, s) d\mu(t), \quad s \in \mathbb{R}^{\aleph_0 *}.$$

It is easily seen that the characteristic function $\hat{\mu}(s)$ has the properties:

(a) $\hat{\mu}(0) = 1$;
(b) $\hat{\mu}(s)$ is continuous;
(c) $\hat{\mu}(s)$ is positive definite on every subspace $\mathbb{R}_J^n \subset \mathbb{R}^{\aleph_0 *}$.

It results from Kolmogorov's theorem and from Bochner's theorem for the group $\mathbb{R}^n$ that every function $g(s)$ on the group $\mathbb{R}^{\aleph_0 *}$ satisfying properties (a)–(c) defines a unique distribution μ_g on the group $\mathbb{R}^{\aleph_0}$ such that $\hat{\mu}_g(s) = g(s)$. Let $A = (\alpha_{ij})_{i,j=1}^{\infty}$ be a symmetric positive semidefinite matrix, i.e., the quadratic form $\langle As, s\rangle = \sum\limits_{i,j=1}^{\infty} \alpha_{ij} s_i s_j$ is nonnegative for all $s = (s_1, \dots, s_n, 0, \dots) \in \mathbb{R}^{\aleph_0 *}$. We can define now the Gaussian distribution on the group $\mathbb{R}^{\aleph_0}$.

DEFINITION 3.10. A distribution μ on the group $\mathbb{R}^{\aleph_0}$ is called *Gaussian* if its characteristic function is represented in the form

$$\hat{\mu}(s) = (t, s) \exp\{-\langle As, s\rangle\}, \quad s \in \mathbb{R}^{\aleph_0 *}, \tag{3.16}$$

where $t \in \mathbb{R}^{\aleph_0}$ and $A = (\alpha_{ij})_{i,j=1}^{\infty}$ is a symmetric positive semidefinite matrix.

Denote by $\Gamma(\mathbb{R}^{\aleph_0})$ the set of Gaussian distributions on the group $\mathbb{R}^{\aleph_0}$. A Gaussian distribution $\gamma \in \Gamma(\mathbb{R}^{\aleph_0})$ is called *symmetric* if $t = 0$ in (3.16). Denote by $\Gamma^s(\mathbb{R}^{\aleph_0})$ the set of symmetric Gaussian distributions on the group $\mathbb{R}^{\aleph_0}$.

We prove now an analogue of Proposition 3.7 for locally compact connected Abelian groups of infinite dimension.

PROPOSITION 3.11. *Let X be a locally compact connected Abelian group of infinite dimension. Then there exists a continuous homomorphism $p : \mathbb{R}^{\aleph_0} \to X$ with the property: for every $\gamma \in \Gamma^s(X)$ there is $\mu \in \Gamma^s(\mathbb{R}^{\aleph_0})$ such that $\gamma = p(\mu)$.*

PROOF. By Theorem 1.15, the group X is topologically isomorphic to a group of the form $\mathbb{R}^m \times K$, where $m \geq 0$ and K is a compact connected Abelian group. First assume $X = K$. Put $D = K^*$. Then by Theorems 1.2 and 1.3, D is a discrete torsion-free Abelian group. By Theorem 1.4, $r_0(D) = \infty$. Choose in D a maximal independent set of elements $\{d_1, \dots, d_l, \dots\}$. Then for every $d \in D$ there exist integers $k, k_1, \dots, k_l$ such that

$$kd = k_1 d_1 + \cdots + k_l d_l.$$

The independence of the set $\{d_1, \dots, d_l\}$ implies that the rational numbers $\{k_j/k\}$ are uniquely determined by d. Inasmuch as D is a torsion-free group, the mapping $\pi : D \to \mathbb{R}^{\aleph_0 *}$ defined by the formula

$$\pi d = (k_1/k, \dots, k_l/k, 0, \dots), \tag{3.17}$$

is a monomorphism.

Define the mapping $p : \mathbb{R}^{\aleph_0} \to K$ by the equality $(pt, d) = (t, \pi d)$ for all $t \in \mathbb{R}^{\aleph_0}$, $d \in D$. It is easy to verify that p is a continuous homomorphism. We observe that if λ is an arbitrary distribution on the group $\mathbb{R}^{\aleph_0}$, then the characteristic function of the distribution $p(\lambda) \in \mathrm{M}^1(K)$ is of the form

$$\widehat{p(\lambda)}(d) = \hat{\lambda}(\pi d), \quad d \in D. \tag{3.18}$$

Note that (3.18) does not follow from Proposition 2.7 but must be proved independently, because the groups $\mathbb{R}^{\aleph_0}$ and $\mathbb{R}^{\aleph_0 *}$ are not locally compact.

Just as in the proof of Proposition 3.7 it is easy to make sure that every non-negative function $\varphi(d)$ on the group D satisfying equation (3.2) is represented in the form

$$\varphi(d) = \langle A\pi d, \pi d \rangle, \quad d \in D, \tag{3.19}$$

where $A = (a_{ij})_{i,j=1}^{\infty}$ is a symmetric positive semidefinite matrix, $a_{ij} = \psi(d_i, d_j)$, $1 \leq i < \infty$, $1 \leq j < \infty$, and πd is defined by (3.17).

Let $\gamma \in \Gamma^s(K)$ and let the function $\varphi(d)$ correspond to the characteristic function $\hat{\gamma}(d)$ in (3.1). We deduce from (3.19) that

$$\hat{\gamma}(d) = \exp\{-\langle A\pi d, \pi d \rangle\}, \quad d \in D. \tag{3.20}$$

Let μ be the Gaussian distribution on the group $\mathbb{R}^{\aleph_0}$ with the characteristic function

$$\hat{\mu}(s) = \exp\{-\langle As, s \rangle\}, \quad s \in \mathbb{R}^{\aleph_0 *}. \tag{3.21}$$

It follows from (3.18) and (3.21) that

$$\widehat{p(\mu)}(d) = \exp\{-\langle A\pi d, \pi d \rangle\}, \quad d \in D. \tag{3.22}$$

We conclude from (3.20) and (3.22) that $\gamma = p(\mu)$.

The general case, when $X \cong \mathbb{R}^m \times K$, where $m \geq 1$ and K is a compact connected Abelian group, $\dim(K) = \infty$, can be considered as in Proposition 3.7. In so doing the monomorphism $\pi : Y \to \mathbb{R}^{\aleph_0 *}$ is defined by the formula

$$\pi y = \pi(s, d) = (s_1, \dots, s_m, k_1/k, \dots, k_l/k, 0, \dots), \quad (s, d) \in Y. \tag{3.23}$$

□

For what follows, we need the following assertion.

LEMMA 3.12. *Let X be a locally compact connected Abelian group with character group Y. Assume X contains no subgroups topologically isomorphic to the circle group $\mathbb{T}$. Then the following statements hold.*

1. *Assume X has finite dimension and $\dim(X) = l$. Let $\pi : Y \to \mathbb{R}^l$ be the monomorphism defined by formula (3.12). Then the subgroup $\pi(Y)$ is dense in $\mathbb{R}^l$.*
2. *Assume $\dim(X) = \infty$. Let $\pi : Y \to \mathbb{R}^{\aleph_0 *}$ be the monomorphism defined by formula (3.23). Then the subgroup $\pi(Y)$ is dense in $\mathbb{R}^{\aleph_0 *}$.*

PROOF. We restrict ourselves to the proof of statement 2. Statement 1 is proved similarly. By Theorem 1.15, the group X is topologically isomorphic to a group of the form $\mathbb{R}^m \times K$, where $m \geq 0$ and K is a compact connected Abelian group, $\dim(K) = \infty$. Then $Y \cong \mathbb{R}^m \times D$, where $D = K^*$. In order not to complicate the notation, suppose $Y = \mathbb{R}^m \times D$. By Theorems 1.2 and 1.3, D is a discrete torsion-free Abelian group. By Theorem 1.4, $r_0(D) = \infty$. Since the restriction of π to $\mathbb{R}^m$ coincides with the identity mapping, it suffices to prove the lemma in the case when $X = K$, $Y = D$. We assume that for all n the group $\mathbb{Z}^n$ is embedded in the natural way in $\mathbb{R}^n$. Moreover, the natural embeddings

$$\mathbb{R} \subset \mathbb{R}^2 \cdots \subset \mathbb{R}^n \subset \cdots \subset \mathbb{R}^{\aleph_0 *}$$

hold. Put $H_n = \overline{\pi(D)} \cap \mathbb{R}^n$. Then H_n is a closed subgroup of $\mathbb{R}^n$ and by Theorem 1.21, $H_n \cong \mathbb{R}^{l_n} \times \mathbb{Z}^{k_n}$. Inasmuch as $\mathbb{Z}^n \subset H_n \subset \mathbb{R}^n$, we have $l_n + k_n = n$. We verify that $k_n = 0$ for all n. Thus, the lemma will be proved.

Assume the contrary, i.e., $k_n > 0$ for some n. We use the following property of closed subgroups of the group $\mathbb{R}^m$: a closed subgroup G of the group $\mathbb{R}^m$ is a direct product of its closed subgroup G_1 and another closed subgroup G_2 if and only if G_1 is an intersection of G with a subspace of $\mathbb{R}^m$ [**16**, Chapter VII, §1]. It follows from $H_n = H_{n+1} \cap \mathbb{R}^n$ that H_n is a topological direct factor of H_{n+1}. Hence the group $\overline{\pi(D)}$ contains a subgroup F topologically isomorphic to $\mathbb{Z}$ as a direct factor, i.e., $\overline{\pi(D)} = F \times G$. Let e be a generator of the group F. Consider an arbitrary element $a_0 = e + g_0 \in (e + G) \cap \pi(D)$. Denote by M the subgroup of $\pi(D)$ generated by a_0. Then $M \cong \mathbb{Z}$ and $\pi(D) = M \times (\pi(D) \cap G)$. Thus, the group $\pi(D)$ contains a subgroup isomorphic to $\mathbb{Z}$ as a direct factor. Hence the group D contains a subgroup isomorphic to $\mathbb{Z}$ as a direct factor. We deduce from this that the group K contains a subgroup topologically isomorphic to the circle group $\mathbb{T}$, contrary to the assumption. □

REMARK 3.13. It was proved in Proposition 3.5 that the support of a symmetric Gaussian distribution on a locally compact Abelian group X is a connected closed subgroup G of X. It results from Propositions 3.7 and 3.11 that for every locally compact connected Abelian group X there exists a symmetric Gaussian distribution $\gamma \in \Gamma^s(X)$ such that $\sigma(\gamma) = X$. Indeed, since on the group $\mathbb{R}^m$ there exists a symmetric Gaussian distribution $\gamma \in \Gamma^s(\mathbb{R}^m)$ such that $\sigma(\gamma) = \mathbb{R}^m$, in view of Theorem 1.15, it suffices to prove this statement for a compact connected Abelian group. So, let us assume $X = K$, where K is a compact connected Abelian group, and let $D = K^*$. Let $\pi : D \to \mathbb{R}^l$ be given by formula (3.7) if $\dim(K) = l < \infty$, and let $\pi : D \to \mathbb{R}^{\aleph_0 *}$ be given by formula (3.17) if $\dim(K) = \infty$. In both cases the required distribution $\gamma \in \Gamma(K)$ is the distribution γ with the characteristic function

$$\hat{\gamma}(d) = \exp\{-\langle \pi d, \pi d \rangle\}, \quad d \in D.$$

This follows from the fact that $\hat{\gamma}(d) = 1$ if and only if $d = 0$ and Corollary 3.6 holds.

The following classical theorems by H. Cramér and J. Marcinkiewicz are well known.

CRAMÉR THEOREM ([**73**, Theorem 6.3.2]). *Let $\mu \in \Gamma(\mathbb{R}^n)$ and let $\mu = \mu_1 * \mu_2$, where $\mu_j \in \mathrm{M}^1(\mathbb{R}^n)$. Then $\mu_j \in \Gamma(\mathbb{R}^n)$, $j = 1, 2$.*

MARCINKIEWICZ THEOREM ([**73**, Theorem 2.5.1]). *Let $\mu \in \mathrm{M}^1(\mathbb{R})$. Assume that the characteristic function $\hat{\mu}(s)$ is of the form*

$$\hat{\mu}(s) = \exp\{\psi(s)\}, \quad s \in \mathbb{R},$$

where $\psi(s)$ is a polynomial. Then $\mu \in \Gamma(\mathbb{R})$.

The group analogues of Cramér's decomposition theorem for the Gaussian distribution and Marcinkiewicz's theorem play an important role in proving characterization theorems on locally compact Abelian groups.

THEOREM 3.14 ([**22**], see also [**36**, Theorem 4.6]). *Let X be a locally compact Abelian group containing no subgroups topologically isomorphic to the circle group $\mathbb{T}$. Let $\mu \in \Gamma(X)$ and let $\mu = \mu_1 * \mu_2$, where $\mu_j \in \mathrm{M}^1(X)$. Then $\mu_j \in \Gamma(X)$, $j = 1, 2$.*

THEOREM 3.15 ([**25**], see also [**36**, Theorem 5.11]). *Let X be a locally compact Abelian group with character group Y. Assume that X contains no subgroups topologically isomorphic to the circle group $\mathbb{T}$. Let $\mu \in \mathrm{M}^1(X)$ and let the characteristic function $\hat{\mu}(y)$ be of the form*

$$\hat{\mu}(y) = \exp\{\psi(y)\}, \quad y \in Y,$$

where $\psi(y)$ is a continuous polynomial. Then $\mu \in \Gamma(X)$.

Notes

The standard definition of the Gaussian distribution on a locally compact Abelian group was first introduced by K.R. Parthasarathy, R. Ranga Rao, and S.R.S. Varadhan in [**86**]. The definition given in [**86**] differs from Definition 3.1, but as proved in [**86**] both definitions are equivalent. In [**86**] the main properties of the Gaussian distributions were also studied. We note that in [**86**] along with other results the representation of the characteristic function of an infinitely divisible distribution on a locally compact Abelian group (the Lévy-Khinchin formula) was obtained. The results of the article [**86**] were included in the monographs [**66**] and [**85**].

Proposition 3.5, Remarks 3.2, 3.3, and 3.13 belong to K.R. Parthasarathy, R. Ranga Rao, and S.R.S. Varadhan [**86**]. We follow [**66**] in the proof of Lemma 3.4. We follow the article [**91**] by V.V. Sazonov and V.N. Tutubalin in the proof of Proposition 3.5. Propositions 3.7 and 3.11 were proved by G.M. Feldman in [**23**] (see also [**28**, §5]).

Gaussian distributions on the finite-dimensional torus $\mathbb{T}^n$ were studied by E. Siebert in [**93**] (see also [**66**, §5.5]). Gaussian distributions γ on the infinite-dimensional torus $\mathbb{T}^{\aleph_0}$ such that diagonal matrices A correspond to γ in the proof of Proposition 3.11 were studied by A. Bendikov in [**5, 6**], E. Siebert in [**93**], and Ch. Berg in [**12**] (see also [**7, 9, 13**]). Gaussian semigroups on infinite-dimensional locally compact groups that need not be Abelian were studied by A. Bendikov and A. Bendikov and L. Saloff-Coste in connection with the theory of potential on these groups constructed by them (see e.g. the monograph [**8**] by A. Bendikov and see also the articles [**10**] and [**11**], where one can find further references).

CHAPTER II

Independent random variables with independent sum and difference

4. Identically distributed random variables

The classical Kac–Bernstein theorem states: If ξ_1 and ξ_2 are independent random variables such that their sum $\xi_1+\xi_2$ and difference $\xi_1-\xi_2$ are also independent, then ξ_j are Gaussian.

Let X be a second countable locally compact Abelian group. In this section we study so-called Gaussian distributions in the sense of Bernstein. These are probability distributions μ on the group X with the following property: if ξ_1 and ξ_2 are independent identically distributed random variables with values in X and distribution μ, then their sum and difference are independent. By the Kac–Bernstein theorem, the class of Gaussian distributions in the sense of Bernstein on the real line coincides with the class of Gaussian distributions. Generally speaking, on the group X the class of Gaussian distributions in the sense of Bernstein is wider than the class of Gaussian distributions. In particular, the Haar distributions on compact Corwin subgroups of X belong to the class of Gaussian distributions in the sense of Bernstein.

In this section, for various classes of locally compact Abelian groups we solve the main problems on Gaussian distributions in the sense of Bernstein. We prove that if the connected component of the zero of a group X contains a finite number of elements of order 2, then every Gaussian distribution in the sense of Bernstein on X is a convolution of a Gaussian distribution, an idempotent distribution, and a signed measure supported in a subgroup of X generated by elements of order 2. Based on this result, we give a complete description of locally compact Abelian groups on which every Gaussian distribution in the sense of Bernstein is represented as a convolution of a Gaussian distribution and an idempotent distribution. These are groups whose connected component of the zero contains at most one element of order 2. Then we describe supports of Gaussian distributions in the sense of Bernstein on an arbitrary locally compact Abelian group X. They are cosets of subgroups of X of the form $\mathbb{R}^m \times K$, where $m \geq 0$ and K is a compact Corwin group. We also prove that if the connected component of the zero of a group X has finite dimension and μ is a Gaussian distribution in the sense of Bernstein on X with a nonvanishing characteristic function, then a zero-one law holds for μ.

DEFINITION 4.1. Let X be a locally compact Abelian group. A probability distribution μ on X is called a *Gaussian distribution in the sense of Bernstein* if μ has the property: if ξ_1 and ξ_2 are independent identically distributed random variables with values in the group X and distribution μ, then the sum $\xi_1 + \xi_2$ and difference $\xi_1 - \xi_2$ are independent.

Denote by $\Gamma_B(X)$ the set of Gaussian distributions in the sense of Bernstein on the group X.

LEMMA 4.2. *Let X be a locally compact Abelian group with character group Y. Let ξ_1 and ξ_2 be independent random variables with values in the group X and distributions μ_1 and μ_2. The sum $\xi_1+\xi_2$ and difference $\xi_1-\xi_2$ are independent if and only if the characteristic functions $\hat\mu_j(y)$ satisfy the equation*

$$\hat\mu_1(u+v)\hat\mu_2(u-v)=\hat\mu_1(u)\hat\mu_2(u)\hat\mu_1(v)\hat\mu_2(-v),\quad u,v\in Y. \tag{4.1}$$

PROOF. We use the fact that if ξ is a random variable with values in X and distribution μ, then $\hat\mu(y)=\mathbf{E}[(\xi,y)]$. Observe that if ξ_1 and ξ_2 are random variables with values in the group X, then ξ_1 and ξ_2 are independent if and only if the equality

$$\mathbf{E}[(\xi_1,u)(\xi_2,v)]=\mathbf{E}[(\xi_1,u)]\mathbf{E}[(\xi_2,v)] \tag{4.2}$$

holds for all $u,v\in Y$. It follows from (4.2) that $\xi_1+\xi_2$ and $\xi_1-\xi_2$ are independent if and only if

$$\mathbf{E}[(\xi_1+\xi_2,u)(\xi_1-\xi_2,v)]=\mathbf{E}[(\xi_1+\xi_2,u)]\mathbf{E}[(\xi_1-\xi_2,v)],\quad u,v\in Y. \tag{4.3}$$

Taking into account that the random variables ξ_1 and ξ_2 are independent, we transform the left-hand side of equality (4.3) as follows

$$\begin{aligned}\mathbf{E}[(\xi_1+\xi_2,u)(\xi_1-\xi_2,v)]&=\mathbf{E}[(\xi_1,u+v)(\xi_2,u-v)]\\&=\mathbf{E}[(\xi_1,u+v)]\mathbf{E}[(\xi_2,u-v)]=\hat\mu_1(u+v)\hat\mu_2(u-v),\quad u,v\in Y.\end{aligned}$$

Analogously, we transform the right-hand side of equality (4.3) as follows

$$\begin{aligned}&\mathbf{E}[(\xi_1+\xi_2,u)]\mathbf{E}[(\xi_1-\xi_2,v)]=\mathbf{E}[(\xi_1,u)(\xi_2,u)]\mathbf{E}[(\xi_1,v)(\xi_2,-v)]\\&=\mathbf{E}[(\xi_1,u)]\mathbf{E}[(\xi_2,u)]\mathbf{E}[(\xi_1,v)]\mathbf{E}[(\xi_2,-v)]=\hat\mu_1(u)\hat\mu_2(u)\hat\mu_1(v)\hat\mu_2(-v),\ u,v\in Y.\end{aligned}$$

□

Equation (4.1) is called the *Kac–Bernstein functional equation.*

COROLLARY 4.3. *Let X be a locally compact Abelian group with character group Y. A distribution μ belongs to the class $\Gamma_B(X)$ if and only if the characteristic function $\hat\mu(y)$ satisfies the equation*

$$\hat\mu(u+v)\hat\mu(u-v)=\hat\mu^2(u)|\hat\mu(v)|^2,\quad u,v\in Y. \tag{4.4}$$

It follows from (4.4) that the class $\Gamma_B(X)$ is a semigroup with respect to convolution.

REMARK 4.4. Let X be a locally compact Abelian group with character group Y. It is obvious that the characteristic function of any Gaussian distribution satisfies equation (4.4). By Corollary 4.3, this implies that $\Gamma(X)\subset\Gamma_B(X)$. Let K be a compact subgroup of the group X. Let us verify that $m_K\in\Gamma_B(X)$ if and only if K is a Corwin group.

Suppose K is a Corwin group. Verify that the characteristic function $\widehat{m}_K(y)$ satisfies equation (4.4). Then by Corollary 4.3, $m_K\in\Gamma_B(X)$. We have

$$\widehat{m}_K(u+v)\widehat{m}_K(u-v)=\widehat{m}_K^2(u)\widehat{m}_K^2(v),\quad u,v\in Y. \tag{4.5}$$

We use representation (2.3) of the characteristic function $\widehat{m}_K(y)$. If $u,v\in A(Y,K)$, then $u\pm v\in A(Y,K)$ and both sides of equation (4.5) are equal to 1. If either $u\in A(Y,K)$, $v\notin A(Y,K)$ or $v\in A(Y,K)$, $u\notin A(Y,K)$, then $u+v\notin A(Y,K)$ and both sides of equation (4.5) are equal to zero. If $u, v \notin A(Y, K)$, then the right-hand side of equation (4.5) is equal to zero. If the left-hand side of equation (4.5) is not equal to zero, we have $u \pm v \in A(Y, K)$. This implies that $2u \in A(Y, K)$. It follows from $K^{(2)} = K$ that then $u \in A(Y, K)$, contrary to the assumption. Hence the left-hand side of equation (4.5) is also equal to zero.

Let us prove the converse assertion. Suppose K is a compact subgroup of the group X and $m_K \in \Gamma_B(X)$. By Corollary 4.3, the characteristic function $\widehat{m}_K(y)$ satisfies equation (4.4) which takes the form (4.5). Substituting $u = v = y$ in (4.5) and taking into account (2.3), we get that if $2y \in A(Y, K)$ then $y \in A(Y, K)$. This implies that K is a Corwin group.

LEMMA 4.5. *Let X be a locally compact Abelian group. Let ξ_1 and ξ_2 be independent random variables with values in the group X and distributions μ_1 and μ_2. If the sum $\xi_1 + \xi_2$ and difference $\xi_1 - \xi_2$ are independent, then μ_j are supported in cosets of a subgroup M of X such that M is topologically isomorphic to a group of the form*

$$\mathbb{R}^m \times K, \tag{4.6}$$

where $m \geq 0$ and K is a compact Corwin group.

PROOF. Denote by Y the character group of the group X. Taking into account Theorem 1.14, we assume that $X = \mathbb{R}^m \times G$, $Y = \mathbb{R}^m \times H$, where $m \geq 0$, $H \cong G^*$, and each of the groups G and H contains a compact open subgroup. Denote by L a compact open subgroup of the group H. Put

$$N_1 = \{y \in Y : \hat{\mu}_1(y) \neq 0\}, \quad N_2 = \{y \in Y : \hat{\mu}_2(y) \neq 0\}, \quad N = N_1 \cap N_2.$$

By Lemma 4.2, the characteristic functions $\hat{\mu}_j(y)$ satisfy equation (4.1). It follows from (4.1) that N is a subgroup of Y. Obviously, N is an open subgroup.

Consider the intersection $B = N \cap L$. Since every open subgroup is closed, B is a compact open subgroup of H. Substituting $u = v = y$ and then $u = y$, $v = -y$ into equation (4.1), we obtain

$$\hat{\mu}_1(2y) = \hat{\mu}_1^2(y)|\hat{\mu}_2(y)|^2, \quad \hat{\mu}_2(2y) = |\hat{\mu}_1(y)|^2\hat{\mu}_2^2(y), \ y \in Y. \tag{4.7}$$

We deduce from (4.7) that for any natural n each of the functions $\hat{\mu}_j(y)$ satisfies the equation

$$|\hat{\mu}_j(2^n y)| = |\hat{\mu}_1(y)\hat{\mu}_2(y)|^{2^{2n-1}}, \quad y \in Y, \ j = 1, 2. \tag{4.8}$$

Let $y \in B$. Inasmuch as B is compact, there exists a convergent subsequence $2^{m_k}y \to y_0 \in B$. We obtain from (4.8) that

$$|\hat{\mu}_j(y_0)| = \lim_{k\to\infty} |\hat{\mu}_j(2^{m_k}y)| = \lim_{k\to\infty} |\hat{\mu}_1(y)\hat{\mu}_2(y)|^{2^{2m_k-1}}, \quad y \in Y, \ j = 1, 2. \tag{4.9}$$

If $|\hat{\mu}_1(y)\hat{\mu}_2(y)| < 1$, then the limit in the right-hand side of (4.9) is equal to zero, contrary to the fact that $y_0 \in B \subset N$. Hence $|\hat{\mu}_1(y)| = |\hat{\mu}_2(y)| = 1$ for all $y \in B$.

Taking into consideration statement (v) of Theorem 2.5, we can replace the distributions μ_j by their shifts λ_j in such a way that $\hat{\lambda}_j(y) = 1$ for all $y \in B$. The characteristic functions $\hat{\lambda}_j(y)$ also satisfy equation (4.1). Applying Proposition 2.10, we get $\sigma(\lambda_j) \subset A(X, B)$. It follows from Theorems 1.6 and 1.9 that $A(X, B) \cong (Y/B)^* = ((\mathbb{R}^m \times H)/B)^* \cong \mathbb{R}^m \times (H/B)^*$. Put $F = (H/B)^*$. Inasmuch as B is an open subgroup of H, the factor-group H/B is discrete. Hence by Theorem 1.2, F is a compact group.

Thus, we reduced the proof of the lemma to the case when $X = \mathbb{R}^m \times F$, $Y = \mathbb{R}^m \times D$, where F is a compact Abelian group, $D \cong F^*$, and μ_j are distributions on X with the characteristic functions satisfying equation (4.1). By Theorem 1.2, D is a discrete Abelian group. Let D_2 be the 2-component of D, i.e., the subgroup of D consisting of all elements $y \in D$ such that the order of y is a power of the number 2. Let $y \in D_2$. Then $2^n y = 0$ for some natural n. We deduce from (4.8) that $|\hat{\mu}_1(y)| = |\hat{\mu}_2(y)| = 1$ for all $y \in D_2$. Applying statement (v) of Theorem 2.5, we can replace the distributions μ_j by their shifts λ_j in such a way that $\hat{\lambda}_j(y) = 1$ for all $y \in D_2$. By Proposition 2.10, $\sigma(\lambda_j) \subset A(X, D_2)$. Put $M = A(X, D_2)$. It is obvious that $M \cong \mathbb{R}^m \times K$, where K is a compact Abelian group. It follows from Theorems 1.8 and 1.9 that $M^* \cong Y/D_2$. It is clear that the factor-group Y/D_2 contains no elements of order 2. By Theorem 1.12, we have $\overline{M^{(2)}} = M$. We conclude from $\overline{M^{(2)}} = M^{(2)}$ that M is a Corwin group. Hence K is also a Corwin group. Returning to the original distributions μ_j we obtain the required statement. □

LEMMA 4.6. *Let X be a locally compact Abelian group X of the form (4.6). Then $X_{(2)} \subset c_X$.*

PROOF. Denote by Y the character group of the group X. Then $Y \cong \mathbb{R}^m \times L$, where $L = K^*$. In order not to complicate the notation, suppose $Y = \mathbb{R}^m \times L$. By Theorem 1.2, L is a discrete Abelian group. In view of Theorem 1.12, it follows from $K^{(2)} = K$ that $L_{(2)} = \{0\}$. This means that the subgroup L contains no elements of order 2. This implies that the subgroup b_L consists of all elements of odd order of the group L. It follows from this that $(b_L)^{(2)} = b_L$ and hence $b_L \subset Y^{(2)}$. Taking into account that $b_Y = b_L$, we conclude from Theorems 1.10 and 1.12 that $c_X = A(X, b_Y) = A(X, b_L) \supset A(X, Y^{(2)}) = X_{(2)}$. □

LEMMA 4.7. *Let X be a locally compact Abelian group with character group Y. Let $\mu \in \Gamma_B(X)$ and assume that the characteristic function $\hat{\mu}(y)$ does not vanish. Then $\hat{\mu}(y)$ can be represented in the form*

$$\hat{\mu}(y) = l(y) \exp\{-\varphi(y)\}, \quad y \in Y, \tag{4.10}$$

where $l(y)$ is a continuous function satisfying the equation

$$l(u+v)l(u-v) = l^2(u), \quad u, v \in Y, \tag{4.11}$$

and the conditions

$$l(-y) = \overline{l(y)}, \quad |l(y)| = 1, \quad y \in Y, \quad l(0) = 1, \tag{4.12}$$

and $\varphi(y)$ is a continuous nonnegative function satisfying equation (3.2).

PROOF. By Corollary 4.3, the characteristic function $\hat{\mu}(y)$ satisfies equation (4.4). We conclude from equation (4.4) that

$$|\hat{\mu}(u+v)||\hat{\mu}(u-v)| = |\hat{\mu}(u)|^2 |\hat{\mu}(v)|^2, \quad u, v \in Y.$$

Take the logarithm of both sides of this equality and put $\varphi(y) = -\ln |\hat{\mu}(y)|$. We get that $\varphi(y)$ is a continuous nonnegative function satisfying equation (3.2) and $|\hat{\mu}(y)| = \exp\{-\varphi(y)\}$. Set $l(y) = \hat{\mu}(y)/|\hat{\mu}(y)|$. The function $l(y)$ is continuous and satisfies conditions (4.12). Moreover, the function $l(y)$ also satisfies equation (4.4). In view of (4.12), this equation is transformed into equation (4.11). Obviously, (4.10) is fulfilled. □

LEMMA 4.8. *Let X be a finite Abelian group with character group Y. Then for any function $f(y)$ on Y there is a complex measure δ on X such that*

$$\hat{\delta}(y) = f(y), \quad y \in Y. \tag{4.13}$$

If all nonzero elements of X have order 2 and $f(y)$ is a real-valued function, then δ is a signed measure.

PROOF. In view of $Y \cong X$, put $X = \{x_1, \dots, x_n\}$, $Y = \{y_1, \dots, y_n\}$. Let δ be a complex measure on X and let $\delta(x_j) = a_j$, $j = 1, 2, \dots, n$. Then the characteristic function $\hat{\delta}(y)$ is of the form

$$\hat{\delta}(y_i) = \sum_{j=1}^{n} a_j (x_j, y_i), \quad i = 1, 2, \dots, n.$$

If numbers $\hat{\delta}(y_i) = f(y_i)$ are given, then we have a system of linear equations to find a_j. Put $a_{ij} = (x_j, y_i)$, $A = (a_{ij})_{i,j=1}^{n}$. This system has a solution because $\det A \neq 0$. Thus, there is a complex measure δ such that (4.13) holds.

If all nonzero elements of X have order 2, then all $(x_j, y_i) = \pm 1$. Taking into account that $f(y)$ is a real-valued function, this implies that all numbers a_j are real, i.e., δ is a signed measure. □

Now we can prove the following decomposition theorem for Gaussian distributions in the sense of Bernstein.

THEOREM 4.9. *Let X be a locally compact Abelian group such that its connected component of the zero c_X contains a finite number of elements of order 2. Let $\mu \in \Gamma_B(X)$. Then*

$$\mu = \gamma * m_K * \delta, \tag{4.14}$$

where $\gamma \in \Gamma(X)$, K is a compact Corwin subgroup of X, and δ is a signed measure on the subgroup $(c_X)_{(2)}$ such that $\delta^{*2} = E_0$.

PROOF. Denote by Y the character group of the group X. By Lemma 4.5, the distribution μ is supported in a coset of a subgroup G of the group X, where G is topologically isomorphic to a group of the form (4.6). Hence we can replace μ by its shift and assume that the group X is of the form (4.6). Then, applying Lemma 4.6, we get that $X_{(2)} \subset c_X$. Hence the subgroup $X_{(2)}$ is finite. Moreover, $\overline{Y^{(2)}} = Y^{(2)}$. By Theorem 1.12, $A(Y, X_{(2)}) = \overline{Y^{(2)}} = Y^{(2)}$. Assume that the number of elements of the subgroup $X_{(2)}$ is n. In view of Theorems 1.9 and 1.12, we have $X_{(2)} \cong (X_{(2)})^* \cong Y/A(Y, X_{(2)}) = Y/Y^{(2)}$. Hence a $Y^{(2)}$-coset decomposition of the group Y is of the form

$$Y = \bigcup_{j=0}^{n-1} (y_j + Y^{(2)}), \quad y_0 = 0. \tag{4.15}$$

Consider the set $E = \{y \in Y : \hat{\mu}(y) \neq 0\}$. By Corollary 4.3, the characteristic function $\hat{\mu}(y)$ satisfies equation (4.4). It follows from (4.4) that the set E is an open subgroup of Y satisfying the condition $E \cap Y^{(2)} = E^{(2)}$. Denote by H a union of cosets in (4.15) for which $E \cap (y_j + Y^{(2)}) \neq \emptyset$. Changing if necessary the numeration of elements y_j, we can assume that

$$H = \bigcup_{j=0}^{l-1}(y_j + Y^{(2)}), \quad y_j \in E,\ j = 0, 1, \dots, l-1.$$

Since E is a subgroup of Y, we conclude that H is also a subgroup of Y. It is easily seen that $\overline{E^{(2)}} = E^{(2)}$ and an $E^{(2)}$-coset decomposition of the group E can be written in the form

$$E = \bigcup_{j=0}^{l-1}(y_j + E^{(2)}). \tag{4.16}$$

By Lemma 4.7, the restriction of the characteristic function $\hat{\mu}(y)$ to E is represented in the form (4.10), where the function $l(y)$ satisfies equation (4.11) on E. Let $u, v \in E$. Swapping u and v in (4.11) we get

$$l(u+v)l(v-u) = l^2(v), \quad u, v \in E. \tag{4.17}$$

On the one hand, multiplying (4.10) and (4.17) we obtain

$$l^2(u+v) = l^2(u)l^2(v), \quad u, v \in E. \tag{4.18}$$

Equality (4.18) means that the function $l^2(y)$ is a character of the group E. On the other hand, substituting $u = v = y$ in equation (4.17), we get

$$l(2y) = l^2(y), \quad y \in E. \tag{4.19}$$

It follows from (4.18) and (4.19) that the restriction of the function $l(y)$ to $E^{(2)}$ is a character of the group $E^{(2)}$. By Theorem 1.9, there exists an element $x_0 \in X$ such that $l(y) = (x_0, y)$ for all $y \in E^{(2)}$. Consider the function $m(y) = (-x_0, y)l(y)$, where $y \in E$. This function also satisfies equation (4.11) on E and the same conditions as the function $l(y)$ does. Moreover,

$$m(2y) = 1, \quad y \in E.$$

Equality (4.19) implies that $m(y) = \pm 1$ for all $y \in E$. Thus, we may conclude from (4.17) that

$$m(u+v)m(u-v) = 1, \quad u, v \in E,$$

and hence $m(u+v) = m(u-v)$ for all $u, v \in E$. This implies that

$$m(u+2v) = m(u), \quad u, v \in E.$$

Since the function $m(y)$ takes a constant value at each coset of the subgroup $E^{(2)}$ in E, it follows from (4.15) and (4.16) that the function $m(y)$ can be extended from E to Y as follows:

$$n(y_j + u) = m(y_j) \text{ for all } u \in Y^{(2)}, \quad j = 0, 1, \dots, l-1,$$

$$n(y_j + u) = 1 \text{ for all } u \in Y^{(2)}, \quad j = l, l+1, \dots, n-1.$$

The extended function $n(y)$ is invariant with respect to the subgroup $Y^{(2)}$. Hence $n(y)$ induces a function on the factor-group $Y/Y^{(2)}$. Note that by Theorems 1.9 and 1.12, we have $(X_{(2)})^* \cong Y/Y^{(2)}$. Taking this into account, it follows by Lemma 4.8 that there is a signed measure δ on $X_{(2)}$ such that $\hat{\delta}(y) = n(y)$ for all $y \in Y$. From what has been said above we get the following representation

$$\hat{\mu}(y) = \begin{cases} (x_0, y)\exp\{-\varphi(y)\}\hat{\delta}(y), & \text{if } y \in E, \\ 0, & \text{if } y \notin E. \end{cases} \tag{4.20}$$

It follows from $\hat{\delta}^2(y) = 1$ for all $y \in Y$ that $\delta^{*2} = E_0$. By Lemma 3.4, we can extend the function $\varphi(y)$ from E to Y retaining its properties. We will keep the notation $\varphi(y)$ for the extended function. Let γ be the Gaussian distribution on the group X with the characteristic function $\hat{\gamma}(y) = (x_0, y)\exp\{-\varphi(y)\}$, $y \in Y$. Thus, it follows from (2.3) and (4.20) that $\hat{\mu}(y) = \hat{\gamma}(y)\widehat{m}_K(y)\hat{\delta}(y)$. This implies that $\mu = \gamma * m_K * \delta$.

Note that the open subgroup E has the property $E \cap Y^{(2)} = E^{(2)}$. This implies that $K = A(X, E)$ is a compact Corwin subgroup. □

REMARK 4.10. Let X be a locally compact Abelian group with character group Y. Let $\mu \in \mathrm{M}^1(X)$. If μ is represented in the form (4.14), then $\mu \in \Gamma_B(X)$. Indeed, taking into account Corollary 4.3, it suffices to verify that the characteristic function $\hat{\mu}(y)$ satisfies equation (4.4). It follows from (3.2) that the characteristic function $\hat{\gamma}(y)$ satisfies equation (4.4). In view of Remark 4.4, the characteristic function $\widehat{m}_K(y)$ also satisfies equation (4.4). Since $\delta^{*2} = E_0$, we have $\hat{\delta}^2(y) = 1$ for all $y \in Y$. Taking into account that δ is concentrated on $X_{(2)}$, this implies that the characteristic function $\hat{\delta}(y)$ satisfies equation (4.4). It follows from what has been said that the characteristic function $\hat{\mu}(y)$ also satisfies equation (4.4) and hence $\mu \in \Gamma_B(X)$.

Based on Theorem 4.9, we can give a complete description of locally compact Abelian groups on which every Gaussian distribution in the sense of Bernstein is represented as a convolution of a Gaussian distribution and an idempotent distribution. To prove this result we need two lemmas.

LEMMA 4.11. *Let X be a locally compact Abelian group. If the connected component of the zero of the group X contains more than one element of order 2, then there exists a compact Corwin subgroup K of the group X such that the factor-group X/K contains a subgroup topologically isomorphic to the 2-dimensional torus $\mathbb{T}^2$.*

PROOF. Assume that for every compact Corwin subgroup G of the group X the factor-group X/G contains no subgroups topologically isomorphic to the 2-dimensional torus $\mathbb{T}^2$. Take a compact Corwin subgroup K of the group X and verify that $K_{(2)}$ is a finite group such that

$$|K_{(2)}| \le 2. \tag{4.21}$$

Put $L = K^*$. By Theorem 1.2, L is a discrete group. It follows from Theorem 1.12 that $A(L, K_{(2)}) = L^{(2)}$. Hence Theorem 1.9 yields that

$$(K_{(2)})^* \cong L/A(L, K_{(2)}) = L/L^{(2)}.$$

If we prove that

$$|L/L^{(2)}| \le 2, \tag{4.22}$$

then it will follow from this that $(K_{(2)})^*$ is a finite group. Hence the group $K_{(2)}$ is also finite and the equality $|K_{(2)}| = |(K_{(2)})^*|$ holds. Thus, (4.21) will be proved.

Inequality (4.22) will be proved if we check that for all $y_1, y_2 \in L$ such that $y_1, y_2 \notin L^{(2)}$ the inclusion $y_1 - y_2 \in L^{(2)}$ is fulfilled. Assume the contrary. Then there exist elements $y_1, y_2 \in L$ such that $y_1, y_2, y_1 - y_2 \notin L^{(2)}$. Denote by M the subgroup of L generated by the elements y_1 and y_2. Inasmuch as $K^{(2)} = K$, by Theorem 1.12, we have that $L_{(2)} = 0$. Hence all nonzero elements of finite order in the group L have odd order. This implies that if $y \in L$, $y \notin L^{(2)}$, then y is an element of infinite order. Since M is a finite generated group, by Theorem 1.27, we have $M = M_1 \times M_2$, where each of the subgroups M_j is isomorphic to either $\mathbb{Z}$ or $\mathbb{Z}(n_j)$. Therefore, there are the following three possibilities: $M \cong \mathbb{Z}$, $M \cong \mathbb{Z} \times \mathbb{Z}(n)$, $M \cong \mathbb{Z}^2$. Consider each of these cases.

1. $M \cong \mathbb{Z}$. We have $y_1 = n_1 e$, $y_2 = n_2 e$, where $e \in L$, e is an element of infinite order, and n_1, n_2 are integers. Inasmuch as $y_1, y_2 \notin L^{(2)}$, the integers n_1 and n_2 are odd. Hence $y_1 - y_2 = (n_1 - n_2)e \in L^{(2)}$, contrary to the assumption. Thus, case 1 is impossible.

2. $M \cong \mathbb{Z} \times \mathbb{Z}(n)$. We have $y_1 = n_1 e + l_1 a$, $y_2 = n_2 e + l_2 a$, where $e, a \in L$, e is an element of infinite order, a is an element of order n, and n_j, l_j are integers. As has been noted above, all nonzero elements of finite order in the group L have odd order. Hence $a = 2b$, $b \in L$. In view of $y_1, y_2, \notin L^{(2)}$, this implies that the integers n_1 and n_2 are odd. Hence $y_1 - y_2 = (n_1 - n_2)e + 2(l_1 - l_2)b \in L^{(2)}$, contrary to the assumption. Thus, case 2 is also impossible.

3. $M \cong \mathbb{Z}^2$. Let $y \in L$ be an element such that $2y \in M$. Then

$$2y = k_1 y_1 + k_2 y_2, \tag{4.23}$$

where k_1, k_2 are integers.

Assume that for each $y \in L$ such that $2y \in M$ the numbers k_1 and k_2 in (4.23) are even. In view of $L_{(2)} = \{0\}$, it follows from (4.23) that $y \in M$. Thus, if $2y \in M$, then $y \in M$. This implies that the annihilator $A(K, M)$ is a compact Corwin subgroup. By Theorems 1.8 and 1.9, $K/A(K, M) \cong M^*$. We have $M^* \cong \mathbb{T}^2$ and $K/A(K, M)$ is a subgroup of $X/A(K, M)$. We see that a factor-group of X by a Corwin subgroup contains a subgroup topologically isomorphic to the 2-dimensional torus $\mathbb{T}^2$, contrary to the assumption. Hence there is an element $y \in L$ such that $2y \in M$ and at least one of the numbers k_j in (4.23) is odd.

Assume $k_1 = 2m_1 + 1$, $k_2 = 2m_2$, where m_j are integers. Then we have $y_1 = 2y - 2m_1 y_1 - 2m_2 y_2 \in L^{(2)}$ contrary to the assumption. Arguing as above we obtain that the case $k_1 = 2m_1$, $k_2 = 2m_2 + 1$ is also impossible. Assume $k_1 = 2m_1 + 1$, $k_2 = 2m_2 - 1$. Then $y_1 - y_2 = 2y - 2m_1 y_1 - 2m_2 y_2 \in L^{(2)}$ contrary to the assumption. Thus, case 3 is also impossible. We proved that inequality (4.22) is fulfilled. Hence inequality (4.21) also holds.

We have by Theorem 1.15, $c_X \cong \mathbb{R}^m \times B$, where B is a compact connected Abelian group. By Theorem 1.13, B is a Corwin group. We deduce from (4.21) that $|B_{(2)}| \leq 2$. Hence $|(c_X)_{(2)}| \leq 2$. The obtained contradiction proves the lemma. □

LEMMA 4.12. *Consider the 2-dimensional torus $\mathbb{T}^2$ with character group $\mathbb{Z}^2$. Then there exists a distribution $\mu \in \Gamma_B(\mathbb{T}^2)$ such that the characteristic function of μ does not vanish and $\mu \notin \Gamma(\mathbb{T}^2)$.*

PROOF. Consider on the group $\mathbb{Z}^2$ the function

$$\psi(m, n) = -\sigma(m^2 + n^2) + i\pi mn, \quad (m, n) \in \mathbb{Z}^2.$$

Choose $\sigma > 0$ in such a way that the inequality

$$\sum_{(m,n)\in\mathbb{Z}^2} \exp\{-\sigma(m^2 + n^2)\} < 2.$$

holds. It follows from this that the inequality

$$f(t, s) = \sum_{(m,n)\in\mathbb{Z}^2} \exp\{\psi(m, n) - i(mt + ns)\} > 0, \quad (t, s) \in \mathbb{R}^2,$$

is fulfilled. It is obvious that

$$\frac{1}{4\pi^2}\int_{-\pi}^{\pi}\int_{-\pi}^{\pi} f(t,s)dtds = 1.$$

Denote by (e^{it}, e^{is}), where $(t,s) \in \mathbb{R}^2$, elements of the group $\mathbb{T}^2$. Let μ be the distribution on the group $\mathbb{T}^2$ with the density $r(e^{it}, e^{is}) = f(t,s)$ with respect to $m_{\mathbb{T}^2}$. Then

$$\hat{\mu}(m,n) = \exp\{\psi(m,n)\}, \quad (m,n) \in \mathbb{Z}^2.$$

Since the function $\exp\{i\pi mn\}$ is not a character, we have $\mu \notin \Gamma(\mathbb{T}^2)$. It is easy to check that the characteristic function $\hat{\mu}(m,n)$ satisfies equation (4.4). By Corollary 4.3, $\mu \in \Gamma_B(\mathbb{T}^2)$. □

REMARK 4.13. Observe that the distribution $\mu \in \Gamma_B(\mathbb{T}^2)$ constructed in Lemma 4.12 illustrates Theorem 4.9. Denote by γ the Gaussian distribution on the 2-dimensional torus $\mathbb{T}^2$ with the characteristic function

$$\hat{\gamma}(m,n) = \exp\{-\sigma(m^2+n^2)\}, \quad (m,n) \in \mathbb{Z}^2.$$

Consider the signed measure δ on $\mathbb{T}^2$ of the form

$$\delta\{(1,1)\} = \delta\{(1,-1)\} = \delta\{(-1,1)\} = \frac{1}{2}, \quad \delta\{(-1,-1)\} = -\frac{1}{2}.$$

Then δ is concentrated on $\mathbb{T}^2_{(2)}$ and $\delta^{*2} = E_{(1,1)}$. It is easy to see that

$$\hat{\delta}(m,n) = \exp\{i\pi mn\}, \quad (m,n) \in \mathbb{Z}^2.$$

Inasmuch as $\hat{\mu}(m,n) = \hat{\gamma}(m,n)\hat{\delta}(m,n)$ for all $(m,n) \in \mathbb{Z}^2$, we have $\mu = \gamma * \delta$.

Put

$$I_B(X) = \Gamma_B(X) \cap I(X).$$

Now we can describe all locally compact Abelian groups such that the equality

$$\Gamma_B(X) = \Gamma(X) * I_B(X) \tag{4.24}$$

holds.

THEOREM 4.14. *Let X be a locally compact Abelian group and let $\mu \in \Gamma_B(X)$. The distribution μ is represented in the form*

$$\mu = \gamma * m_K,$$

where $\gamma \in \Gamma(X)$ and K is a compact Corwin subgroup of X, if and only if the connected component of the zero of the group X contains no more than one element of order 2.

PROOF. Denote by Y the character group of the group X. Necessity. Assume that the connected component of the zero of the group X contains more than one element of order 2. Then by Lemma 4.11, there exists a compact Corwin subgroup K of the group X such that the factor-group X/K contains a subgroup F topologically isomorphic to the 2-dimensional torus $\mathbb{T}^2$. Therefore, we can consider the distribution $\mu \in \Gamma_B(\mathbb{T}^2)$, $\mu \notin \Gamma(\mathbb{T}^2)$, constructed in Lemma 4.12, as a distribution on the factor-group X/K. We also denote this distribution by μ. We may suppose that the characteristic function $\hat{\mu}(y)$ is defined on the annihilator $A(Y, K)$ because by Theorem 1.9, $(X/K)^* \cong A(Y, K)$. Consider on the group Y the function

$$h(y) = \begin{cases} \hat{\mu}(y) & \text{if} \quad y \in A(Y, K), \\ 0 & \text{if} \quad y \notin A(Y, K). \end{cases}$$

Since $A(Y, K)$ is a subgroup and $\hat{\mu}(y)$ is a positive definite function, by Proposition 2.9 $h(y)$ is also a positive definite function. By Theorem 1.11, the annihilator $A(Y, K)$ is an open subgroup, because K is a compact subgroup. Hence the function $h(y)$ is continuous. By Bochner's theorem, there exists a distribution $\lambda \in \mathrm{M}^1(X)$ such that $\hat{\lambda}(y) = h(y)$.

We will check that $\lambda \in \Gamma_B(X)$. In view of Corollary 4.3, it suffices to show that the characteristic function $\hat{\lambda}(y)$ satisfies equation (4.4). Consider 3 cases:

1. $u, v \in A(Y, K)$. Then (4.4) holds, because the function $\hat{\mu}(y)$ satisfies equation (4.4).

2. Either $u \in A(Y, K)$, $v \notin A(Y, K)$ or $v \in A(Y, K)$, $u \notin A(Y, K)$. Then both sides of equation (4.4) are equal to zero.

3. $u, v \notin A(Y, K)$. Then the right-hand side of equation (4.4) is equal to zero. If the left-hand side of equation (4.4) is not equal to zero, then $u \pm v \in A(Y, K)$. Hence $2u \in A(Y, K)$. Since K is a compact Corwin group, this implies that $u \in A(Y, K)$, that contradicts the assumption. Thus, the left-hand side of equation (4.4) is also equal to zero.

Thus, we proved that equation (4.4) becomes an equality for all $u, v \in Y$. Hence $\lambda \in \Gamma_B(X)$. Inasmuch as $\hat{\mu}(m, n) \neq 0$ for all $(m, n) \in \mathbb{Z}^2$ and $\mu \notin \Gamma(\mathbb{T}^2)$, it follows from this that $\lambda \notin \Gamma(X) * I(X)$. The necessity is proved.

Sufficiency. Since the subgroup c_X contains no more than one element of order 2, by Theorem 4.9, $\mu = \gamma * m_K * \delta$, where γ is a Gaussian distribution, K is a compact Corwin subgroup of X, and δ is a signed measure on $(c_X)_{(2)}$ such that $\delta^{*2} = E_0$. We have $\hat{\delta}^2(y) = 1$ for all $y \in Y$. Hence $\hat{\delta}(y) = \pm 1$ for all $y \in Y$. It is obvious that $\hat{\delta}(0) = 1$. By the condition of the theorem, either $(c_X)_{(2)} = \{0\}$ or $(c_X)_{(2)} \cong \mathbb{Z}(2)$. If $(c_X)_{(2)} = \{0\}$, then $\delta = E_0$. Let $(c_X)_{(2)} \cong \mathbb{Z}(2)$. Taking into account that δ is concentrated on $(c_X)_{(2)}$, this implies that $\hat{\delta}(y)$ is a character of the group Y. Hence δ is a degenerate distribution. □

EXAMPLE 4.15. An example of a locally compact Abelian group for which Theorem 4.14 holds is a locally compact Abelian group with the connected component of the zero of dimension 1. Indeed, it follows from Theorem 1.15 that if $\dim(c_X) = 1$, then either $c_X \cong \mathbb{R}$ or c_X is compact. Assume that c_X is compact. Denote by H the character group of the group c_X. By Theorems 1.2–1.4, H is a discrete torsion-free Abelian group of torsion-free rank 1. Observe also that a torsion-free Abelian group of torsion-free rank 1 is isomorphic to a subgroup of the group $\mathbb{Q}$. Hence either $H \cong \mathbb{Z}$ or $H \cong H_{\boldsymbol{a}}$ for some $\boldsymbol{a} = (a_0, a_1, \dots, a_n, \dots)$. In view of Theorem 1.1, this implies that either $c_X \cong \mathbb{T}$ or $c_X \cong \Sigma_{\boldsymbol{a}}$. In each of these cases the subgroup c_X contains no more then one element of order 2.

We supplement Theorem 4.14 with the following statement.

PROPOSITION 4.16. *Let X be a locally compact Abelian group. Every Gaussian distribution in the sense of Bernstein on X with a nonvanishing characteristic* *function is a Gaussian distribution if and only if the group X has no subgroups topologically isomorphic to the 2-dimensional torus $\mathbb{T}^2$.*

PROOF. The necessity follows from Lemma 4.12.

Sufficiency. Denote by Y the character group of the group X. By Lemma 4.7, the characteristic function $\hat{\mu}(y)$ is represented in the form (4.10). Put $\nu = \mu * \bar{\mu}$. Then $\hat{\nu}(y) = \exp\{-2\varphi(y)\}$ and hence $\nu \in \Gamma^s(X)$. Proposition 3.5 yields that $\sigma(\nu)$ is a closed connected subgroup of the group X. Since μ is a factor of ν, applying Proposition 2.1 we can assume that the group X is connected. There are two possibilities: either the group X contains no subgroups topologically isomorphic to the circle group $\mathbb{T}$ or X contains such subgroup.

1. Let the group X contain no subgroups topologically isomorphic to the circle group $\mathbb{T}$. Inasmuch as μ is a factor of ν, it follows from Theorem 3.14 that $\mu \in \Gamma(X)$. In this case the proposition is proved.

2. Let the group X contain a subgroup F topologically isomorphic to the circle group $\mathbb{T}$. By Theorem 1.22, the subgroup F is a topological direct factor of the group X, i.e., $X = F \times G$, where G is a connected Abelian group. By the assumption G has no subgroups topologically isomorphic to the circle group $\mathbb{T}$. By Theorem 1.15, $G \cong \mathbb{R}^m \times K$, where $m \geq 0$ and K is a compact connected Abelian group. We restrict ourselves to the case $m = 0$. The general case can be considered similarly. We assume for definiteness that $\dim(K) = \infty$. In the case when $\dim(K) < \infty$ the proof is simplified. We can suppose $Y = L \times D$, where $L \cong \mathbb{Z}$ and $D \cong K^*$. Applying Theorems 1.2 and 1.3, we conclude that D is a discrete torsion-free Abelian group. By Theorem 1.4, $r_0(D) = \infty$.

By Lemma 4.7, the characteristic function $\hat{\mu}(y)$ can be represented in the form (4.10). The proposition will be proved if we verify that the function $l(y)$ in (4.10) is a character of the group Y, i.e., the function $l(y)$ satisfies the equation

$$l(u+v) = l(u)l(v), \quad u, v \in Y. \tag{4.25}$$

Denote by $y = (n, d)$, where $n \in \mathbb{Z}$, $d \in D$, elements of the group Y. Substituting $u = v = y$ in equation (4.11) and taking into account that $l(0) = 1$, we obtain

$$l(2y) = l^2(y), \quad y \in Y. \tag{4.26}$$

It follows from (4.26) and (4.11) by induction that $l(ny) = l^n(y)$ for all $y \in Y$, $n \in \mathbb{Z}$. Hence the function $l(n, 0)$, $n \in \mathbb{Z}$, satisfies equation (4.25). By Corollary 4.3, the characteristic function $\hat{\mu}(y)$ satisfies equation (4.4). Consider the restriction of equation (4.4) to the subgroup D. By Bochner's theorem, $\hat{\mu}(0, d)$, $d \in D$, is the characteristic function of a distribution $\lambda \in \mathrm{M}^1(K)$. Since the function $\hat{\mu}(0, d)$ also satisfies equation (4.4), by Corollary 4.3, $\lambda \in \Gamma_B(K)$. Taking into account that the group K contains no subgroups topologically isomorphic to the circle group $\mathbb{T}$, it follows from case 1 that $\lambda \in \Gamma(K)$. This yields that the function $l(0, d)$, $d \in D$, satisfies equation (4.25).

Consider on the group Y the function

$$p(n, d) = l(n, 0)l(0, d), \quad (n, d) \in Y.$$

Taking into account what has been said above, the function $p(n, d)$ satisfies equation (4.25). Put

$$q(n, d) = l(n, d)/p(n, d), \quad (n, d) \in Y.$$

It is obvious that the function $q(n,d)$ satisfies equation (4.11). We verify that $q(n,d) = 1$ for all $(n,d) \in Y$. Thus, the proposition will be proved.

We have $q(n,0) = q(0,d) = 1$ for all $n \in \mathbb{Z}$, $d \in D$. Substituting $u = (n,0)$, $v = (n,d)$ in equation (4.10), we get

$$q(2n,d)q(0,-d) = q^2(n,0).$$

We deduce from this that $q(2n,d) = 1$ for all $(n,d) \in Y$. In particular, $q(2n,2d) = 1$ for all $(n,d) \in Y$. Inasmuch as the function $q(n,d)$ satisfies equation (4.11), the function $q(n,d)$ satisfies equation (4.26), i.e., $q(2n,2d) = q^2(n,d)$. This implies that $q(n,d) = \pm 1$ for all $(n,d) \in Y$. Let us prove that $q(n,d) = 1$ for all $(n,d) \in Y$.

Assume that there exists an element $(n_0, d_0) \in Y$ such that $q(n_0,d_0) = -1$. Choose in D a maximal independent set of elements $\{d_1, \dots, d_l, \dots\}$ and let $\pi : D \to \mathbb{R}^{\aleph_0 *}$ be the homomorphism defined by (3.17). Since the group K contains no subgroups topologically isomorphic to the circle group $\mathbb{T}$, there is a subgroup B of D of finite rank, say m, such that $d_0 \in B$ and the subgroup $\overline{\pi(B)}$ is topologically isomorphic to $\mathbb{R}^m$. Denote by $\tau : \overline{\pi(B)} \to \mathbb{R}^m$ this topological isomorphism. Put $\theta = \tau\pi$. Then $\overline{\theta(B)} = \mathbb{R}^m$.

Denote by $\|t\|$ a norm of the vector $t \in \mathbb{R}^m$. Take $\varepsilon > 0$. Consider the point $s_0 = (\theta d_0)/2 \in \mathbb{R}^m$ and choose an element $a \in B$ such that

$$\|s_0 - \theta a\| < \varepsilon/2.$$

Substitute $u = (n_0, d_0 - a)$, $v = (0,a)$ in equation (4.11). Taking into account that $q^2(n,d) = 1$ for all $(n,d) \in B$, we obtain

$$q(n_0,d_0)q(n_0,d_0 - 2a) = 1.$$

It follows from this that $q(n_0, d_0 - 2a) = -1$. Moreover,

$$\|\theta(d_0 - 2a)\| = \|\theta d_0 - 2\theta a\| = 2\|s_0 - \theta a\| < \varepsilon.$$

Set $b = d_0 - 2a$. Thus, we proved that for any $\varepsilon > 0$ there exists an element $(n_0, b) \in L \times B$ such that $q(n_0,b) = -1$ and $\|\theta b\| < \varepsilon$.

Since $p(n,d)$ is a character of the group Y, the function

$$q(n,d)\exp\{-\varphi(n,d)\} = \hat{\mu}(y)/p(n,d)$$

is a positive definite function on Y. Denote by $g(n,d)$ its restriction to the subgroup $L \times B$. Then $g(n,d)$ is also a positive definite function. As follows from the proof of Proposition 3.7, the function $\varphi(n,d)$ on the subgroup $L \times B$ is represented in the form

$$\varphi(n,d) = \sigma n^2 + 2n\langle t, \theta d\rangle + \langle A\theta d, \theta d\rangle, \tag{4.27}$$

where $\sigma \geq 0$, $t \in \mathbb{R}^m$, and $A = (\alpha_{i,j})_{i,j=1}^m$ is a symmetric positive semidefinite matrix.

By Bochner's theorem, $g(n,d)$ is a characteristic function. Applying statement (vii) of Theorem 2.5 to the function $g(n,d)$, we get

$$|g(n_0,b) - g(n_0,0)|^2 \leq 2(1 - g(0,b)).$$

Taking into account (4.27), we find from this the following inequality

$$\begin{aligned}(4.28)\quad &\left|-\exp\{-\sigma n_0^2 - 2n_0\langle t,\theta b\rangle - \langle A\theta b,\theta b\rangle\} - \exp\{-\sigma n_0^2\}\right|^2 \\ &= \exp\{-2\sigma n_0^2\}\left|\exp\{-2n_0\langle t,\theta b\rangle - \langle A\theta b,\theta\rangle\} + 1\right|^2 \\ &\le 2(1-\exp\{-\langle A\theta b,\theta b\rangle\}).\end{aligned}$$

But inequality (4.28) is impossible if the norm $\|\theta b\|$ is small enough. The obtained contradiction proves that $q(n,d)=1$ for all $(n,d)\in Y$. We conclude from this that the function $l(n,d)=p(n,d)$ satisfies equation (4.25). □

Let us now prove a theorem describing supports of Gaussian distributions in the sense of Bernstein on an arbitrary locally compact Abelian group. The proof is based on Theorem 4.9 and on the following lemma.

LEMMA 4.17. *Let*

$$X=\mathbb{R}^l\times\mathbb{T}^m\times F,$$

where $l\ge 0$, $m\ge 0$, and F is a finite Corwin group. Let a distribution μ on the group X be represented in the form

$$\mu=\gamma*m_K*\delta, \tag{4.29}$$

where $\gamma\in\Gamma^s(\mathbb{R}^l\times\mathbb{T}^m)$, K is a compact Corwin subgroup of X, δ is a signed measure on $X_{(2)}$ such that $\delta^{*2}=E_0$. Assume X is the minimal closed subgroup containing $\sigma(\mu)$. Then $\sigma(\mu)=X$.

PROOF. We first make the following remark. By Proposition 3.5, the support $\sigma(\gamma)$ is a closed connected subgroup of X. Inasmuch as $\sigma(m_K)=K$, we have $\sigma(\gamma*m_K)=\sigma(\gamma)+K$. Since F is a finite Corwin group, F has no elements of order 2. This implies that $X_{(2)}\subset\mathbb{T}^m$. Since X is the minimal closed subgroup containing $\sigma(\mu)$, it follows from this that $\sigma(\gamma)+K=X$. Hence $\sigma(\gamma*m_K)=X$. Generally speaking, if the support of a distribution is the whole group, then the support of the convolution of this distribution with a signed measure can have "holes", i.e., the support of the convolution need not be the whole group. We prove that it is impossible in our case.

Note that K is a subgroup of $\mathbb{T}^m\times F$. This implies that K can be represented as a direct product of two groups

$$K=A\times B,$$

where $A\subset\mathbb{T}^m$, $A\cong\mathbb{T}^a$, $a\ge 0$, and B is a finite group. We have $m_K=m_A\times m_B=m_A*m_B$. Represent the group $\mathbb{T}^m$ in the form $\mathbb{T}^m=\mathbb{T}^{m-a}\times\mathbb{T}^a$. Observe now that since $A\cong\mathbb{T}^a$, by Theorem 1.22, the subgroup A is a direct factor of $\mathbb{T}^m$. For this reason without loss of generality, we can assume that $A=\mathbb{T}^a$ and hence

$$m_K=m_{\mathbb{T}^a}*m_B.$$

Denote by Y the character group of the group X and by $(s_1,\dots,s_l,n_1,\dots,n_m)$, where $s_j\in\mathbb{R}$, $n_i\in\mathbb{Z}$, elements of the group $\mathbb{R}^l\times\mathbb{Z}^m\cong(\mathbb{R}^l\times\mathbb{T}^m)^*$. In view of (2.3), the characteristic function $\widehat{m}_{\mathbb{T}^a}(y)$ is of the form

$$\widehat{m}_{\mathbb{T}^a}(y)=\begin{cases}1 & \text{if}\quad y\in A(Y,\mathbb{T}^a),\\ 0 & \text{if}\quad y\notin A(Y,\mathbb{T}^a).\end{cases}$$

This implies that $\gamma * m_{\mathbb{T}^a} = \gamma_1 * m_{\mathbb{T}^a}$, where γ_1 is a symmetric Gaussian distribution on the group $\mathbb{R}^l \times \mathbb{T}^{m-a}$ with the characteristic function

$$\hat{\gamma}_1(s_1, \dots, s_l, n_1, \dots, n_{m-a}) = \hat{\gamma}(s_1, \dots, s_l, n_1, \dots, n_{m-a}, 0, \dots, 0).$$

Hence $\gamma * m_K = \gamma_1 * m_K$. For this reason we can assume that in (4.29) $\gamma \in \Gamma^s(\mathbb{R}^l \times \mathbb{T}^{m-a})$.

Since F is a finite Abelian group, by Theorem 1.27, we have $F \cong \mathbb{Z}(k_1) \times \cdots \times \mathbb{Z}(k_n)$. In order not to complicate the notation, suppose

$$F = \mathbb{Z}(k_1) \times \cdots \times \mathbb{Z}(k_n).$$

Embed the group F into the group $\mathbb{T}^n$ in the natural way and consider the group X as a subgroup of the group $\mathbb{R}^l \times \mathbb{T}^m \times \mathbb{T}^n$. Denote by $(t_1, \dots, t_l, u_1, \dots, u_m, v_1, \dots, v_n)$ and $(t_1, \dots, t_l, e^{iu_1}, \dots, e^{iu_m}, e^{iv_1}, \dots, e^{iv_n})$, where $t_j, u_j, v_j \in \mathbb{R}$, elements of the groups $\mathbb{R}^l \times \mathbb{R}^m \times \mathbb{R}^n$ and $\mathbb{R}^l \times \mathbb{T}^m \times \mathbb{T}^n$ respectively. Let p be the continuous homomorphism

$$p : \mathbb{R}^l \times \mathbb{R}^m \times \mathbb{R}^n \to \mathbb{R}^l \times \mathbb{T}^m \times \mathbb{T}^n,$$

defined by the formula

$$p(t_1, \dots, t_l, u_1, \dots, u_m, v_1, \dots, v_n) = (t_1, \dots, t_l, e^{iu_1}, \dots, e^{iu_m}, e^{iv_1}, \dots, e^{iv_n}).$$

Let $\alpha \in \Gamma^s(\mathbb{R}^l \times \mathbb{R}^{m-a})$ be the Gaussian distribution such that

$$\gamma = p(\alpha).$$

Put $L = \sigma(\alpha)$. Then L is a linear subspace in $\mathbb{R}^l \times \mathbb{R}^{m-a} \subset \mathbb{R}^l \times \mathbb{R}^m \times \mathbb{R}^n$. Hence L is a subgroup of $\mathbb{R}^l \times \mathbb{R}^m \times \mathbb{R}^n$.

Consider in $\mathbb{R}^a \subset \mathbb{R}^m$ the a-cube

$$C = \{(u_1, \dots, u_a) : \ 0 \le u_j < 2\pi, \ j = 1, 2, \dots, a\}.$$

Denote by Λ the Lebesgue measure on the a-cube C normalized in such a way that $\Lambda(C) = 1$. It is obvious that

$$m_{\mathbb{T}^a} = p(\Lambda).$$

Inasmuch as B is a finite Abelian group, by Theorem 1.27, we have

$$B = B_1 \times \cdots \times B_k,$$

where $B_j \cong \mathbb{Z}(l_j)$, $j = 1, 2, \dots, k$. Then there exist subgroups $T_j \subset \mathbb{T}^m \times \mathbb{T}^n$ such that $B_j \subset T_j$, $T_j \cong \mathbb{T}$, $j = 1, 2, \dots, k$. We have

$$m_B = m_{B_1} \times \cdots \times m_{B_k} = m_{B_1} * \cdots * m_{B_k}.$$

Denote by R_j the connected component of the zero of the group $p^{-1}(T_j)$. It is obvious that $R_j \cong \mathbb{R}$. Put $Z_j = R_j \cap p^{-1}(B_j)$, $j = 1, 2, \dots, k$. Then $Z_j \cong \mathbb{Z}$. Let e_j be a generatrix of the group Z_j. Consider the set $D_j = \{0, e_j, 2e_j, \dots, (l_j - 1)e_j\}$. Denote by ω_j the uniform distribution on D_j. It is easy to see that $m_{B_j} = p(\omega_j)$. Put

$$\omega = \omega_1 * \cdots * \omega_k.$$

Then

$$m_B = p(\omega).$$

Denote by P the subgroup of $\mathbb{R}^l \times \mathbb{R}^m \times \mathbb{R}^n$ generated by the subgroups L, $\mathbb{R}^a$, and $Z_1, \dots, Z_k$. Then the subgroup P is topologically isomorphic to a group of the form $\mathbb{R}^{\widetilde{l}} \times \mathbb{Z}^{\widetilde{m}}$, where $\widetilde{l} \ge 0$, $\widetilde{m} \ge 0$. Put $U = P \cap \operatorname{Ker} p$, $V = P/U$. Let τ be the natural homomorphism

$$\tau : P \to V.$$

It is obvious that $V \cong \mathbb{R}^p \times \mathbb{T}^d \times G$, where G is a finite Abelian group. Observe also that G is a Corwin group. In order not to complicate the notation, suppose $V = \mathbb{R}^p \times \mathbb{T}^d \times G$. It is easy to see that we can replace the group V by a topologically isomorphic group (this means the change of variables in the natural notation of elements of V) in such a way that

$$V = \mathbb{R}^p \times \mathbb{T}^q \times \mathbb{T}^a \times G.$$

Moreover, $\sigma(\tau(\alpha)) = \mathbb{R}^p \times \mathbb{T}^q$ and $m_{\mathbb{T}^a} = \tau(\Lambda)$. Put $\lambda = \tau(\alpha)$. The Gaussian distribution λ and a Haar measure $m_{\mathbb{R}^p\times\mathbb{T}^q}$ are mutually absolutely continuous. Denote by $(t_1, \dots, t_p, e^{is_1}, \dots, e^{is_q}, e^{iu_1}, \dots, e^{iu_a})$, where $t_j, s_j, u_j \in \mathbb{R}$, elements of the group $\mathbb{R}^p \times \mathbb{T}^q \times \mathbb{T}^a$. Let $f(t_1, \dots, t_p, e^{is_1}, \dots, e^{is_q})$ be the density of λ with respect to $m_{\mathbb{R}^p\times\mathbb{T}^q}$. It is not difficult to verify that $f(t_1, \dots, t_p, e^{is_1}, \dots, e^{is_q})$ is an entire function in variables t_i, s_j. It is easy to see that the distribution $\lambda * m_{\mathbb{T}^a}$ has the density

$$g(t_1, \dots, t_p, e^{is_1}, \dots, e^{is_q}, e^{iu_1}, \dots, e^{iu_a}) = f(t_1, \dots, t_p, e^{is_1}, \dots, e^{is_q}) \tag{4.30}$$

with respect to $m_{\mathbb{R}^p\times\mathbb{T}^q\times\mathbb{T}^a} = m_{\mathbb{R}^p\times\mathbb{T}^q} \times m_{\mathbb{T}^a}$ and this density is an entire function in variables t_i, s_j, u_k. It follows from (4.30) that the function g depends only on variables t_i, s_j.

It is not difficult to make sure that $\tau(\omega) = m_S$, where S is a finite subgroup of V, and the distribution $\lambda * m_{\mathbb{T}^a} * m_S$ and a Haar measure m_V are mutually absolutely continuous.

Define the mapping

$$\theta : V \to \mathbb{R}^l \times \mathbb{T}^m \times \mathbb{T}^n$$

by the formula $\theta = p\tau^{-1}$. The mapping θ is well defined and θ is a continuous monomorphism. Since $\theta(V) \subset X$, we will consider θ as a continuous monomorphism from V to X.

Put $M = \theta(V) \subset X$. Inasmuch as any subgroup of $X_{(2)}$ is a direct factor of $X_{(2)}$, we have $X_{(2)} = M_{(2)} \times N$, where N is a subgroup of $X_{(2)}$. Thus, any element of the group $X_{(2)}$ can be uniquely represented in the form $x_{ij} = a_i + b_j$, $a_i \in M_{(2)}$, $b_j \in N$. Represent the signed measure δ in the form

$$\delta = \sum_{i,j} c_{ij} E_{x_{ij}} = \sum_{i,j} c_{ij} E_{a_i} * E_{b_j} = \sum_j \delta_j * E_{b_j},$$

where

$$\delta_j = \sum_i c_{ij} E_{a_i} \tag{4.31}$$

are some signed measures on $M_{(2)}$. It follows from (4.29) and (4.31) that

$$\mu = \sum_j \gamma * m_K * \delta_j * E_{b_j}. \tag{4.32}$$

On the one hand, the signed measures $\gamma * m_K * \delta_j * E_{b_j}$ are concentrated on disjointed cosets $M + b_j$. Since μ is a measure, this implies that all signed measures $\gamma * m_K * \delta_j$ are also measures. On the other hand, since $\delta^{*2} = E_0$, we have $\mu^{*2} = \gamma^{*2} * m_K$. Hence the distribution μ^{*2} is concentrated on M. This implies that in (4.32) all measures $\gamma * m_K * \delta_j$ except one must be equal to zero. Thus, we have

$$\mu = \gamma * m_K * \delta = \gamma * m_K * \left(\sum_i c_{ij_0} E_{a_i}\right) * E_{b_{j_0}}. \tag{4.33}$$

Assume first that $b_{j_0} = 0$ in (4.33), i.e.,

$$\mu = \gamma * m_K * \delta_{j_0}. \tag{4.34}$$

Consider on V the signed measure $\rho = \theta^{-1}(\delta_{j_0})$. Inasmuch as $\gamma = \theta(\lambda)$, $m_{\mathbb{T}^a} = \theta(m_{\mathbb{T}^a})$, $m_B = \theta(m_S)$, $\delta_{j_0} = \theta(\rho)$, it follows from (4.29) and (4.34) that

$$\mu = \gamma * m_K * \delta_{j_0} = \gamma * m_{\mathbb{T}^a} * m_B * \delta_{j_0} = \theta(\lambda * m_{\mathbb{T}^a} * m_S * \rho). \tag{4.35}$$

Put $c = |S|$. Each element of the group S is represented in the form $s_{ij} = c_i + d_j$, where $c_i \in \mathbb{T}^q \times \mathbb{T}^a$, $d_j \in G$. We have

$$m_S = \frac{1}{c}\sum_{i,j} E_{s_{ij}} = \frac{1}{c}\sum_{i,j} E_{c_i} * E_{d_j} = \sum_j \kappa * E_{d_j},$$

where

$$\kappa = \frac{1}{c}\sum_i E_{c_i}.$$

This implies that

$$\lambda * m_{\mathbb{T}^a} * m_S * \rho = \sum_j \lambda * m_{\mathbb{T}^a} * \kappa * \rho * E_{d_j}.$$

Since θ is a monomorphism, it follows from (4.35) that the signed measure $\lambda * m_{\mathbb{T}^a} * m_S * \rho$ is actually a measure. Note that the signed measure $\lambda * m_{\mathbb{T}^a} * \kappa * \rho$ is concentrated on the subgroup $\mathbb{R}^p \times \mathbb{T}^q \times \mathbb{T}^a$. Hence the signed measures $\lambda * m_{\mathbb{T}^a} * \kappa * \rho * E_{d_j}$ are concentrated on disjointed cosets $\mathbb{R}^p \times \mathbb{T}^q \times \mathbb{T}^a + d_j$. This implies that the signed measure $\lambda * m_{\mathbb{T}^a} * \kappa * \rho$ is also a measure. We find from (4.30) that the measure $\lambda * m_{\mathbb{T}^a} * \kappa * \rho$ has the density with respect to m_V and this density is an entire function in variables t_i, s_j, u_k. Denote by r this density. Inasmuch as r is an entire function, we have

$$m_{\mathbb{R}^{p+q+a}}\{(t_1, \dots, t_p, s_1, \dots, s_q, u_1, \dots, u_a) \in \mathbb{R}^{p+q+a} :$$
$$r(t_1, \dots, t_p, e^{is_1}, \dots, e^{is_q}, e^{iu_1}, \dots, e^{iu_a}) = 0\} = 0.$$

Hence the distribution $\lambda * m_{\mathbb{T}^a} * m_S * \rho$ and m_V are mutually absolutely continuous. On the one hand, taking into account (4.35), we conclude that $\sigma(\mu) = \overline{\theta(V)}$. On the other hand, the fact that X is the minimal closed subgroup containing $\sigma(\mu)$ implies that $\sigma(\mu) = X$. The lemma is proved in the case when in (4.33) $b_{j_0} = 0$.

If in (4.33) $b_{j_0} \neq 0$, apply the reasoning given above in the case when $b_{j_0} = 0$ to the distribution $\nu = \mu * E_{b_{j_0}}$. We get that $\sigma(\nu) = \overline{\theta(V)}$. This implies the equality

$$\sigma(\mu) = \overline{\theta(V)} + b_{j_0}. \tag{4.36}$$

If $\overline{\theta(V)} \neq X$, then the form of the group X and (4.36) imply that the minimal closed subgroup of X containing $\sigma(\mu)$ is a property subgroup of X, contrary to the condition of the lemma. Thus, $\overline{\theta(V)} = X$. Hence $\sigma(\mu) = X$. □

We can now prove the following statement (compare with Lemma 4.5).

THEOREM 4.18. *Let X be a locally compact Abelian group and let $\mu \in \Gamma_B(X)$. Then $\sigma(\mu)$ is a coset of a subgroup of X topologically isomorphic to a group of the form* (4.6).

PROOF. Denote by Y the character group of the group X. Put

$$B = \{y \in Y : |\hat{\mu}(y)| = 1\}.$$

We note that by studying the support of μ we can assume that $B = \{0\}$. Indeed, by Corollary 2.11, B is a closed subgroup. It follows from statement (v) of Theorem 2.5 that there exists an element $x \in X$ such that $\hat{\mu}(y) = (x, y)$ for all $y \in B$. Set $\nu = \mu * E_{-x}$. Then $\nu \in \Gamma_B(X)$ and ν has the property

$$B = \{y \in Y : \hat{\nu}(y) = 1\}.$$

Put $G = A(X, B)$, $H = G^*$. By Proposition 2.10, we have $\sigma(\nu) \subset G$. Moreover,

$$\{h \in H : |\hat{\nu}(h)| = 1\} = \{0\},$$

because

$$\{h \in H : \hat{\nu}(h) = 1\} = \{h \in H : |\hat{\nu}(h)| = 1\}.$$

This remark and Lemma 4.5 show that we can suppose that the group X is of the form (4.6) and the distribution μ satisfies the condition

$$\{y \in Y : |\hat{\mu}(y)| = 1\} = \{0\}. \tag{4.37}$$

The theorem will be proved if we show that $\mu(U) > 0$ for any open subset $U \subset X$. Assume that $\mu(U) = 0$ for an open subset $U \subset X$. Take an open subset U_0 in U and a neighborhood of the zero V in X such that $U_0 + V \subset U$. The intersection $K \cap V$ is a neighborhood of the zero in the group K. By Theorem 1.18, there exists a closed subgroup $S \subset K \cap V$ such that $K/S \cong \mathbb{T}^m \times F$, where $m \geq 0$ and F is a finite Abelian group. Obviously, F is a Corwin group. We have as a result $X/S \cong M$, where

$$M = \mathbb{R}^l \times \mathbb{T}^m \times F.$$

Denote by τ this topological isomorphism between X/S and M. Put $p = \tau\pi$, where π is the natural homomorphism $\pi : X \to X/S$. Thus,

$$p : X \to M.$$

Set $N = M^*$. It is obvious that $p(\mu) \in \Gamma_B(M)$. Moreover, it follows from (4.37) that

$$\{y \in N : |\widehat{p(\mu)}(y)| = 1\} = \{0\}. \tag{4.38}$$

Since the connected component of the zero of the group M contains a finite number of elements of order 2, by Theorem 4.9, the distribution $p(\mu)$ can be represented in the form

$$p(\mu) = \gamma * m_K * \delta * E_{x_0},$$

where $\gamma \in \Gamma^s(\mathbb{R}^l \times \mathbb{T}^m)$, K is a compact Corwin subgroup of M, δ is a signed measure on $M_{(2)}$ such that $\delta^{*2} = E_0$, and $x_0 \in M$. Put $\alpha = p(\mu) * E_{-x_0}$. Then $\alpha \in \Gamma_B(M)$. It follows from (4.38) that

$$\{y \in N : |\hat{\alpha}(y)| = 1\} = \{0\}.$$

In particular, this implies that the minimal closed subgroup of M containing the support $\sigma(\alpha)$ coincides with M. On the one hand, applying Lemma 4.17 to the distribution α we get $\sigma(\alpha) = M$. Hence

$$\sigma(p(\mu)) = M. \tag{4.39}$$

On the other hand, we have

$$p(\mu)(p(U_0)) = \mu\{p^{-1}(p(U_0))\} = \mu\{U_0 + S\} \leq \mu\{U_0 + V\} \leq \mu\{U\} = 0.$$

This contradicts (4.39) because, by Theorem 1.23, p is an open mapping and hence $p(U_0)$ is an open set. Thus, $\mu(U) > 0$ and the theorem is proved.

It should be noted that if X is an arbitrary locally compact Abelian group of the form (4.6), then there exists $\mu \in \Gamma_B(X)$ such that $\sigma(\mu) = X$. □

To conclude this section, we prove a zero-one law for Gaussian distributions in the sense of Bernstein. For this we need the following lemma.

LEMMA 4.19. *Let $X = \mathbb{R}^n \times \mathbb{T}^m$, where $n \geq 0$, $m \geq 0$. Let $\gamma \in \Gamma(X)$ and let γ be absolutely continuous with respect to a Haar measure m_X. Let δ be a signed measure on X concentrated on a finite number of points and let $\mu = \gamma * \delta \in \mathrm{M}^1(X)$. Then μ and γ are mutually absolutely continuous.*

PROOF. Consider the group $\mathbb{T}^m$ with character group $\mathbb{Z}^m$. Denote by $k = (k_1, \dots, k_m)$ elements of the group $\in \mathbb{Z}^m$. We prove the lemma only for the group $\mathbb{T}^m$. The general case can be considered similarly. We also suppose γ is a symmetric Gaussian distribution. The characteristic function $\hat{\gamma}(k)$ can be represented in the form

$$\hat{\gamma}(k) = \exp\left\{-\frac{1}{2}\langle Ak, k\rangle\right\}, \quad k \in \mathbb{Z}^m,$$

where A is a symmetric positive semi-definite matrix. Since γ is absolutely continuous with respect to $m_{\mathbb{T}^m}$, we have $\det A \neq 0$. It is obvious that the Gaussian distribution γ and $m_{\mathbb{T}^m}$ are mutually absolutely continuous. Moreover, μ is absolutely continuous with respect to γ. We prove that γ is absolutely continuous with respect to μ. It is easy to see that the density $r(e^{it})$ of the Gaussian distribution γ with respect to $m_{\mathbb{T}^m}$ can be represented in the form

$$r(e^{it}) = \sqrt{\frac{(2\pi)^m}{\det A}} \sum_{k \in \mathbb{Z}^m} \exp\left\{-\frac{1}{2}\langle A^{-1}(t + 2\pi k), t + 2\pi k\rangle\right\}, \tag{4.40}$$

where $t = (t_1, \dots, t_m) \in \mathbb{R}^m$, $e^{it} = (e^{it_1}, \dots, e^{it_m}) \in \mathbb{T}^m$. In particular, when $m = 1$, if we write the characteristic function $\hat{\gamma}(k)$ in the form

$$\hat{\gamma}(k) = \exp\left\{-\frac{\sigma^2 k^2}{2}\right\}, \quad k \in \mathbb{Z},$$

then we have

$$r(e^{it}) = \frac{\sqrt{2\pi}}{\sigma} \sum_{n=-\infty}^{\infty} \exp\left\{-\frac{1}{2\sigma^2}(t + 2\pi n)^2\right\}, \quad t \in \mathbb{R}. \tag{4.41}$$

We note that the function $r(e^{it})$ may be extended by formula (4.40) to $\mathbb{C}^m$ as an entire function in t. Indeed, let $f(t)$ be an arbitrary function in $t = (t_1, \dots, t_m) \in \mathbb{R}^m$. Then $f(t)$ can be extended to $\mathbb{C}^m$ as an entire function in t if and only if the function $g(s) = f(t)$, where $s = Tt$ and T is an invertible matrix, can be extended. Taking into account this remark we can assume that A in (4.40) is a diagonal matrix. Hence the statement that the function $r(e^{it})$ may be extended to $\mathbb{C}^m$ as an entire function in t, is reduced to the one-dimensional case. Obviously, if $m = 1$, it follows from (4.41) that the function $r(e^{it})$ can be extended by formula (4.41) to $\mathbb{C}$ as an entire function in t.

Let $\delta = \sum c_j E_{x_j}$, where $c_j \in \mathbb{R}$, $x_j = e^{ia_j} \in \mathbb{T}^m$. Then μ has the density

$$g(e^{it}) = \sum c_j r(e^{i(t-a_j)}),$$

with respect to $m_{\mathbb{T}^m}$. This density is also an entire function in t. It follows from this that

$$m_{\mathbb{R}^m}\{t \in \mathbb{R}^m : g(e^{it}) = 0\} = 0.$$

Hence $m_{\mathbb{T}^m}$ is absolutely continuous with respect to μ. So, γ is also absolutely continuous with respect to μ. $\square$

THEOREM 4.20. *Let X be a locally compact Abelian group such that the connected component of the zero of X has finite dimension. Let $\mu \in \Gamma_B(X)$. Assume that the characteristic function of μ does not vanish. Then for any Borel subgroup G of the group X either $\mu(G) = 0$ or $\mu(G) = 1$.*

PROOF. Observe that if c_X has finite dimension, then the group c_X contains a finite number of elements of order 2. Indeed, by Theorem 1.15, we have $c_X \cong \mathbb{R}^p \times K$, where $p \geq 0$ and K is a compact connected Abelian group. Therefore, we may assume that the group c_X is compact. Put $H = (c_X)^*$. By Theorems 1.2 and 1.3, H is a discrete torsion-free Abelian group. By Theorem 1.4, the dimension of c_X is equal to the torsion-free rank $r_0(H)$. Hence $r_0(H) < \infty$. By Theorem 1.12, $A(H, (c_X)_{(2)}) = H^{(2)}$. It follows from this that $((c_X)_{(2)})^* \cong H/H^{(2)}$. Taking into account that all nonzero elements of the factor-group $H/H^{(2)}$ have order 2, we can consider the factor-group $H/H^{(2)}$ as a linear space under the field $\mathbb{Z}(2)$ and its dimension does not exceed $r_0(H)$. This implies that $((c_X)_{(2)})^*$ is a finite group. Hence $(c_X)_{(2)}$ is also a finite group.

Since the group c_X contains a finite number of elements of order 2, by Theorem 4.9, the distribution μ can be represented in the form

$$\mu = \gamma * \delta * E_x, \tag{4.42}$$

where $\gamma \in \Gamma^s(c_X)$, δ is a signed measure on $(c_X)_{(2)}$ such that $\delta^{*2} = E_0$, and $x \in X$. By Proposition 3.7, $\gamma = p(M)$, where

$$p : \mathbb{R}^l \to c_X$$

is a continuous homomorphism and $M \in \Gamma^s(\mathbb{R}^l)$. We may also suppose that $\sigma(M) = \mathbb{R}^l$. Then the Gaussian distribution M and a Haar measure $m_{\mathbb{R}^l}$ are mutually absolutely continuous. Put $L = \operatorname{Ker} p$. Then L is a closed subgroup of $\mathbb{R}^l$. Let

$$\tau : \mathbb{R}^l \to \mathbb{R}^l/L$$

be the natural homomorphism. We have $\mathbb{R}^l/L \cong \mathbb{R}^n \times \mathbb{T}^m$. In order not to complicate the notation, suppose $\mathbb{R}^l/L = \mathbb{R}^n \times \mathbb{T}^m$.

Define the mapping

$$q : \mathbb{R}^n \times \mathbb{T}^m \to X$$

by the formula $q = p\tau^{-1}$. The mapping q is well defined and q is a continuous monomorphism. We have $\gamma = q(N)$, where $N = \tau(M) \in \Gamma^s(\mathbb{R}^n \times \mathbb{T}^m)$.

Note that the Gaussian distribution N and a Haar measure $m_{\mathbb{R}^n \times \mathbb{T}^m}$ are mutually absolutely continuous. Set $F = q(\mathbb{R}^n \times \mathbb{T}^m)$, $A = (c_X)_{(2)} \cap F$. By Suslin's theorem, F is a Borel subgroup of X. Since every subgroup of $(c_X)_{(2)}$ is a direct factor of $(c_X)_{(2)}$, we have $(c_X)_{(2)} = A \times B$, where B is a subgroup of $(c_X)_{(2)}$. Then any element of the subgroup $(c_X)_{(2)}$ can be uniquely represented in the form $x_{ij} = a_i + b_j$, where $a_i \in A$, $b_j \in B$.

Assume that the signed measure δ is of the form

$$\delta = \sum_{i,j} c_{ij} E_{x_{ij}} = \sum_{i,j} c_{ij} E_{a_i} * E_{b_j} = \sum_j \left(\sum_i c_{ij} E_{a_i} \right) * E_{b_j}.$$

where $c_{ij} \in \mathbb{R}$. This implies that

$$\mu = \left(\sum_j \gamma * \omega_j * E_{b_j} \right) * E_x, \tag{4.43}$$

where

$$\omega_j = \sum_i c_{ij} E_{a_i}$$

are some signed measures on F. We note that the Gaussian distribution γ is also concentrated on F. Inasmuch as μ is a measure, all signed measures $\gamma * \omega_j$ are actually measures, because the signed measures $\gamma * \omega_j * E_{b_j}$ are concentrated on the disjoint cosets $F + b_j$.

We use now that $\delta^{*2} = E_0$. It follows from (4.42) that $\mu^{*2} = \gamma^{*2} * E_{2x}$. Hence the distribution μ^{*2} is concentrated on the set $F + 2x$.

Assume that at least two measures $\gamma * \omega_{j_1}$ and $\gamma * \omega_{j_2}$ are nonzero. Then (4.43) implies that

$$\mu^{*2}(F + 2x + b_{j_1} + b_{j_2}) > 0.$$

Taking into account that F and $F + b_{j_1} + b_{j_2}$ are disjoint sets, this inequality contradicts the fact that the distribution μ^{*2} is concentrated on the set $F + 2x$. Thus, all convolutions $\gamma * \omega_j$ except only one are equal to zero. Since $\hat{\gamma}(y) \neq 0$ for all $y \in Y$, we get that $\omega_j = 0$ for all j except one, say $j = j_0$. So,

$$\delta = \left(\sum_i c_{ij_0} E_{a_i} \right) * E_{b_{j_0}}. \tag{4.44}$$

It follows from (4.42) and (4.44) that

$$\mu = \gamma * \left(\sum_i c_{ij_0} E_{a_i} \right) * E_{x + b_{j_0}}. \tag{4.45}$$

Put $\tilde{x} = x + b_{j_0}$. There are two possibilities for $\tilde{x}$: either $\tilde{x} \in F$ or $\tilde{x} \notin F$.

1. $\tilde{x} \in F$. It follows from (4.44) that the signed measure $\delta * E_x$ is concentrated on F. This implies that the distribution $\mu = \gamma * \delta * E_x$ is also concentrated on F. Put $\lambda = q^{-1}(\delta * E_x)$. Then λ is a signed measure on $\mathbb{R}^n \times \mathbb{T}^m$. We have $\mu = q(\nu)$, where $\nu = N * \lambda$. Let G be a Borel subgroup of X and let $\mu(G) > 0$. Then $\nu(q^{-1}(G)) > 0$. Lemma 4.19 implies that the distributions ν and N are mutually absolutely continuous. Therefore, $N(q^{-1}(G)) > 0$. Inasmuch as the Gaussian distribution N is absolutely continuous with respect to $m_{\mathbb{R}^n \times \mathbb{T}^m}$, we have $m_{\mathbb{R}^n \times \mathbb{T}^m}(q^{-1}(G)) > 0$. Since $q^{-1}(G)$ is a Borel subgroup of $\mathbb{R}^n \times \mathbb{T}^m$ and $m_{\mathbb{R}^n \times \mathbb{T}^m}(q^{-1}(G)) > 0$, the subgroup $q^{-1}(G)$ contains a neighborhood of the zero in the group $\mathbb{R}^n \times \mathbb{T}^m$. Taking into account that the group $\mathbb{R}^n \times \mathbb{T}^m$ is connected, we obtain $q^{-1}(G) = \mathbb{R}^n \times \mathbb{T}^m$.

Hence $G = F$. It follows from this that $\mu(G) = \mu(F) = 1$. So, in case 1 the theorem is proved.

2. $\tilde{x} \notin F$. Rewrite (4.45) in the form

$$\mu = \gamma * \omega_{j_0} * E_{\tilde{x}}. \tag{4.46}$$

It is easy to see that $F = p(\mathbb{R}^l)$. There are two possibilities: either $n\tilde{x} \in F$ for some natural $n \geq 2$ or $n\tilde{x} \notin F$ for all natural n.

2a. Assume that $n_0\tilde{x} = p(t_0) \in F$ for some natural $n_0 \geq 2$ and also assume that n_0 is minimal having this property. Set $x_0 = \tilde{x} - p\left(\frac{t_0}{n_0}\right)$. Then $n_0 x_0 = 0$. Extend p to a homomorphism

$$\bar{p} : \mathbb{R}^l \times \mathbb{Z}(n_0) \to X$$

by the formula

$$\bar{p}(t, k) = p(t) + kx_0, \quad t \in \mathbb{R}^l, \ k = 0, 1, \ldots, n_0 - 1.$$

Observe that $L = \operatorname{Ker} p = \operatorname{Ker} \bar{p}$. Taking this into account we will keep the notation τ for the natural homomorphism

$$\tau : \mathbb{R}^l \times \mathbb{Z}(n_0) \to (\mathbb{R}^l \times \mathbb{Z}(n_0))/L.$$

It is easy to see that $(\mathbb{R}^l \times \mathbb{Z}(n_0))/L \cong \mathbb{R}^n \times \mathbb{T}^m \times \mathbb{Z}(n_0)$. In order not to complicate the notation, suppose $(\mathbb{R}^l \times \mathbb{Z}(n_0))/L = \mathbb{R}^n \times \mathbb{T}^m \times \mathbb{Z}(n_0)$. Put

$$S = \mathbb{R}^n \times \mathbb{T}^m \times \mathbb{Z}(n_0).$$

Define the mapping

$$\bar{q} : S \to X$$

by the formula $\bar{q} = \bar{p}\tau^{-1}$. The mapping $\bar{q}$ is well defined and $\bar{q}$ is a continuous monomorphism. We have $\gamma * E_{\tilde{x}} = \bar{q}(P)$, where $P = \tau\left(M * E_{\left(\frac{t_0}{n_0}, 1\right)}\right) \in \Gamma(S)$. We note that the Gaussian distribution P is absolutely continuous with respect to a Haar measure m_S. Set $\kappa = \bar{q}^{-1}(\omega_{j_0})$. Then κ is a signed measure on $\mathbb{R}^n \times \mathbb{T}^m$. It follows from (4.46) that $\mu = \bar{q}(\pi)$, where $\pi = P * \kappa$.

Let G be a Borel subgroup of X and let $\mu(G) > 0$. Then $\pi(\bar{q}^{-1}(G)) > 0$. Lemma 4.19 implies that the distributions π and P are mutually absolutely continuous. Hence $P(\bar{q}^{-1}(G)) > 0$. It follows from this that $m_S(\bar{q}^{-1}(G)) > 0$. Since $\bar{q}^{-1}(G)$ is a Borel subgroup of S and $m_S(\bar{q}^{-1}(G)) > 0$, the subgroup $\bar{q}^{-1}(G)$ contains a neighborhood of the zero in the group S. Taking into account that $\mathbb{R}^n \times \mathbb{T}^m$ is a locally compact connected Abelian group, we get $\bar{q}^{-1}(G) \supset \mathbb{R}^n \times \mathbb{T}^m$. Therefore, $G \supset F$. It follows from (4.45) that μ is concentrated on the coset $\tilde{x} + F$. So, if $\mu(G) > 0$, then $\{\tilde{x} + F\} \cap G \neq \emptyset$. Inasmuch as $G \supset F$, this implies that $G \supset \{\tilde{x} + F\}$. Hence $\mu(G) \geq \mu(\tilde{x} + F) = 1$. In case 2a the theorem is also proved.

2b. If $n\tilde{x} \notin F$ for all natural n, we reason as above. The only distinction is that we extend p to a homomorphism

$$\bar{p} : \mathbb{R}^l \times \mathbb{Z} \to X$$

by the formula

$$\bar{p}(t, k) = p(t) + k\tilde{x}, \quad t \in \mathbb{R}^l, \ k \in \mathbb{Z}.$$

$\square$

5. General case

Let X be a second countable locally compact Abelian group. In this section we study the following problem. Let ξ_1 and ξ_2 be independent random variables with values in X and distributions μ_1 and μ_2. Assume that the sum $\xi_1+\xi_2$ and difference $\xi_1-\xi_2$ are independent. What can we say about the distributions μ_j? In contrast to Section 4, we do not suppose that the random variables ξ_1 and ξ_2 are identically distributed. We prove a decomposition theorem for μ_j assuming that the connected component of the zero of the group X contains a finite number of elements of order 2. It is interesting to observe that in this case, not only elements of order 2 play an important role, but also elements of order 4. Based on this result, we give a complete description of locally compact Abelian groups on which μ_j are represented as convolutions of Gaussian distributions and idempotent distributions. These are groups whose connected component of the zero contains no elements of order 2.

The proof of the main result of this section is based on the following lemma. In what follows we also use this statement.

LEMMA 5.1. *Let Y be a locally compact Abelian group. Let $\Phi(y)$ be a continuous function on Y satisfying the equation*

$$\Delta_{2h}\Delta_h^2\Phi(y)=0, \quad h,y\in Y \tag{5.1}$$

and the conditions $\Phi(-y)=\Phi(y)$, $\Phi(0)=0$. Let

$$Y=\bigcup_\alpha(y_\alpha+\overline{Y^{(2)}})$$

be the $\overline{Y^{(2)}}$-coset decomposition of the group Y. Then the function $\Phi(y)$ can be represented in the form

$$\Phi(y)=\varphi(y)+r_\alpha, \quad y\in y_\alpha+\overline{Y^{(2)}}, \tag{5.2}$$

where $\varphi(y)$ is a continuous function on Y satisfying equation (3.2).

PROOF. It should be noted that every function $\varphi(y)$ on the subgroup $Y^{(2)}$ satisfying equation (3.2) can be extended to a function $\widetilde{\varphi}(y)$ on the group Y by the formula

$$\widetilde{\varphi}(y)=\frac{1}{4}\varphi(2y), \quad y\in Y. \tag{5.3}$$

The function $\widetilde{\varphi}(y)$ also satisfies equation (3.2). Analogously, every additive function $l(y)$, i.e., a function satisfying Cauchy's functional equation

$$l(u+v)=l(u)+l(v) \tag{5.4}$$

on the subgroup $Y^{(2)}$ can be extended to an additive function $\widetilde{l}(y)$ on the group Y by the formula

$$\widetilde{l}(y)=\frac{1}{2}l(2y), \quad y\in Y. \tag{5.5}$$

Observe that every continuous polynomial $f(y)$ on Y of degree ≤ 2 can be represented as a sum

$$f(y)=\varphi(y)+l(y)+c,$$

where $\varphi(y)$ is a continuous function satisfying equation (3.2), $l(y)$ is a continuous function satisfying equation (5.4), and c is a constant.

As noted in Remark 3.2, formulas (3.4) and (3.5) establish a one-to-one correspondence between functions $\varphi(y)$ satisfying equation (3.2) and 2-additive functions $\Phi(u,v)$. Consider an arbitrary coset $y_\alpha + \overline{Y^{(2)}}$ of the subgroup $\overline{Y^{(2)}}$ in Y. Fix an element y_α. The function $\Phi(y_\alpha + y)$, as a function in y, also satisfies equation (5.1). For this reason its restriction to the subgroup $\overline{Y^{(2)}}$ is a polynomial of degree ≤ 2. Hence we have the following representation

$$\Phi(y_\alpha + y) = \varphi_\alpha(y) + l_\alpha(y) + c_\alpha, \quad y \in \overline{Y^{(2)}}, \tag{5.6}$$

where the continuous function $\varphi_\alpha(y)$ satisfies equation (3.2), the continuous function $l_\alpha(y)$ satisfies equation (5.4), and c_α is a constant. Extend the functions $\varphi_\alpha(y)$ and $l_\alpha(y)$ from the subgroup $Y^{(2)}$ to the functions $\widetilde{\varphi}_\alpha(y)$ and $\widetilde{l}_\alpha(y)$ on the group Y by formulas (5.3) and (5.5) respectively. Passing from the function $\widetilde{\varphi}_\alpha(y)$ to the corresponding 2-additive function $\psi_\alpha(u,v)$ we find from (5.6)

$$\begin{aligned} \Phi(-y_\alpha - y) &= \Phi(y_\alpha + (-2y_\alpha - y)) = \widetilde{\varphi}_\alpha(-2y_\alpha - y) + \widetilde{l}_\alpha(-2y_\alpha - y) + c_\alpha \\ &= 4\psi_\alpha(y_\alpha, y_\alpha) + 4\psi_\alpha(y_\alpha, y) + \psi_\alpha(y,y) - 2\widetilde{l}_\alpha(y_\alpha) - \widetilde{l}_\alpha(y) + c_\alpha, \quad y \in \overline{Y^{(2)}}. \end{aligned} \tag{5.7}$$

Inasmuch as

$$\Phi(y_\alpha + y) = \psi_\alpha(y,y) + \widetilde{l}_\alpha(y) + c_\alpha, \quad y \in \overline{Y^{(2)}}, \tag{5.8}$$

and $\Phi(-y) = \Phi(y)$, we conclude from equalities (5.7) and (5.8) that

$$2\psi_\alpha(y_\alpha, y) = \widetilde{l}_\alpha(y), \quad y \in \overline{Y^{(2)}}. \tag{5.9}$$

It follows from (5.8) and (5.9) that

$$\begin{aligned} \Phi(y_\alpha + y) &= \psi_\alpha(y_\alpha + y, y_\alpha + y) - 2\psi_\alpha(y_\alpha, y) - \psi_\alpha(y_\alpha, y_\alpha) \\ &\quad + \widetilde{l}_\alpha(y) + c_\alpha = \widetilde{\varphi}_\alpha(y_\alpha + y) + r_\alpha, \quad y \in \overline{Y^{(2)}}, \end{aligned} \tag{5.10}$$

where r_α is a constant.

From equation (5.1), we obtain that the function $\Phi(y)$ satisfies the equation

$$\Phi(y + 4h) - 2\Phi(y + 3h) + 2\Phi(y + h) - \Phi(y) = 0, \quad y, h \in Y. \tag{5.11}$$

Substituting $y = 0$, $h = y_\alpha + 2u$, $u \in Y$, in equation (5.11) we arrive at

$$\Phi(4y_\alpha + 8u) - 2\Phi(3y_\alpha + 6u) + 2\Phi(y_\alpha + 2u) = 0, \quad u \in Y.$$

Taking into account (5.10) we infer that

$$\widetilde{\varphi}_0(4y_\alpha + 8u) - 2\widetilde{\varphi}_\alpha(3y_\alpha + 6u) + 2\widetilde{\varphi}_\alpha(y_\alpha + 2u) = 0, \quad u \in Y.$$

Passing here from the functions $\widetilde{\varphi}_0(y)$ and $\widetilde{\varphi}_\alpha(y)$ to the corresponding 2-additive functions $\psi_0(u,v)$ and $\psi_\alpha(u,v)$, we obtain

$$\begin{aligned} 4(\psi_0(u,u) - \psi_\alpha(u,u)) &+ 4(\psi_0(y_\alpha, u) - \psi_\alpha(y_\alpha, u)) \\ &+ (\psi_0(y_\alpha, y_\alpha) - \psi_\alpha(y_\alpha, y_\alpha)) = 0, \quad u \in Y. \end{aligned}$$

This implies that $\widetilde{\varphi}_\alpha(y) = \widetilde{\varphi}_0(y)$ for all $y \in Y$ and representation (5.2) follows from (5.10). □

The main result of this section is the following theorem.

THEOREM 5.2. *Let X be a locally compact Abelian group. Assume that the connected component of the zero of X contains a finite number of elements of order 2. Let ξ_1 and ξ_2 be independent random variables with values in the group X and distributions μ_1 and μ_2. Assume that the sum $\xi_1 + \xi_2$ and difference $\xi_1 - \xi_2$ are independent. Then there exists a subgroup $G \subset X_{(2)}$ such that distributions $p(\mu_k)$ (p is the natural homomorphism $p : X \to X/G$) are of the form*

$$p(\mu_k) = \gamma * m_V * \pi_k * E_{x_k},$$

where $\gamma \in \Gamma^s(X/G)$, V is a compact Corwin subgroup of the factor-group X/G, π_k are signed measures on $(X/G)_{(4)}$, $x_k \in X/G$.

PROOF. By Lemma 4.5, the distributions μ_j are supported in cosets of a subgroup M of X such that M is topologically isomorphic to a group of the form (4.6). It follows from this that we may assume that $X = \mathbb{R}^m \times K$, where $m \geq 0$ and K is a compact Corwin group. Then by Lemma 4.6, we have $X_{(2)} \subset c_X$. Hence $X_{(2)}$ is a finite subgroup.

Denote by Y the character group of the group X. Since K is a compact Corwin group, it is easy to see that Y is topologically isomorphic to a group of the form $\mathbb{R}^m \times D$, where D is a discrete Abelian group with no elements of order 2. Note that $\overline{Y^{(2)}} = Y^{(2)}$. Let $|X_{(2)}| = n$. It follows from Theorems 1.9 and 1.12 that $|X_{(2)}| = |Y/Y^{(2)}|$. Let

$$Y = \bigcup_{j=0}^{n-1} (y_j + Y^{(2)}) \tag{5.12}$$

be a $Y^{(2)}$-coset decomposition of the group Y. Put

$$N_1 = \{y \in Y : \hat{\mu}_1(y) \neq 0\}, \quad N_2 = \{y \in Y : \hat{\mu}_2(y) \neq 0\}, \quad N = N_1 \cap N_2.$$

By Lemma 4.2, the characteristic functions $\hat{\mu}_j(y)$ satisfy equation (4.1). It follows from equation (4.1) that N is an open subgroup of the group Y satisfying the following conditions:

(i) if $2y \in N$, then $y \in N$;
(ii) $N \cap Y^{(2)} = N^{(2)}$.

Equation (4.1) also implies that

$$|\hat{\mu}_1(u+v)||\hat{\mu}_2(u-v)| = |\hat{\mu}_1(u-v)||\hat{\mu}_2(u+v)|$$

for all $u, v \in Y$. It follows from this that for arbitrary elements a and b from a given coset $y_j + Y^{(2)}$ we have the equality

$$|\hat{\mu}_1(a)||\hat{\mu}_2(b)| = |\hat{\mu}_1(b)||\hat{\mu}_2(a)|. \tag{5.13}$$

Denote by H a union of cosets $y_j + Y^{(2)}$ such that $N \cap (y_j + Y^{(2)}) \neq \emptyset$. Since N is a subgroup of Y, we conclude that H is also a subgroup of Y. Changing the numeration if it is necessary, we can assume that

$$H = \bigcup_{j=0}^{l-1} (y_j + Y^{(2)}). \tag{5.14}$$

Moreover, we may suppose that $y_j \in N$ in (5.14). Taking this into account (ii) implies that

$$N = \bigcup_{j=0}^{l-1} (y_j + N^{(2)}). \tag{5.15}$$

Put $G = A(X, H)$. By Theorems 1.8 and 1.12, we have $H = A(Y, G)$ and $H^* \cong X/G$. It is clear that $G \subset X_{(2)}$. Assume that $N \cap (y_j + Y^{(2)}) = \emptyset$ for some j. If $a \in N_1 \cap (y_j + Y^{(2)})$, then $a \notin N_2 \cap (y_j + Y^{(2)})$ and the right-hand side of (5.13) vanishes. Hence $\hat{\mu}_2(b) = 0$ for any $b \in y_j + Y^{(2)}$.

Thus, we proved the following assertion. If a locally compact Abelian group X is topologically isomorphic to a group of the form (4.6), then either $\hat{\mu}_1(y) \equiv 0$ or $\hat{\mu}_2(y) \equiv 0$ for each coset $y_j + Y^{(2)}$ which is disjoint with $A(Y, G)$. This reasoning also shows that if $N \cap (y_j + Y^{(2)}) \neq \emptyset$, then $N_1 \cap (y_j + Y^{(2)}) = N_2 \cap (y_j + Y^{(2)})$.

Consider the restriction of equation (4.1) to the subgroup N. The functions $|\hat{\mu}_1(y)|$ and $|\hat{\mu}_2(y)|$ also satisfy equation (4.1). Put

$$f_k(y) = -\ln|\hat{\mu}_k(y)|, \quad y \in N,\ k = 1, 2.$$

It follows from (4.1) that the functions $f_k(y)$ satisfy the equation

$$f_1(u + v) + f_2(u - v) = A(u) + A(v), \quad u, v \in N, \tag{5.16}$$

where $A(u) = f_1(u) + f_2(u)$. Apply the finite difference method to solve equation (5.16). Let h be an arbitrary element of N. Substitute $u + h$ for u and $v + h$ for v in equation (5.16). Subtracting equation (5.16) from the resulting equation we obtain

$$\Delta_{2h} f_1(u + v) = \Delta_h A(u) + \Delta_h A(v), \quad u, v \in N. \tag{5.17}$$

Put $v = 0$ in (5.17) and subtract from (5.17) the obtained equation. We get

$$\Delta_v \Delta_{2h} f_1(u) = \Delta_h A(v) + \Delta_h A(0), \quad u, v \in N.$$

It follows from this that

$$\Delta_v^2 \Delta_{2h} f_1(u) = 0, \quad u, v, h \in N.$$

The function $f_2(y)$ satisfies the same equation. Applying Lemma 5.1, we obtain the representations

$$f_k(y) = \varphi_k(y) + p_{k,j}, \quad y \in y_j + N^{(2)},\ k = 1, 2.$$

Since equation (4.1) implies that $|\hat{\mu}_1(2y)| = |\hat{\mu}_2(2y)|$ for all $y \in Y$, the functions $\varphi_1(y)$ and $\varphi_2(y)$ coincide on the subgroup $N^{(2)}$. Hence they also coincide on the subgroup N, i.e.,

$$\varphi_1(y) = \varphi_2(y) = \varphi(y), \quad y \in N.$$

Taking this into account, we get from equation (4.1) that $p_{1,j} = -p_{2,j} = p_j$, $j = 0, 1, \ldots, l - 1$. It is obvious that $\varphi(y) \geq 0$.

Put

$$l_k(y) = \hat{\mu}_k(y)/|\hat{\mu}_k(y)|, \quad y \in N,\ k = 1, 2.$$

The functions $l_1(y)$ and $l_2(y)$ satisfy the equation

$$l_1(u + v) l_2(u - v) = l_1(u) l_2(u) l_1(v) l_2(-v), \quad u, v \in N. \tag{5.18}$$

Moreover, $|l_k(u)| = 1$, $l_k(-u) = \overline{l_k(u)}$ for all $u \in N$, $l_k(0) = 1$, $k = 1, 2$. Putting $v = u$ and then $v = -u$ in (5.18), we find

$$l_k(2u) = l_k^2(u), \quad u \in N, \ k = 1, 2. \tag{5.19}$$

Replacing u by v in equation (5.18) we get

$$l_1(u+v)l_2(v-u) = l_1(v)l_2(v)l_1(u)l_2(-u), \quad u, v \in N. \tag{5.20}$$

Multiplying (5.18) and (5.20) we find

$$l_1^2(u+v) = l_1^2(u)l_1^2(v), \quad u, v \in N. \tag{5.21}$$

It follows from (5.19) and (5.21) that the restriction of the function $l_1(y)$ to $N^{(2)}$ is a character of the subgroup $N^{(2)}$. The same reasoning shows that the restriction of the function $l_2(y)$ to $N^{(2)}$ is also a character of the subgroup $N^{(2)}$. By Theorem 1.12, there exist elements $x_k \in X/G$ such that $l_k(y) = (x_k, y)$ for all $y \in N^{(2)}$. Put $m_k(y) = (-x_k, y)l_k(y)$ for all $y \in H$. Then

$$m_k(2y) = 1, \quad y \in N, \ k = 1, 2,$$

and (5.19) implies that $m_k(y) = \pm 1$ for all $y \in N$, $k = 1, 2$. Hence

$$m_k(y) = m_k(-y), \quad y \in N, \ k = 1, 2, \tag{5.22}$$

and the functions $m_1(y)$ and $m_2(y)$ satisfy the equation

$$m_1(u+v)m_2(u-v) = m_1(u)m_2(u)m_1(v)m_2(v), \quad u, v \in N. \tag{5.23}$$

This implies that for any elements a and b from a given coset $y_j + N^{(2)}$ the equality

$$m_1(a)m_2(b) = m_1(b)m_2(a) \tag{5.24}$$

holds. Fix an element $a \in y_j + N^{(2)}$. It follows from (5.24) that for all $b \in y_j + N^{(2)}$ either $m_1(b) = m_2(b)$ or $m_1(b) = -m_2(b)$. Consider the subgroup

$$L = N^{(2)} \cup (y_j + N^{(2)}),$$

where $y_j \in N$. Take into account that $m_1(y) = m_2(y)$ for all $y \in N^{(2)}$. Moreover, either $m_1(y) = m_2(y)$ or $m_1(y) = -m_2(y)$ for all $y \in y_j + N^{(2)}$. In view of (5.23), we easily get that

$$m_k(u+v)m_k(u-v) = 1, \quad u, v \in L, \ k = 1, 2. \tag{5.25}$$

We conclude now from (5.15) and (5.25) that

$$m_k(u+4v) = m_k(u), \quad u, v \in N, \ k = 1, 2,$$

i.e., the functions $m_k(y)$ are invariant with respect to the subgroup $N^{(4)}$.

It follows from representation (5.12) that

$$Y^{(2)} = \bigcup_{j=0}^{n-1} (2y_j + Y^{(4)}).$$

Taking into consideration (5.14), this implies that

$$H = \bigcup_{0 \le i \le n-1, \ 0 \le j \le l-1} (2y_i + y_j + Y^{(4)}). \tag{5.26}$$

Similarly, (5.15) implies that

$$N = \bigcup_{0 \le i, j \le l-1} (2y_i + y_j + N^{(4)}).$$

Taking into account (5.26), extend the functions $m_k(y)$ from N to some functions $n_k(y)$ on H by the formulas

$$n_k(2y_i + y_j + u) = m_k(2y_i + y_j), \quad u \in Y^{(4)},\ 0 \le i, j \le l-1,$$

$$n_k(2y_i + y_j + u) = 1, \quad u \in Y^{(4)},\ l \le i \le n-1,\ 0 \le j \le l-1, \quad k = 1, 2.$$

It follows from (5.22) and (5.26) that the functions $n_k(y)$ also satisfy the conditions

$$n_k(y) = n_k(-y), \quad y \in H,\ k = 1, 2. \tag{5.27}$$

The functions $n_k(y)$ are invariant with respect to the subgroup $Y^{(4)}$. In particular each of them satisfies the equation

$$n_k(u + 4v) = n_k(u), \quad u, v \in H, \quad k = 1, 2, \tag{5.28}$$

and hence each of them defines a function on the factor-group $H/Y^{(4)}$.

Set $F = A(X/G, Y^{(4)})$. Then $F \subset (X/G)_{(4)}$ and note that $F \cong (H/Y^{(4)})^*$. Taking into account that $H/Y^{(4)}$ is a finite group and applying Lemma 4.8, we conclude that there exist complex measures δ_k on X/G such that they are supported in the group F and $\hat\delta_k(y) = n_k(y)$ for all $y \in H$. It is obvious that $F \cong (\mathbb{Z}(4))^m$ for some m. An arbitrary character of the group $(\mathbb{Z}(4))^m$ is of the form

$$((k_1, \dots, k_m), (l_1, \dots, l_m)) = \exp\left\{\frac{i\pi}{2}\sum_{j=1}^m k_j l_j\right\},$$

$$(k_1, \dots, k_m) \in (\mathbb{Z}(4))^m, \quad (l_1, \dots, l_m) \in (\mathbb{Z}(4))^m.$$

The complex measures δ_k, considered as complex measures on F, are defined by the formulas

$$\delta_k\{(k_1, \dots, k_m)\} = \frac{1}{4^m}\sum_{(l_1,\dots,l_m)\in(\mathbb{Z}(4))^m} n_k(l_1, \dots, l_m)\exp\left\{\frac{i\pi}{2}\sum_{j=1}^m k_j l_j\right\}, \quad k = 1, 2. \tag{5.29}$$

Observe that (5.27) and (5.28) imply that

$$n_k(3y) = n_k(y), \quad y \in H,\ k = 1, 2. \tag{5.30}$$

We also note that $\exp\left\{\frac{i\pi}{2}\sum_{j=1}^m k_j l_j\right\} = \pm i$ if and only if $\sum_{j=1}^m k_j l_j$ is an odd number. Inasmuch as (5.30) implies that

$$n_k(l_1, \dots, l_m) = n_k(3l_1, \dots, 3l_m)$$

and the numbers $\exp\left\{\frac{i\pi}{2}\sum_{j=1}^m k_j l_j\right\}$ and $\exp\left\{\frac{i\pi}{2}\sum_{j=1}^m 3k_j l_j\right\}$ are complex conjugate, it follows from (5.29) that all numbers $\delta_k\{(k_1, \dots, k_m)\}$ are real, i.e., actually δ_k are signed measures.

Return to the representations of the functions $f_k(y)$ on N. We have

$$f_1(y) = \varphi(u) + p_j, \quad f_2(y) = \varphi(y) - p_j, \quad y \in y_j + N^{(2)},\ j = 0, 1, \dots, l-1.$$

By Lemma 4.8, there exist signed measures ϵ_1 and ϵ_2 on $(X/G)_{(2)}$ such that

$$\hat\epsilon_1(y) = e^{p_j}, \quad \hat\epsilon_2(y) = e^{-p_j}, \quad y \in y_j + Y^{(2)},\ j = 0, 1, \dots, l-1. \tag{5.31}$$

Put $\pi_k = \delta_k * \epsilon_k$, $k = 1, 2$. Then π_k are signed measures on $(X/G)_{(4)}$.

By Lemma 3.4, we can extend the function $\varphi(y)$ from N to H retaining its properties. We will keep the notation $\varphi(y)$ for the extended function. Let γ be a symmetric Gaussian distribution on the factor-group X/G with the characteristic function $\hat{\gamma}(y) = \exp\{-\varphi(y)\}$ for all $y \in H$. Set $V = A(X/G, N)$. It is easily seen that V is a compact Corwin subgroup. Thus, we obtain that the restriction to H of the characteristic functions of the distributions μ_k can be represented in the form

$$\hat{\mu}_k(y) = \begin{cases} \exp\{-\varphi(y)\}\hat{\pi}_k(y)(x_k, y), & \text{if} \quad y \in N, \\ 0, & \text{if} \quad y \notin N, \quad k = 1, 2. \end{cases}$$

The desired representation $p(\mu_k) = \gamma * m_V * \pi_k * E_{x_k}$, $k = 1, 2$, results now from (2.3) and Corollary 2.8. □

Taking into account Lemmas 4.5 and 4.6, Theorem 5.2 implies the following statement.

COROLLARY 5.3. *Let X be a locally compact Abelian group such that its connected component of the zero has no elements of order 2. Let ξ_1 and ξ_2 be independent random variables with values in the group X and distributions μ_1 and μ_2. Assume that the sum $\xi_1 + \xi_2$ and difference $\xi_1 - \xi_2$ are independent. Then $\mu_j \in \Gamma(X) * I(X)$, $j = 1, 2$.*

Corollary 5.3 fails if the connected component of the zero of a group X contains an element of order 2. For the proof, we need the following lemmas.

LEMMA 5.4. *Let X be a locally compact Abelian group. If the connected component of the zero of the group X contains an element of order 2, then there exists a compact Corwin subgroup K of the group X such that the factor-group X/K contains a subgroup topologically isomorphic to the circle group $\mathbb{T}$.*

PROOF. Assume that for every compact Corwin subgroup G of the group X the factor-group X/G contains no subgroups topologically isomorphic to the circle group $\mathbb{T}$.

Take a compact Corwin subgroup K of the group X and verify that K contains no elements of order 2. Put $L = K^*$. By Theorem 1.2, L is a discrete group. It follows from Theorem 1.12 that $A(L, K_{(2)}) = L^{(2)}$. Hence if we prove that $L^{(2)} = L$, i.e., L is a Corwin group, then we prove that K has no elements of order 2.

Suppose that there is an element $y_0 \in L$ such that $y_0 \notin L^{(2)}$. Denote by M the subgroup of L generated by the element y_0. Since $K^{(2)} = K$, it follows from Theorem 1.12 that $L_{(2)} = 0$. Hence all nonzero elements of finite order in the group L have odd order. This implies that if $y \in L$ and $y \notin L^{(2)}$, then y is an element of infinite order. Thus, $M \cong \mathbb{Z}$. Take $y \in L$ such that $2y \in M$. Then

$$2y = ky_0. \tag{5.32}$$

Assume k in (5.32) is even. In view of $L_{(2)} = \{0\}$, we have that $y \in M$. Thus, if $2y \in M$, then $y \in M$. This implies that $A(K, M)$ is a compact Corwin group. By Theorems 1.8 and 1.9, $K/A(K, M) \cong M^*$. We have $M^* \cong \mathbb{T}$ and $K/A(K, M)$ is a subgroup of $X/A(K, M)$. This contradicts the fact that for every compact Corwin subgroup G of the group X the factor-group X/G has no subgroups topologically isomorphic to the circle group $\mathbb{T}$. Thus, the case when k is even is impossible.

Assume that k in (5.32) is odd. Then we have $2y = (2l - 1)y_0$. This implies that $y_0 \in L^{(2)}$. The obtained contradiction shows that the case when k is odd is also impossible. This means that L is a Corwin group.

By Theorem 1.15, we have $c_X \cong \mathbb{R}^m \times B$, where $m \geq 0$ and B is a compact connected Abelian group. By Theorem 1.13, B is a Corwin group. We proved that B contains no elements of order 2. The obtained contradiction proves the lemma. □

LEMMA 5.5. *Consider the circle group $\mathbb{T}$ with character group $\mathbb{Z}$. There exist independent random variables ξ_1 and ξ_2 with values in $\mathbb{T}$ and distributions μ_1 and μ_2 with nonvanishing characteristic functions such that the sum $\xi_1 + \xi_2$ and difference $\xi_1 - \xi_2$ are independent, whereas $\mu_j \notin \Gamma(\mathbb{T})$, $j = 1, 2$.*

PROOF. Consider on the group $\mathbb{Z}$ the functions

$$g_1(n) = \begin{cases} \exp\{-an^2\} & \text{if } \ n \in \mathbb{Z}^{(2)}, \\ b\exp\{-an^2\} & \text{if } \ n \notin \mathbb{Z}^{(2)}, \end{cases} \qquad g_2(n) = \begin{cases} \exp\{-an^2\} & \text{if } \ n \in \mathbb{Z}^{(2)}, \\ b^{-1}\exp\{-an^2\} & \text{if } \ n \notin \mathbb{Z}^{(2)}. \end{cases}$$

Choose real numbers a and b in such a way that

$$\sum_{n\in\mathbb{Z}} g_j(n) < 2, \quad j = 1, 2.$$

It follows from this that

$$f_j(t) = \sum_{n\in\mathbb{Z}} g_j(n)\exp\{-int\} > 0, \quad t \in \mathbb{R}, \ j = 1, 2.$$

It is also obvious that

$$\frac{1}{2\pi}\int_{-\pi}^{\pi} f_j(t)dt = 1, \quad j = 1, 2.$$

Let μ_j be the distribution on the circle group $\mathbb{T}$ with the density $r_j(e^{it}) = f_j(t)$ with respect to $m_{\mathbb{T}}$. Then $\hat{\mu}_j(n) = g_j(n)$, $n \in \mathbb{Z}$, $j = 1, 2$. Let ξ_1 and ξ_2 be independent random variables with values in X and distributions μ_1 and μ_2. It is obvious that $\mu_j \notin \Gamma(\mathbb{T})$ and $\hat{\mu}_j(n) \neq 0$ for all $n \in \mathbb{Z}$. We will verify that $\xi_1 + \xi_2$ and $\xi_1 - \xi_2$ are independent. Thus, the lemma will be proved.

By Lemma 4.2, it suffices to show that the characteristic functions $\hat{\mu}_j(n)$ satisfy equation (4.1). By considering separately the cases:

(i) $m, n \in \mathbb{Z}^{(2)}$;
(ii) either $m \in \mathbb{Z}^{(2)}$, $n \notin \mathbb{Z}^{(2)}$ or $m \notin \mathbb{Z}^{(2)}, n \in \mathbb{Z}^{(2)}$;
(iii) $m, n \notin \mathbb{Z}^{(2)}$;

we are convinced of this. □

PROPOSITION 5.6. *Let X be a locally compact Abelian group. If the connected component of the zero of the group X contains an element of order 2, then there exist independent random variables ξ_1 and ξ_2 with values in X and distributions λ_1 and λ_2 such that the sum $\xi_1 + \xi_2$ and difference $\xi_1 - \xi_2$ are independent, whereas $\lambda_j \notin \Gamma(X) * I(X)$, $j = 1, 2$.*

PROOF. Denote by Y the character group of the group X. Assume that the connected component of the zero of X contains an element of order 2. Then by Lemma 5.4, there exists a compact Corwin subgroup K of the group X such that the factor-group X/K contains a subgroup F topologically isomorphic to the circle group $\mathbb{T}$. Therefore, we can consider the distributions μ_j, constructed in Lemma 5.5, as distributions on the factor-group X/K. We also denote these distributions by μ_j.

Since by Theorem 1.9, $(X/K)^* \cong A(Y,K)$, we may suppose that the characteristic functions $\hat{\mu}_j(y)$ are defined on the annihilator $A(Y,K)$.

Consider on the group Y the functions

$$h_j(y) = \begin{cases} \hat{\mu}_j(y) & \text{if} \quad y \in A(Y,K), \\ 0 & \text{if} \quad y \notin A(Y,K), \quad j = 1,2. \end{cases}$$

Inasmuch as $A(Y,K)$ is a subgroup and $\hat{\mu}_j(y)$ are positive definite functions, by Proposition 2.9, $h_j(y)$ are also positive definite functions. By Theorem 1.11, the annihilator $A(Y,K)$ is an open subgroup, because K is a compact subgroup. Hence the functions $h_j(y)$ are continuous. By Bochner's theorem, there exist distributions $\lambda_j \in \mathrm{M}^1(X)$ such that $\hat{\lambda}_j(y) = h_j(y)$ for all $y \in Y$.

Let ξ_1 and ξ_2 be independent random variables with values in X and distributions λ_j. We will check that the sum $\xi_1+\xi_2$ and difference $\xi_1-\xi_2$ are independent. In view of Lemma 4.2, it suffices to show that the characteristic functions $\hat{\lambda}_j(y)$ satisfy equation (4.1). Consider three cases:

1. $u, v \in A(Y,K)$. Since the characteristic functions $\hat{\mu}_j(y)$ satisfy equation (4.1), equation (4.1) holds.

2. Either $u \in A(Y,K)$, $v \notin A(Y,K)$ or $v \in A(Y,K)$, $u \notin A(Y,K)$. Then both sides of equation (4.1) are equal to zero.

3. $u, v \notin A(Y,K)$. Then the right-hand side of equation (4.1) is equal to zero. If the left-hand side of equation (4.1) is not equal to zero, then $u \pm v \in A(Y,K)$. Hence $2u \in A(Y,K)$. Inasmuch as K is a compact Corwin group, this implies that $u \in A(Y,K)$, which contradicts the assumption. Thus, the left-hand side of equation (4.1) is also equal to zero.

We see that for all $u, v \in Y$ equation (4.1) becomes the equality. □

Corollary 5.3 and Proposition 5.6 imply directly the following statement.

THEOREM 5.7. *Let X be a locally compact Abelian group. Let ξ_1 and ξ_2 be independent random variables with values in the group X and distributions μ_1 and μ_2. The independence of the sum $\xi_1+\xi_2$ and difference $\xi_1-\xi_2$ implies that $\mu_j \in \Gamma(X) * I(X)$, $j=1,2$, if and only if the connected component of the zero of the group X contains no elements of order 2.*

Let X be a locally compact Abelian group topologically isomorphic to a group of the form (4.6). Assume also that the group X contains only one element of order 2, i.e., $X_{(2)} \cong \mathbb{Z}(2)$. Then we can strengthen Theorem 5.2. The following statement holds.

THEOREM 5.8. *Let X be a locally compact Abelian group topologically isomorphic to a group of the form (4.6). Suppose X contains only one element of order 2. Let p be the natural homomorphism $p: X \to X/X_{(2)}$. Assume that ξ_1 and ξ_2 are independent random variables with values in the group X and distributions μ_1 and μ_2 such that the sum $\xi_1+\xi_2$ and difference $\xi_1-\xi_2$ are independent. Then either*

$$\mu_k = \gamma * m_V * \epsilon_k * E_{z_k}, \quad k=1,2,$$

where $\gamma \in \Gamma^s(X)$, V is a compact Corwin subgroup of the group X, ϵ_k are signed measures on $X_{(2)}$ such that $\epsilon_1 * \epsilon_2 = E_0$, $z_k \in X$ or

$$p(\mu_k) = \gamma * m_V * E_{x_k}, \quad k=1,2,$$

where $\gamma \in \Gamma^s(X/X_{(2)})$, V *is a compact Corwin subgroup of the factor-group* $X/X_{(2)}$, $x_k \in X/X_{(2)}$ *and at least one of the distributions* μ_k *is represented in the form*

$$\mu_k = \lambda * m_U * E_{z_k},$$

where $\lambda \in \Gamma^s(X)$, U *is a compact Corwin subgroup of* X, $z_k \in X$.

PROOF. We argue as in the proof of Theorem 5.2 and retain the same notation. Since $X_{(2)}$ is a finite group and by Theorem 1.12, $Y^{(2)} = A(Y, X_{(2)})$, we have $X_{(2)} \cong (X_{(2)})^* \cong Y/Y^{(2)}$. Hence a $Y^{(2)}$-coset decomposition of the group Y is of the form

$$Y = Y^{(2)} \cup (y_1 + Y^{(2)}).$$

There are two possibilities for the subgroup H:

1. $H = Y$ and then $G = A(X, Y) = \{0\}$;
2. $H = Y^{(2)}$ and then $G = A(X, Y^{(2)}) = X_{(2)}$.

1. Assume $H = Y$. Then we have

$$N = N^{(2)} \cup (y_1 + N^{(2)}). \tag{5.33}$$

Since $m_1(y) = m_2(y)$ for all $y \in N^{(2)}$ and $m_1(y) = \pm m_2(y) = \pm 1$ for all $y \in y_1 + N^{(2)}$, it follows from (5.33) that the functions $m_k(y)$ are characters of the subgroup N. By Theorem 1.9, there exist elements $t_k \in X$ such that $m_k(y) = (t_k, y)$ for all $y \in N$. Set $z_k = x_k + t_k$. We obtain as a result the representation

$$\mu_k = \gamma * m_V * \epsilon_k * E_{z_k}, \quad k = 1, 2,$$

where $\gamma \in \Gamma^s(X)$, V is a compact Corwin subgroup of the group X, ϵ_k are signed measures on $X_{(2)}$, $z_k \in X$. Moreover, it follows from (5.31) that $\hat{\epsilon}_1(y)\hat{\epsilon}_2(y) = 1$ for all $y \in Y$. Hence $\epsilon_1 * \epsilon_2 = E_0$.

2. Assume $H = Y^{(2)}$. Then $N \subset Y^{(2)}$ and it follows from $N \cap Y^{(2)} = N^{(2)}$ that

$$N = N^{(2)}. \tag{5.34}$$

There are two possibilities for the subgroup N. Namely, either $N \neq \{0\}$ or $N = \{0\}$.

2a. Let $N \neq \{0\}$. Put $W = N^*$. It follows from (5.34) that the group W has no elements of order 2. Consider the restriction of equation (4.1) to the subgroup N and apply Corollary 5.3 to the group W. We get that the restrictions of the characteristic functions $\hat{\mu}_k(y)$ to N are the characteristic functions of some Gaussian distributions. It follows easily from this that

$$p(\mu_k) = \gamma * m_V * E_{x_k}, \quad k = 1, 2, \tag{5.35}$$

where $\gamma \in \Gamma^s(X/X_{(2)})$, V is a compact Corwin subgroup of the factor-group $X/X_{(2)}$, $x_k \in X/X_{(2)}$. Moreover, since either $\hat{\mu}_1(y) = 0$ or $\hat{\mu}_2(y) = 0$ for all $y \in y_1 + Y^{(2)}$, it is not difficult to prove taking into account (5.35) that at least one of the distributions μ_k is represented in the form

$$\mu_k = \lambda * m_U * E_{z_k},$$

where $\lambda \in \Gamma^s(X)$, U is a compact Corwin subgroup of X, $z_k \in X$.

2b. Let $N = \{0\}$. It is possible only if X is a compact group. We have $V = A(X/X_{(2)}, N) = X/X_{(2)}$. Taking into account that the equalities

$$N_1 \cap Y^{(2)} = N_2 \cap Y^{(2)} = N \cap Y^{(2)}$$

always hold, we get the representation

$$p(\mu_k) = m_{X/X_{(2)}}, \quad k = 1, 2. \tag{5.36}$$

Moreover, since either $\hat\mu_1(y) = 0$ or $\hat\mu_2(y) = 0$ for all $y \in y_1 + Y^{(2)}$, it follows from (5.36) that $\mu_k = m_X$ at least for one of the distributions μ_k. For definiteness, suppose that $\mu_1 = m_X$. Returning to the random variables ξ_k, this means that the random variables ξ_1 and $2\xi_2$ are identically distributed with distribution m_X. □

EXAMPLE 5.9. We illustrate Theorem 5.8 by the following example. Consider the circle group $\mathbb{T}$ with character group $\mathbb{Z}$. We have $\mathbb{T}_{(2)} \cong \mathbb{Z}(2)$. Let a subgroup G be the same as in Theorem 5.8. Then there are two possibilities for G: either $G = \{0\}$ or $G = \mathbb{T}_{(2)}$.

1. $G = \{0\}$. We observe that all compact Corwin subgroups V of the circle group $\mathbb{T}$ are of the form: either $V = \mathbb{T}$ or V is the subgroup of mth roots of 1, where m is odd, i.e., $V \cong \mathbb{Z}(m)$. If $V = \mathbb{T}$, then by Theorem 5.8, we have

$$\mu_1 = \mu_2 = m_{\mathbb{T}}.$$

Let $V \cong \mathbb{Z}(m)$. Then by Theorem 5.8,

$$\mu_k = \gamma * m_V * \epsilon_k * E_{z_k}, \tag{5.37}$$

where $\gamma \in \Gamma^s(\mathbb{T})$, ϵ_k are signed measures on $\mathbb{T}_{(2)}$ such that $\epsilon_1 * \epsilon_2 = E_1$, $z_k \in \mathbb{T}$, $k = 1, 2$. It follows from $V \cong \mathbb{Z}(m)$ that $N = \mathbb{Z}^{(m)}$. Then (5.37) implies that the characteristic functions of distributions μ_k are represented in the form

$$\hat\mu_1(n) = \begin{cases} \exp\{-\sigma n^2 + int_1\}, & \text{if } \ n \in \mathbb{Z}^{(2m)}, \\ \exp\{-\sigma n^2 + int_1 + q\}, & \text{if } \ n \in \mathbb{Z}^{(m)} \backslash \mathbb{Z}^{(2m)}, \\ 0, & \text{if } \ n \notin \mathbb{Z}^{(m)}, \end{cases}$$

$$\hat\mu_2(n) = \begin{cases} \exp\{-\sigma n^2 + int_2\}, & \text{if } \ n \in \mathbb{Z}^{(2m)}, \\ \exp\{-\sigma n^2 + int_2 - q\}, & \text{if } \ n \in \mathbb{Z}^{(m)} \backslash \mathbb{Z}^{(2m)}, \\ 0, & \text{if } \ n \notin \mathbb{Z}^{(m)}, \end{cases}$$

where $\sigma \geq 0$, t_k, $q \in \mathbb{R}$, $k = 1, 2$.

2. $G = \mathbb{T}_{(2)}$. It follows from (5.34) that $N = \{0\}$. Hence $V = A(\mathbb{T}/\mathbb{T}_{(2)}, N) = \mathbb{T}/\mathbb{T}_{(2)}$. We get by Theorem 5.8, $p(\mu_k) = m_{\mathbb{T}/\mathbb{T}_{(2)}}$, $k = 1, 2$, i.e., $\hat\mu_1(2n) = \hat\mu_2(2n) = 0$, $n \in \mathbb{Z}$, $n \neq 0$. Moreover, either $\hat\mu_1(2n+1) = 0$ or $\hat\mu_2(2n+1) = 0$ for all $n \in \mathbb{Z}$. This implies at least one of the distributions μ_k, say μ_1, such that $\mu_1 = m_{\mathbb{T}}$. As regards to the second distribution, we know only that $\hat\mu_2(2n) = 0$ for all $n \in \mathbb{Z}$, $n \neq 0$. Returning to the random variables ξ_k this means that ξ_1 and $2\xi_2$ are identically distributed random variables with the distribution $m_{\mathbb{T}}$.

The obtained description of possible distributions μ_k for the circle group $\mathbb{T}$ is the main content of the article [**4**]. Similarly, it is possible to get from Theorem 5.8 the description of possible distributions μ_k for the group $\mathbb{R} \times \mathbb{T}$ and for the $\boldsymbol{a}$-adic solenoids $\Sigma_{\boldsymbol{a}}$ found in [**76**].

EXAMPLE 5.10. Consider the 2-dimensional torus $\mathbb{T}^2$ with character group $\mathbb{Z}^2$. Denote by (e^{it}, e^{is}), where $(t, s) \in \mathbb{R}^2$, elements of the group $\mathbb{T}^2$. We will construct independent random variables ξ_1 and ξ_2 with values in $\mathbb{T}^2$ and distributions μ_1 and μ_2 such that the sum $\xi_1 + \xi_2$ and difference $\xi_1 - \xi_2$ are independent. In so doing $\mu_k = \gamma * \pi_k$, $k = 1, 2$, where $\gamma \in \Gamma^s(\mathbb{T}^2)$, and π_k are signed measures such that they are supported in $\mathbb{T}^2_{(4)}$ but not in $\mathbb{T}^2_{(2)}$. Thus, we will show that the statement in Theorem 5.2 that π_k are signed measures on $\left(\mathbb{T}^2/G\right)_{(4)}$, generally speaking cannot be strengthened to the statement that π_k are signed measures on $\left(\mathbb{T}^2/G\right)_{(2)}$ (compare with Theorem 5.8).

Consider on the group $\mathbb{Z}^2$ the functions

$$l_1(m,n) = \begin{cases} 1, & \text{if } (m,n) \in \{(\mathbb{Z}^2)^{(2)},\ (1,0)+(\mathbb{Z}^2)^{(4)},\ (3,0)+(\mathbb{Z}^2)^{(4)}, \\ & \qquad (0,1)+(\mathbb{Z}^2)^{(4)},\ (0,3)+(\mathbb{Z}^2)^{(4)}, \\ & \qquad (1,1)+(\mathbb{Z}^2)^{(4)},\ (3,3)+(\mathbb{Z}^2)^{(4)}\}, \\ -1, & \text{if } (m,n) \in \{(1,2)+(\mathbb{Z}^2)^{(4)},\ (3,2)+(\mathbb{Z}^2)^{(4)}, \\ & \qquad (2,1)+(\mathbb{Z}^2)^{(4)},\ (2,3)+(\mathbb{Z}^2)^{(4)}, \\ & \qquad (1,3)+(\mathbb{Z}^2)^{(4)},\ (3,1)+(\mathbb{Z}^2)^{(4)}\} \end{cases}$$

and

$$l_2(m,n) = \begin{cases} 1, & \text{if } (m,n) \in \{(\mathbb{Z}^2)^{(2)},\ (1,0)+(\mathbb{Z}^2)^{(4)},\ (3,0)+(\mathbb{Z}^2)^{(4)}, \\ & \qquad (0,1)+(\mathbb{Z}^2)^{(4)},\ (0,3)+(\mathbb{Z}^2)^{(4)}, \\ & \qquad (1,3)+(\mathbb{Z}^2)^{(4)},\ (3,1)+(\mathbb{Z}^2)^{(4)}\}, \\ -1, & \text{if } (m,n) \in \{(1,2)+(\mathbb{Z}^2)^{(4)},\ (3,2)+(\mathbb{Z}^2)^{(4)}, \\ & \qquad (2,1)+(\mathbb{Z}^2)^{(4)},\ (2,3)+(\mathbb{Z}^2)^{(4)}, \\ & \qquad (1,1)+(\mathbb{Z}^2)^{(4)},\ (3,3)+(\mathbb{Z}^2)^{(4)}\}. \end{cases}$$

Take $\sigma > 0$ such that

$$\sum_{(m,n)\in\mathbb{Z}^2} \exp\{-\sigma(m^2+n^2)\} < 2.$$

It follows from this that the inequalities

$$(5.38)\quad f_k(t,s) = \sum_{(m,n)\in\mathbb{Z}^2} \exp\{-\sigma(m^2+n^2) - i(mt+ns)\} l_k(m,n) > 0,$$
$$(t,s) \in \mathbb{R}^2,\ k=1,2,$$

hold. It is obvious that

$$\frac{1}{4\pi^2}\int_{-\pi}^{\pi}\int_{-\pi}^{\pi} f_k(t,s)dtds = 1, \quad k=1,2.$$

Let μ_k be the distributions on the 2-dimensional torus $\mathbb{T}^2$ with the densities $r_k(e^{it},e^{is}) = f_k(t,s)$ with respect to the Haar distribution $m_{\mathbb{T}^2}$. Let ξ_k be independent random variables with values in $\mathbb{T}^2$ and distributions μ_k, $k=1,2$. It is obvious that

$$(5.39)\qquad \hat\mu_k(m,n) = \exp\{-\sigma(m^2+n^2)\} l_k(m,n), \quad (m,n)\in\mathbb{Z}^2,\ k=1,2.$$

We can check directly that the functions $l_k(m,n)$ satisfy equation (4.1). This implies that the characteristic functions $\hat\mu_k(m,n)$ also satisfy equation (4.1). By Lemma 4.2, the sum $\xi_1+\xi_2$ and difference $\xi_1-\xi_2$ are independent.

By the construction, the functions $l_k(m,n)$ are invariant with respect to the subgroup $(\mathbb{Z}^2)^{(4)}$. Therefore, they define some functions $\widetilde{l}_k(m,n)$ on the factor-group $\mathbb{Z}^2/(\mathbb{Z}^2)^{(4)} \cong (\mathbb{Z}(4))^2$. By Lemma 4.8, there exist complex measures π_k on the group $\mathbb{T}^2$ supported in $\mathbb{T}^2_{(4)}$ such that

$$(5.40)\qquad \hat\pi_k(m,n) = l_k(m,n), \quad (m,n)\in\mathbb{Z}^2,\ k=1,2.$$

Since the functions $\widetilde{l}_k(m,n)$ satisfy condition (5.30), arguing as in the proof of Theorem 5.2, we get that the complex measures π_k actually are signed measures. Inasmuch as the characteristic functions $\hat{\pi}_k(m,n)$ are not invariant with respect to the subgroup $(\mathbb{Z}^2)^{(2)}$, the signed measures π_k are not supported in $\mathbb{T}^2_{(2)}$.

Let γ be the symmetric Gaussian distribution on the group $\mathbb{T}^2$ with the characteristic function

$$\hat{\gamma}(m,n) = \exp\{-\sigma(m^2+n^2)\}, \quad (m,n) \in \mathbb{Z}^2. \tag{5.41}$$

It follows from (5.39)–(5.41) that $\mu_k = \gamma * \pi_k$, $k = 1, 2$.

REMARK 5.11. Assume that under conditions of Theorem 5.2 the independent random variables ξ_1 and ξ_2 are identically distributed, i.e., $\mu_1 = \mu_2 = \mu \in \Gamma_B(X)$. By Theorem 4.9, this implies that

$$\mu = \gamma * m_K * \pi,$$

where $\gamma \in \Gamma(X)$, K is a compact Corwin subgroup of the group X, π is a signed measure on $X_{(2)}$ such that $\pi^{*2} = E_0$. Taking into account Theorem 5.2 and Example 5.10, we see that even if we suppose $G = \{0\}$ in Theorem 5.2, the distributions μ_k have, generally speaking, more complicated decomposition than in the case of identically distributed random variables. If the connected component of the zero of a group X contains more than one element of order 2, in this case an important role also play elements of order 4.

Notes

The Kac–Bernstein theorem was discovered independently by M. Kac in [**68**] and by S.N. Bernstein in [**14**]. This result is one of the most celebrated characterization theorems in mathematical statistics.

Theorem 4.14, where the complete description of locally compact Abelian groups X for which every Gaussian distribution in the sense of Bernstein is represented as a convolution of a Gaussian distribution and an idempotent distribution, i.e., equality (4.24) holds, was proved by G.M. Feldman in [**24**], see also [**36**, Theorem 9.9]. It should be noted that A. Rukhin in [**90**] proved equality (4.24) assuming that Y is a Corwin group. H. Heyer and Ch. Rall in [**67**] proved equality (4.24) for groups X satisfying the following conditions: X is a Corwin group and either $Y = \overline{Y^{(2)}}$ or $Y/\overline{Y^{(2)}} \cong \mathbb{Z}(2)$. Lemma 4.5 and Proposition 4.16 were proved by G.M. Feldman in [**24**]. Theorems 4.9, 4.18, 4.20 and Lemma 4.17 were proved by G.M. Feldman in [**39**].

We note that Gaussian distributions in the sense of Bernstein on arbitrary topological separable metric Abelian groups, need not be locally compact, were studied by T. Byczkowski in [**18, 19**] and by H. Byczkowska and T. Byczkowski in [**17**]. In particular for a wide class of such groups X in [**18, 19**] it was proved that if μ is a Gaussian distribution in the sense of Bernstein on X without nontrivial idempotent factors and B is a Borel subgroup of X, then either $\mu(B) = 0$ or $\mu(B) = 1$ (a zero-one low).

Let ξ_1 and ξ_2 be independent random variables with values in a locally compact Abelian group X and distributions μ_1 and μ_2 such that their sum and difference are independent. A.L. Rukhin in [**89, 90**] obtained a sufficient condition on X under which distributions μ_1 and μ_2 were represented as convolutions of Gaussian distributions and idempotent distributions. He proved that this is the case if both

X and Y are Corwin groups. The complete descriptions of such groups has been obtained by G.M. Feldman in [**24**]. These are groups with the connected component of the zero containing no elements of order 2. If the connected component of the zero of a group X contains an element of order 2, then the situation is much more complicated. This was shown by Y. Baryshnikov, B. Eisenberg and W. Stadje in [4] for the circle group $\mathbb{T}$ and by M.V. Myronyuk in [**76**] for the group $\mathbb{R} \times \mathbb{T}$ and $\boldsymbol{a}$-adic solenoids $\Sigma_{\boldsymbol{a}}$. The results of Section 5, with the exception of Lemma 5.1, were obtained by G.M. Feldman in [**43**]. Lemma 5.1 was proved in [**38**].

We make some remarks on non-Abelian groups. Let X be a second countable non-Abelian topological group, but it need not be locally compact. The operation in X we write as multiplication. Let ξ_1 and ξ_2 be independent identically distributed random variables with values in X and distribution μ. In the article [**83**] D. Neuenschwander suggested in the non-Abelian case instead of the condition:

(i) $\xi_1\xi_2$ and $\xi_1\xi_2^{-1}$ are independent,

to consider the condition:

(ii) $(\xi_1\xi_2,\ \xi_2\xi_1)$ and $(\xi_1\xi_2^{-1},\ \xi_2^{-1}\xi_1)$ are independent.

We note that if X is an Abelian group, then conditions (i) and (ii) are equivalent. In the case of the Heisenberg group $\mathcal{H}$, D. Neuenschwander proved that it follows from condition (ii) that μ is a Gaussian distribution concentrated on an Abelian subgroup of $\mathcal{H}$ [**83**]. Then D. Neuenschwander, B. Roynette, and R. Schott generalized this result to simply connected nilpotent Lie groups [**84**], see also [**57**]. P. Graczyk and J.-J. Loeb proved in [**61**] that if X is an arbitrary nilpotent group, then condition (ii) implies that the distribution μ is concentrated on an Abelian subgroup of X. In contrast to [**84**] the proof by P. Graczyk–J.-J. Loeb uses neither the fact that X is a Lie group nor the Campbell-Hausdorff formula. In [**61**] they also proved that if X is either a discrete group or a compact group, or X is in a class of solvable groups (e.g. the "$ax + b$" group), then condition (ii) implies that the distribution μ is concentrated on an Abelian subgroup of X. As a corollary, they show that the Gaussian distributions on Riemannian symmetric spaces do not satisfy condition (ii).

CHAPTER III

Characterization of probability distributions through the independence of linear forms

6. General characterization theorems

The Kac–Bernstein theorem was the first among characterization theorems where independent linear forms of independent random variables ξ_j under different restrictions on ξ_j were studied. These studies were completed with the following Skitovich–Darmois theorem: Let ξ_j, $j = 1, 2, \dots, n$, $n \geq 2$, be independent random variables and let α_j, β_j be nonzero real numbers. If the linear forms $L_1 = \alpha_1\xi_1 + \cdots + \alpha_n\xi_n$ and $L_2 = \beta_1\xi_1 + \cdots + \beta_n\xi_n$ are independent, then all random variables ξ_j are Gaussian. The Skitovich–Darmois theorem was generalized by S.G. Ghurye and I. Olkin to the multivariable case when instead of random variables, random vectors ξ_j in the space $\mathbb{R}^m$ are considered and coefficients of the linear forms L_1 and L_2 are nonsingular $m \times m$ matrices. In this case the independence of L_1 and L_2 implies that all random vectors ξ_j are Gaussian. It should be noted that nonsingular $m \times m$ matrices are topological automorphisms of the group $\mathbb{R}^m$.

Let X be a second countable locally compact Abelian group. In this section, we begin the study of group analogues of the Skitovich–Darmois theorem. Let ξ_j, $j = 1, 2, \dots, n$, $n \geq 2$, be independent random variables with values in X and distributions μ_j with nonvanishing characteristic functions. First we prove that if the group X contains no subgroups topologically isomorphic to the circle group $\mathbb{T}$, then the independence of the linear forms $L_1 = \alpha_1\xi_1 + \cdots + \alpha_n\xi_n$ and $L_2 = \beta_1\xi_1 + \cdots + \beta_n\xi_n$, where $\alpha_j, \beta_j \in \mathrm{Aut}(X)$, implies that all distributions μ_j are Gaussian. This theorem fails if X contains a subgroup topologically isomorphic to the circle group $\mathbb{T}$. We also prove that for an arbitrary locally compact Abelian group X the description of distributions which are characterized by the independence of L_1 and L_2 is reduced to the case when X is a connected group. Next, based on this result, we prove that this problem is reduced to the case when the group X is topologically isomorphic to a group of the form $\mathbb{R}^{\mathfrak{n}} \times \mathbb{T}^{\mathfrak{m}}$, where $\mathfrak{n} \leq \aleph_0$, $\mathfrak{m} \leq \aleph_0$. It should be noted that the group $\mathbb{R}^{\aleph_0}$ is not locally compact, although its structure is simple enough. For two independent random variables we prove that if a group X contains no subgroups topologically isomorphic to the 2-dimensional torus $\mathbb{T}^2$, then the independence of L_1 and L_2 implies that μ_j are either Gaussian distributions or convolutions of Gaussian distributions and signed measures supported in a subgroup of X generated by an element of X of order 2.

LEMMA 6.1. *Let X be a locally compact Abelian group with character group Y. Let α_j, β_j, $j = 1, 2, \dots, n$, $n \geq 2$, be continuous endomorphisms of X. Let ξ_j be independent random variables with values in the group X and distributions μ_j. The linear forms $L_1 = \alpha_1\xi_1 + \cdots + \alpha_n\xi_n$ and $L_2 = \beta_1\xi_1 + \cdots + \beta_n\xi_n$ are independent if* *and only if the characteristic functions $\hat\mu_j(y)$ satisfy the equation*

$$\prod_{i=1}^n \hat\mu_j(\widetilde\alpha_j u + \widetilde\beta_j v) = \prod_{i=1}^n \hat\mu_j(\widetilde\alpha_j u)\prod_{i=1}^n \hat\mu_j(\widetilde\beta_j v), \quad u, v \in Y. \tag{6.1}$$

PROOF. We note that the linear forms L_1 and L_2 are independent if and only if the equality

$$\begin{aligned}(6.2)\quad &\mathbf{E}[(\alpha_1\xi_1 + \cdots + \alpha_n\xi_n, u)(\beta_1\xi_1 + \cdots + \beta_n\xi_n, v)] \\ &\qquad = \mathbf{E}[(\alpha_1\xi_1 + \cdots + \alpha_n\xi_n, u)]\mathbf{E}[(\beta_1\xi_1 + \cdots + \beta_n\xi_n, v)]\end{aligned}$$

holds for all $u, v \in Y$. Taking into account that the random variables ξ_j are independent and $\hat\mu_j(y) = \mathbf{E}[(\xi_j, y)]$, we transform the left-hand side of equality (6.2) as follows

$$\begin{aligned}\mathbf{E}[(\alpha_1\xi_1 + \cdots + \alpha_n\xi_n, u)(\beta_1\xi_1 + \cdots + \beta_n\xi_n, v)] &= \mathbf{E}\left[\prod_{j=1}^n (\xi_j, \widetilde\alpha_j u + \widetilde\beta_j v)\right] \\ = \prod_{j=1}^n \mathbf{E}[(\xi_j, \widetilde\alpha_j u + \widetilde\beta_j v)] &= \prod_{j=1}^n \hat\mu_j(\widetilde\alpha_j u + \widetilde\beta_j v), \quad u, v \in Y.\end{aligned}$$

We transform similarly the right-hand side of equality (6.2) as follows

$$\begin{aligned}\mathbf{E}[(\alpha_1\xi_1 + \cdots + \alpha_n\xi_n, u)]\mathbf{E}[(\beta_1\xi_1 + \cdots + \beta_n\xi_n, v)] &= \mathbf{E}\left[\prod_{j=1}^n (\xi_j, \widetilde\alpha_j u)\right]\mathbf{E}\left[\prod_{j=1}^n (\xi_j, \widetilde\beta_j v)\right] \\ = \prod_{j=1}^n \mathbf{E}[(\xi_j, \widetilde\alpha_j u)]\prod_{j=1}^n \mathbf{E}[(\xi_j, \widetilde\beta_j v)] &= \prod_{j=1}^n \hat\mu_j(\widetilde\alpha_j u)\prod_{j=1}^n \hat\mu_j(\widetilde\beta_j v), \quad u, v \in Y.\end{aligned}$$

□

Equation (6.1) is called the *Skitovich–Darmois functional equation.*

COROLLARY 6.2. *Let X be a locally compact Abelian group. Let α_j, β_j, $j = 1, 2, \dots, n$, $n \geq 2$, be continuous endomorphisms of X. Let ξ_j be independent random variables with values in the group X and distributions μ_j. Assume that the linear forms $L_1 = \alpha_1\xi_1 + \cdots + \alpha_n\xi_n$ and $L_2 = \beta_1\xi_1 + \cdots + \beta_n\xi_n$ are independent. Let λ_j be shifts of μ_j and let η_j be independent random variables with values in the group X and distributions λ_j. Then the linear forms $M_1 = \alpha_1\eta_1 + \cdots + \alpha_n\eta_n$ and $M_2 = \beta_1\eta_1 + \cdots + \beta_n\eta_n$ are also independent.*

THEOREM 6.3. *Let X be a locally compact Abelian group containing no subgroups topologically isomorphic to the circle group $\mathbb{T}$. Let α_j, β_j, $j = 1, 2, \dots, n$, $n \geq 2$, be topological automorphisms of X. Let ξ_j be independent random variables with values in the group X and distributions μ_j with nonvanishing characteristic functions. If the linear forms $L_1 = \alpha_1\xi_1 + \cdots + \alpha_n\xi_n$ and $L_2 = \beta_1\xi_1 + \cdots + \beta_n\xi_n$ are independent, then $\mu_j \in \Gamma(X)$, $j = 1, 2, \dots, n$.*

PROOF. Denote by Y the character group of the group X. Let ξ be a random variable with values in X and distribution μ. Let $\alpha \in \mathrm{Aut}(X)$. By Proposition 2.7, the characteristic function of the random variable $\alpha\xi$ is equal to $\hat\mu(\widetilde\alpha y)$. Taking into account Definition 3.1, it follows from this that $\mu \in \Gamma(X)$ if and only if $\alpha(\mu) \in \Gamma(X)$. Therefore, we can put $\zeta_j = \alpha_j\xi_j$ and reduce the proof of the theorem to the case when $L_1 = \xi_1 + \dots + \xi_n$ and $L_2 = \delta_1\xi_1 + \dots + \delta_n\xi_n$, where $\delta_j \in \mathrm{Aut}(X)$. By Lemma 6.1, the characteristic functions $\hat{\mu}_j(y)$ satisfy equation (6.1) which takes the form

$$\prod_{j=1}^{n} \hat{\mu}_j(u + \widetilde{\delta}_j v) = \prod_{j=1}^{n} \hat{\mu}_j(u) \prod_{j=1}^{n} \hat{\mu}_j(\widetilde{\delta}_j v), \quad u, v \in Y. \tag{6.3}$$

Put $\nu_j = \mu_j * \bar{\mu}_j$. Then $\hat{\nu}_j(y) = |\hat{\mu}_j(y)|^2 > 0$ for all $y \in Y$. The characteristic functions $\hat{\nu}_j(y)$ also satisfy equation (6.3). If we prove that $\nu_j \in \Gamma(X)$, then applying Theorem 3.14 we get that $\mu_j \in \Gamma(X)$. Therefore, we can assume that $\hat{\mu}_j(y) > 0$ for all $y \in Y$, $j = 1, 2, \dots, n$.

Put $\varphi_j(y) = -\ln \hat{\mu}_j(y)$. It follows from (6.3) that the functions $\varphi_j(y)$ satisfy the equation

$$\sum_{j=1}^{n} \varphi_j(u + \widetilde{\delta}_j v) = P(u) + Q(v), \quad u, v \in Y, \tag{6.4}$$

where

$$P(u) = \sum_{j=1}^{n} \varphi_j(u), \quad Q(v) = \sum_{j=1}^{n} \varphi_j(\widetilde{\delta}_j v). \tag{6.5}$$

We use the finite difference method to solve equation (6.4). Let h_1 be an arbitrary element of the group Y. Set $k_1 = -\widetilde{\delta}_n^{-1} h_1$. Then $h_1 + \widetilde{\delta}_n k_1 = 0$. Substitute $u + h_1$ for u and $v + k_1$ for v in equation (6.4). Subtracting equation (6.4) from the resulting equation we obtain

$$\sum_{j=1}^{n-1} \Delta_{l_{1j}} \varphi_j(u + \widetilde{\delta}_j v) = \Delta_{h_1} P(u) + \Delta_{k_1} Q(v), \quad u, v \in Y, \tag{6.6}$$

where $l_{1j} = h_1 + \widetilde{\delta}_j k_1 = (\widetilde{\delta}_j - \widetilde{\delta}_n) k_1$, $j = 1, 2, \dots, n-1$. Let h_2 be an arbitrary element of the group Y. Put $k_2 = -\widetilde{\delta}_{n-1}^{-1} h_2$. Then $h_2 + \widetilde{\delta}_{n-1} k_2 = 0$. Substitute $u + h_2$ for u and $v + k_2$ for v in equation (6.6). Subtracting equation (6.6) from the resulting equation we obtain

$$\sum_{j=1}^{n-2} \Delta_{l_{2j}} \Delta_{l_{1j}} \varphi_j(u + \widetilde{\delta}_j v) = \Delta_{h_2} \Delta_{h_1} P(u) + \Delta_{k_2} \Delta_{k_1} Q(v), \quad u, v \in Y,$$

where $l_{2j} = h_2 + \widetilde{\delta}_j k_2 = (\widetilde{\delta}_j - \widetilde{\delta}_{n-1}) k_2$, $j = 1, 2, \dots, n-2$. Arguing as above we obtain the equation

$$\begin{aligned} \Delta_{l_{n-1,1}} \Delta_{l_{n-2,1}} \dots \Delta_{l_{11}} \varphi_1(u + \widetilde{\delta}_1 v) \\ = \Delta_{h_{n-1}} \Delta_{h_{n-2}} \dots \Delta_{h_1} P(u) + \Delta_{k_{n-1}} \Delta_{k_{n-2}} \dots \Delta_{k_1} Q(v), \quad u, v \in Y, \end{aligned} \tag{6.7}$$

where h_m is an arbitrary element of Y, $k_m = -\widetilde{\delta}_{n-m+1}^{-1} h_m$, $m = 1, 2, \dots, n-1$, $l_{mj} = h_m + \widetilde{\delta}_j k_m = (\widetilde{\delta}_j - \widetilde{\delta}_{n-m+1}) k_m$, $j = 1, 2, \dots, n-m$. Let h_n be an arbitrary element of the group Y. Put $k_n = -\widetilde{\delta}_1^{-1} h_n$. Then $h_n + \widetilde{\delta}_1 k_n = 0$. Substitute $u + h_n$ for u and $v + k_n$ for v in equation (6.7). Subtracting equation (6.7) from the resulting equation we obtain

$$\Delta_{h_n} \Delta_{h_{n-1}} \dots \Delta_{h_1} P(u) + \Delta_{k_n} \Delta_{k_{n-1}} \dots \Delta_{k_1} Q(v) = 0, \quad u, v \in Y. \tag{6.8}$$

Let h be an arbitrary element of the group Y. Substitute $u+h$ for u in equation (6.8) and subtract equation (6.8) from the resulting equation. We get

$$\Delta_h \Delta_{h_n} \Delta_{h_{n-1}} \dots \Delta_{h_1} P(u) = 0, \quad u \in Y. \tag{6.9}$$

We note that u, h, and h_m, $m = 1, 2, \dots, n$ are arbitrary elements of the group Y. Put $h_1 = \dots = h_n = h$ in (6.9). We have

$$\Delta_h^{n+1} P(u) = 0, \quad u, h \in Y. \tag{6.10}$$

Thus, $P(y)$ is a continuous polynomial on Y. Set $\gamma = \mu_1 * \dots * \mu_n$. Then

$$\hat{\gamma}(y) = \prod_{j=1}^{n} \hat{\mu}_j(y), \quad y \in Y.$$

In view of (6.5), we have

$$\hat{\gamma}(y) = e^{-P(y)}.$$

Since the group X contains no subgroups topologically isomorphic to the circle group $\mathbb{T}$ and $P(y)$ is a continuous polynomial, by Theorem 3.15, $\gamma \in \Gamma(X)$. Applying Theorem 3.14, we conclude that all μ_j are Gaussian distributions. □

REMARK 6.4. Theorem 6.3 fails if a locally compact Abelian group X contains a subgroup topologically isomorphic to the circle group $\mathbb{T}$. This immediately follows from Lemma 5.5.

Theorem 6.3 and Lemma 5.5 imply directly the following statement (compare with Proposition 4.16).

PROPOSITION 6.5. *Let X be a locally compact Abelian group. Let ξ_1 and ξ_2 be independent random variables with values in X and distributions μ_1 and μ_2 with nonvanishing characteristic functions. The independence of the sum $\xi_1 + \xi_2$ and difference $\xi_1 - \xi_2$ implies that $\mu_j \in \Gamma(X)$, $j = 1, 2$, if and only if the group X has no subgroups topologically isomorphic to the circle group $\mathbb{T}$.*

PROPOSITION 6.6. *Let X be a locally compact Abelian group. Let α_j, β_j, $j = 1, 2, \dots, n$, $n \geq 2$, be topological automorphisms of X. Let ξ_j be independent random variables with values in the group X and distributions μ_j with nonvanishing characteristic functions. If the linear forms $L_1 = \alpha_1\xi_1 + \dots + \alpha_n\xi_n$ and $L_2 = \beta_1\xi_1 + \dots + \beta_n\xi_n$ are independent, then each distribution μ_j is supported in a coset of the subgroup c_X.*

PROOF. We note that c_X is a characteristic subgroup. For this reason we can assume that $L_1 = \xi_1 + \dots + \xi_n$ and $L_2 = \delta_1\xi_1 + \dots + \delta_n\xi_n$, where $\delta_j \in \mathrm{Aut}(X)$. Put $\nu_j = \mu_j * \bar{\mu}_j$, $\varphi_j(y) = -\ln \hat{\nu}_j(y)$, $P(y) = \sum_{j=1}^n \varphi_j(y)$. As follows from the proof of Theorem 6.3, the function $P(y)$ satisfies equation (6.10). By Proposition 1.30, $P(y) = 0$ for all $y \in b_Y$. This yields that $\hat{\nu}_j(y) = 1$ for all $y \in b_Y$. By Proposition 2.10, the inclusions $\sigma(\nu_j) \subset A(X, b_Y)$, $j = 1, 2, \dots, n$, hold. By Theorem 1.10, we have $c_X = A(X, b_Y)$. Hence $\sigma(\nu_j) \subset c_X$. It follows from Proposition 2.1 that each distribution μ_j is supported in a coset of the subgroup c_X. □

Let X be a locally compact Abelian group. Taking into account that c_X is a characteristic subgroup of the group X, it follows from Corollary 6.2 and Proposition 6.6 that the description of distributions which are characterized by the independence of linear forms L_1 and L_2 is reduced to the case when the group X is connected. Now we will take the next step and prove that this problem can be reduced to the case when the group X is topologically isomorphic to a group of the form $\mathbb{R}^{\mathfrak{n}} \times \mathbb{T}^{\mathfrak{m}}$, where $\mathfrak{n} \leq \aleph_0$, $\mathfrak{m} \leq \aleph_0$. It should be noted that the group $\mathbb{R}^{\aleph_0}$ is not locally compact although its structure is simple enough.

LEMMA 6.7. *A locally compact connected Abelian group X is topologically isomorphic to a group of the form $\mathbb{R}^a \times K \times \mathbb{T}^{\mathfrak{m}}$, where $a \geq 0$, K is a compact connected Abelian group containing no subgroups topologically isomorphic to the circle group $\mathbb{T}$, and $\mathfrak{m} \leq \aleph_0$.*

PROOF. By Theorem 1.15, $X \cong \mathbb{R}^a \times F$, where $a \geq 0$ and F is a compact connected Abelian group. Put $H = F^*$. By Theorems 1.2 and 1.3, H is a countable discrete torsion-free Abelian group. By Theorem 1.26, the group H is represented in the form $H = N \times M$, where M is a free Abelian group, i.e., $M \cong \mathbb{Z}^{\mathfrak{m}*}$, $\mathfrak{m} \leq \aleph_0$ and N is a torsion-free Abelian group without free factor-groups. Put $K = N^*$. By Theorems 1.2, 1.3, and 1.9 K is a compact connected Abelian group containing no subgroups topologically isomorphic to the circle group $\mathbb{T}$. By Pontryagin's duality theorem, $F \cong K \times \mathbb{T}^{\mathfrak{m}}$. □

THEOREM 6.8. *Let X be a locally compact Abelian group such that its connected component of the zero is topologically isomorphic to a group of the form $\mathbb{R}^a \times K \times \mathbb{T}^{\mathfrak{m}}$, where $a \geq 0$, K is a compact connected Abelian group containing no subgroups topologically isomorphic to the circle group $\mathbb{T}$, $\mathfrak{m} \leq \aleph_0$. Assume K has finite dimension and $\dim(K) = l$. Let $\alpha_j, \beta_j \in \mathrm{Aut}(X)$, $j = 1, 2, \dots, n$, $n \geq 2$. Let ξ_j be independent random variables with values in the group X and distributions μ_j with nonvanishing characteristic functions. If the linear forms $L_1 = \alpha_1\xi_1 + \cdots + \alpha_n\xi_n$ and $L_2 = \beta_1\xi_1 + \cdots + \beta_n\xi_n$ are independent, then there exist a continuous monomorphism $p : G \to X$, where $G = \mathbb{R}^b \times \mathbb{T}^{\mathfrak{m}}$, $b = a + l$, and $x_j \in X$, $j = 1, 2, \dots, n$, such that all distributions $\mu_j * E_{x_j}$ are concentrated on the subgroup $p(G)$. Furthermore, $p(G)$ is a characteristic subgroup.*

PROOF. Denote by Y the character group of the group X. It follows from Corollary 6.2 and Proposition 6.6 that we can assume that X is a connected group, i.e., $X = c_X$. Assume for definiteness that $\mathfrak{m} = \aleph_0$. Then $X = \mathbb{R}^a \times K \times \mathbb{T}^{\aleph_0}$, and $Y \cong \mathbb{R}^a \times D \times \mathbb{Z}^{\aleph_0*}$, where $D = K^*$. In order not to complicate the notation, suppose $Y = \mathbb{R}^a \times D \times \mathbb{Z}^{\aleph_0*}$. Since K is a compact connected Abelian group, by Theorems 1.2 and 1.3, D is a countable discrete torsion-free Abelian group. Since K contains no subgroups topologically isomorphic to the circle group $\mathbb{T}$, by Theorem 1.9, the group D has no free factor-groups. Since the group K has dimension l, by Theorem 1.4, $r_0(D) = l$. Put $H = G^*$. Then $H \cong \mathbb{R}^b \times \mathbb{Z}^{\aleph_0*}$. We also assume $H = \mathbb{R}^b \times \mathbb{Z}^{\aleph_0*}$. Denote by $y = (s, d, m)$, where $s = (s_1, \dots, s_a) \in \mathbb{R}^a$, $d \in D$, $m = (m_1, \dots, m_q, 0, \dots,) \in \mathbb{Z}^{\aleph_0*}$, elements of the group Y and by (w, m), where $w \in \mathbb{R}^b$, $m \in \mathbb{Z}^{\aleph_0*}$, elements of the group H. As in the proof of Proposition 3.7, consider the natural embedding of the group Y into the group H. Choose in D a maximal independent set of elements $\{d_1, \dots, d_l\}$. Then for every $d \in D$ there exist integers $k, k_1, \dots, k_l$, such that (3.6) is fulfilled. Moreover, the rational numbers $\{k_j/k\}$ are uniquely determined by d. Define the mapping $\pi : Y \to H$ as follows

$$\pi y = \pi(s, d, m) = (s_1, \dots, s_a, k_1/k, \dots, k_l/k, m_1, \dots, m_q, 0, \dots,), \tag{6.11}$$
$$s = (s_1, \dots, s_a) \in \mathbb{R}^a,\ d \in D,\ m = (m_1, \dots, m_q, 0, \dots,) \in \mathbb{Z}^{\aleph_0*}.$$

Since D is a torsion-free group, π is a continuous monomorphism of the group Y to H. Inasmuch as the group K has no subgroups topologically isomorphic to the circle group $\mathbb{T}$, it follows from Lemma 3.12 that $\overline{\pi(\mathbb{R}^a \times D)} = \mathbb{R}^b$. Then $\overline{\pi(Y)} = H$. Put $p = \widetilde{\pi}$. Since $\overline{\pi(Y)} = H$, it follows from statement (b) of Theorem 1.20 that $p : G \to X$ is a continuous monomorphism.

By Lemma 6.1, the characteristic functions $\hat{\mu}_j(y)$ satisfy equation (6.1). Consider the distributions $\nu_j = \mu_j * \bar{\mu}_j$, $j = 1, 2, \ldots, n$. Then $\hat{\nu}_j(y) = |\hat{\mu}_j(y)|^2 > 0$ for all $y \in Y$. The characteristic functions $\hat{\nu}_j(y)$ also satisfy equation (6.1).

Since $\mathbb{R}^a$ is the connected component of the zero of the group Y, this implies that $\mathbb{R}^a$ is a characteristic subgroup of the group Y. For this reason $K \times \mathbb{T}^{\aleph_0} = A(X, \mathbb{R}^a)$ is a characteristic subgroup of the group X. It follows from Theorem 1.26 that D is a characteristic subgroup of the group $D \times \mathbb{Z}^{\aleph_0 *}$. This implies that $\mathbb{T}^{\aleph_0} = A(K \times \mathbb{T}^{\aleph_0}, D)$ is a characteristic subgroup of the group $K \times \mathbb{T}^{\aleph_0}$. Hence $\mathbb{T}^{\aleph_0}$ is a characteristic subgroup of the group X. It follows from this that $\mathbb{R}^a \times D = A(Y, \mathbb{T}^{\aleph_0})$ is a characteristic subgroup of the group Y. Therefore, we can consider the restriction of equation (6.1) for the characteristic functions $\hat{\nu}_j(y)$ to this subgroup. Since $\mathbb{R}^a \times D \cong (\mathbb{R}^a \times K)^*$ and the group $\mathbb{R}^a \times K$ contains no subgroups topologically isomorphic to the circle group $\mathbb{T}$, it follows from Lemma 6.1 and Theorem 6.3 that these restrictions are the characteristic functions of some Gaussian distributions on the group $\mathbb{R}^a \times K$. Taking into account that $\hat{\nu}_j(y) > 0$, we have the representations

$$\hat{\nu}_j(y) = \exp\{-\varphi_j(y)\}, \quad y \in \mathbb{R}^a \times D, \ j = 1, 2, \ldots, n, \tag{6.12}$$

where $\varphi_j(y)$ are continuous nonnegative functions on $\mathbb{R}^a \times D$ satisfying equation (3.2). As shown in the proof of Proposition 3.7, there exist symmetric positive semidefinite $b \times b$ matrices A_j, such that

$$\varphi_j(y) = \langle A_j \pi y, \pi y \rangle, \quad y \in \mathbb{R}^a \times D, \quad j = 1, 2, \ldots, n. \tag{6.13}$$

Assume that $(w, m) = \pi y$ for some $y \in Y$. Consider on the subgroup $\pi(Y)$ the functions $h_j(w, m) = \hat{\nu}_j(\pi^{-1}(w, m))$, $j = 1, 2, \ldots, n$. It follows from (6.12) and (6.13) that

$$h_j(w, 0) = \exp\{-\langle A_j w, w \rangle\}, \quad (w, 0) \in \pi(\mathbb{R}^a \times D), \ j = 1, 2, \ldots, n. \tag{6.14}$$

Taking into account statement (vii) of Theorem 2.5, it follows from (6.14) that the positive definite functions $h_j(w, m)$ are uniformly continuous on the subgroup $\pi(Y)$ in topology induced on $\pi(Y)$ by the standard topology of H. Since $\overline{\pi(Y)} = H$, the functions $h_j(w, m)$ can be extended by continuity to some continuous functions $\bar{h}_j(w, m)$, $w \in \mathbb{R}^b$, $m \in \mathbb{Z}^{\aleph_0 *}$, on the group H. Obviously, $\bar{h}_j(w, m)$ are also positive definite functions. By Bochner's theorem, there exist distributions λ_j on G such that $\hat{\lambda}_j(w, m) = \bar{h}_j(w, m)$, $w \in \mathbb{R}^b$, $m \in \mathbb{Z}^{\aleph_0 *}$. Inasmuch as $\hat{\lambda}_j(\pi y) = \hat{\nu}_j(y)$ for all $y \in Y$, it follows from Proposition 2.7 that $\nu_j = p(\lambda_j)$. We see that the distributions ν_j are concentrated on $p(G)$. By Suslin's theorem, $p(G)$ is a Borel subgroup. By Proposition 2.1, this implies that there exist elements $x_j \in X$, $j = 1, 2, \ldots, n$, such that all distributions $\mu_j * E_{x_j}$ are concentrated on $p(G)$.

It remains to prove that $p(G)$ is a characteristic subgroup. Let $\alpha \in \mathrm{Aut}(X)$. Put $T_\alpha = \pi \widetilde{\alpha} \pi^{-1}$. Then T_α is an algebraic automorphism of the subgroup $\pi(Y)$. The automorphism $\widetilde{\alpha}$ is determined by its restriction to the subgroup $\mathbb{R}^a$ and its values on the maximal independent system elements of the subgroup $D \times \mathbb{Z}^{\aleph_0 *}$. For this reason the automorphism $\widetilde{\alpha}$ determines, in the natural way, a matrix $A(\alpha)$. It is easily seen that if $\pi y = (s, q_1/q, \dots, q_l/q, m)$, $y = (s, d, m) \in Y$, then $T_\alpha \pi y = A(\alpha)\pi y$, where the expression in the right-hand side of this equality is the product of the matrix $A(\alpha)$ and the vector πy. It is clear that the automorphism T_α can be uniquely extended to the topological automorphism of the group H. Denote by $\overline{T}_\alpha$ the extended automorphism. We have $\widetilde{\alpha} y = \pi^{-1} T_\alpha \pi y$, $T_\alpha \pi y = \overline{T}_\alpha \pi y$ for all $y \in Y$. Take $g \in G$, $y \in Y$. We have

$$(\alpha p g, y) = (pg, \widetilde{\alpha} y) = (pg, \pi^{-1} T_\alpha \pi y) = (g, T_\alpha \pi y) = (g, \overline{T}_\alpha \pi y) = (p\widetilde{\overline{T}}_\alpha g, y).$$

It follows from this that $\alpha p g = p\widetilde{\overline{T}}_\alpha g$. Hence $p(G)$ is a characteristic subgroup. □

REMARK 6.9. Let $X = \mathbb{R}^a \times K \times \mathbb{T}^{\mathfrak{m}}$, where $a \geq 0$, K is a compact connected Abelian group containing no subgroups topologically isomorphic to the circle group $\mathbb{T}$, $\mathfrak{m} \leq \aleph_0$. Assume K has finite dimension and $\dim(K) = l$. Denote by L the arcwise connected component of the zero of the group X. We verify that $p(G) = L$. By Dixmier's theorem L is the union of all one-parametric subgroups of the group X [**1**, §8.19]. Let $q : \mathbb{R} \to X$ be a continuous homomorphism. Put $f = \widetilde{q}$. Taking into account Dixmier's theorem the required statement will be proved if we check that $qt \in p(G)$ for all $t \in \mathbb{R}$. To this end it suffices to show that there exists an element $g_t \in G$ such that $(qt, y) = (pg_t, y)$ for all $y \in Y$. Let $\{e_j\}_{j=1}^a$ be a natural basis in $\mathbb{R}^a$ and let $\{b_j\}$ be a natural basis in $\mathbb{Z}^{\mathfrak{m}*}$. Taking into account (3.6), we obtain

$$fy = f(s, d, m) = \sum_{j=1}^{a} s_j f e_j + \sum_{j=1}^{l} \frac{k_j}{k} f d_j + \sum_j m_j f b_j,$$

$$s = (s_1, \dots, s_a) \in \mathbb{R}^a,\ d \in D,\ m = (m_1, \dots, m_k, 0 \dots) \in \mathbb{Z}^{\mathfrak{m}*}.$$

This implies that

$$(qt, y) = (t, fy) = \exp\left\{ it \left(\sum_{j=1}^{a} s_j f e_j + \sum_{j=1}^{l} \frac{k_j}{k} f d_j + \sum_j m_j f b_j \right) \right\}. \tag{6.15}$$

Put

$$g_t = (tfe_1, \dots, tfe_a, tfd_1, \dots, tfd_l, e^{itfb_1}, e^{itfb_2}, \dots) \in G.$$

Let $y = (s_1, \dots, s_a, d, m_1, \dots, m_k, 0, \dots) \in Y$. Taking into account (6.11) we get

$$(pg_t, y) = (g_t, fy) = \exp\left\{ it \left(\sum_{j=1}^{a} s_j f e_j + \sum_{j=1}^{l} \frac{k_j}{k} f d_j + \sum_j m_j f b_j \right) \right\}. \tag{6.16}$$

The required statement follows from (6.15) and (6.16).

In Theorem 6.8 we considered the case when the group K has finite dimension. Let us study now the case when the group K has infinite dimension. We need some properties of nonlocally compact Abelian groups $\mathbb{R}^{\aleph_0 *} \times \mathbb{Z}^{\mathfrak{m}*}$ and $\mathbb{R}^{\aleph_0} \times \mathbb{T}^{\mathfrak{m}}$, $\mathfrak{m} \leq \aleph_0$. The group $\mathbb{R}^{\aleph_0 *} \times \mathbb{Z}^{\mathfrak{m}*}$ is nuclear [**2**, (7.8), (7.10)]. For such groups Bochner's theorem holds. The factor-group of a nuclear group with respect to a closed subgroup is also nuclear. The groups $\mathbb{R}^{\aleph_0 *} \times \mathbb{Z}^{\mathfrak{m}*}$ and $\mathbb{R}^{\aleph_0} \times \mathbb{T}^{\mathfrak{m}}$ are strongly reflexive [**2**, (17.3), (17.8)]. For these groups the Pontryagin duality theorem holds. They have, in particular, some properties analogous to those of locally compact Abelian groups. We note that the definitions of the character, the character group, and the annihilator for these groups are the same as for locally compact Abelian groups. Some properties of the groups $\mathbb{R}^{\aleph_0 *} \times \mathbb{Z}^{\mathfrak{m}*}$ and $\mathbb{R}^{\aleph_0} \times \mathbb{T}^{\mathfrak{m}}$ are formulated below as a lemma (for the proof see [**2**]).

LEMMA 6.10. *Consider the groups* $\mathbb{R}^{\aleph_0} \times \mathbb{T}^{\mathfrak{m}}$ *and* $\mathbb{R}^{\aleph_0 *} \times \mathbb{Z}^{\mathfrak{m}*}$, *where* $\mathfrak{m} \le \aleph_0$. *Let* P *be a closed subgroup of* $\mathbb{R}^{\aleph_0} \times \mathbb{T}^{\mathfrak{m}}$. *The following statements hold.*

1. *The character group of the group* $\mathbb{R}^{\aleph_0} \times \mathbb{T}^{\mathfrak{m}}$ *is topologically isomorphic to the group* $\mathbb{R}^{\aleph_0 *} \times \mathbb{Z}^{\mathfrak{m}*}$ *and the character group of the group* $\mathbb{R}^{\aleph_0 *} \times \mathbb{Z}^{\mathfrak{m}*}$ *is topologically isomorphic to the group* $\mathbb{R}^{\aleph_0} \times \mathbb{T}^{\mathfrak{m}}$.
2. *There is a one-to-one correspondence between the family of all continuous positive definite functions on the group* $\mathbb{R}^{\aleph_0 *} \times \mathbb{Z}^{\mathfrak{m}*}$ *and the family of all regular finite measures on the group* $\mathbb{R}^{\aleph_0} \times \mathbb{T}^{\mathfrak{m}}$.
3. *For any* $g \in \mathbb{R}^{\aleph_0} \times \mathbb{T}^{\mathfrak{m}}$, $g \notin P$ *there is a character* $h \in A(\mathbb{R}^{\aleph_0 *} \times \mathbb{Z}^{\mathfrak{m}*}, P)$ *such that* $(g, h) \neq 1$.
4. *Every continuous character of the subgroup* P *has the form* $g \to (g, h)$ *for some* $h \in \mathbb{R}^{\aleph_0 *} \times \mathbb{Z}^{\mathfrak{m}*}$.
5. *The natural homomorphisms* $(\mathbb{R}^{\aleph_0 *} \times \mathbb{Z}^{\mathfrak{m}*})/A(\mathbb{R}^{\aleph_0 *} \times \mathbb{Z}^{\mathfrak{m}*}, P) \to P^*$ *and* $(\mathbb{R}^{\aleph_0} \times \mathbb{T}^{\mathfrak{m}}/P)^* \to A(\mathbb{R}^{\aleph_0 *} \times \mathbb{Z}^{\mathfrak{m}*}, P)$ *are topological isomorphisms.*

THEOREM 6.11. *Let* X *be a locally compact Abelian group such that its connected component of the zero is topologically isomorphic to a group of the form* $\mathbb{R}^a \times K \times \mathbb{T}^{\mathfrak{m}}$, *where* $a \ge 0$, K *is a compact connected Abelian group containing no subgroups topologically isomorphic to the circle group* $\mathbb{T}$, $\mathfrak{m} \le \aleph_0$. *Assume that* K *has infinite dimension. Let* $\alpha_j, \beta_j \in \mathrm{Aut}(X)$, $j = 1, 2, \dots, n$, $n \ge 2$. *Let* ξ_j *be independent random variables with values in the group* X *and distributions* μ_j *with nonvanishing characteristic functions. If the linear forms* $L_1 = \alpha_1\xi_1 + \cdots + \alpha_n\xi_n$ *and* $L_2 = \beta_1\xi_1 + \cdots + \beta_n\xi_n$ *are independent, then there exist a continuous monomorphism* $p : G \to X$, *where* $G = \mathbb{R}^{\aleph_0} \times \mathbb{T}^{\mathfrak{m}}$, *and* $x_j \in X$, $j = 1, 2, \dots, n$, *such that all distributions* $\mu_j * E_{x_j}$ *are concentrated on the subgroup* $p(G)$. *Furthermore,* $p(G)$ *is a characteristic subgroup.*

PROOF. We prove Theorem 6.11 in an analogy with the proof of Theorem 6.8. We assume $X = \mathbb{R}^a \times K \times \mathbb{T}^{\mathfrak{m}}$, where K is a compact connected Abelian group containing no subgroups topologically isomorphic to the circle group $\mathbb{T}$ and $Y = \mathbb{R}^a \times D \times \mathbb{Z}^{\mathfrak{m}*}$, where $D = K^*$ is a countable discrete torsion-free Abelian group containing no free factor-groups. Since the group K has infinite dimension, by Theorem 1.4, $r_0(D) = \infty$. Put $G = \mathbb{R}^{\aleph_0} \times \mathbb{T}^{\mathfrak{m}}$. Denote by H the character group of the group G. By Lemma 6.10, $H \cong \mathbb{R}^{\aleph_0 *} \times \mathbb{Z}^{\mathfrak{m}*}$. In order not to complicate the notation, suppose $H = \mathbb{R}^{\aleph_0 *} \times \mathbb{Z}^{\mathfrak{m}*}$. As in the proof of Proposition 3.11 consider the natural embedding of the group Y into the group H. Choose in D a maximal independent set of elements $\{d_1, \dots, d_l, \dots\}$. In the same way as in Theorem 6.8, we construct the continuous homomorphism $\pi : Y \to H$. Since the group K has no subgroups topologically isomorphic to the circle group $\mathbb{T}$, by Lemma 3.12, we have $\overline{\pi(\mathbb{R}^a \times D)} = \mathbb{R}^{\aleph_0 *}$. This implies that $\overline{\pi(Y)} = H$. Put $p = \widetilde{\pi}$. Inasmuch as G is a strongly reflexive group, it follows from $\overline{\pi(Y)} = H$ that the homomorphism $p : G \to X$ is a monomorphism.

Consider the distributions $\nu_j = \mu_j * \bar{\mu}_j$ and argue as in the proof of Theorem 6.8. We show that the restrictions of the characteristic functions $\hat{\nu}_j(y)$ to the subgroup $\mathbb{R}^a \times D$ are the characteristic functions of some Gaussian distributions on the group $\mathbb{R}^a \times K$. This implies that the representations (6.12) and (6.13) hold, where A_j are infinite symmetric positive semidefinite matrices. Next, arguing as in the proof of Theorem 6.8, we come to the continuous positive definite functions $\bar{h}_j(w,k)$, $w \in \mathbb{R}^{\aleph_0*}$, $m \in \mathbb{Z}^{\mathfrak{m}*}$, on the group H. Applying Lemma 6.10, we can correspond to each function $\bar{h}_j(w,m)$ a distribution λ_j on the group G such that $\hat{\lambda}_j(w,m) = \bar{h}_j(w,m)$, $w \in \mathbb{R}^{\aleph_0*}$, $m \in \mathbb{Z}^{\mathfrak{m}*}$. We have $\hat{\lambda}_j(\pi y) = \hat{\nu}_j(y)$, $y \in Y$. This implies that $\nu_j = p(\lambda_j)$. The final part of the proof of the theorem is the same as in the proof of Theorem 6.8. □

REMARK 6.12. Arguing as in Remark 6.9, we see that if in Theorem 6.11 the subgroup K has infinite dimension, then the subgroup $p(G)$ is the arcwise connected component of the zero of the group X.

In what follows, we need the following general statement.

PROPOSITION 6.13. *Let X and G be complete separable metric Abelian groups, $\alpha_j, \beta_j \in \mathrm{Aut}(X)$, $j = 1, 2, \dots, n$, $n \geq 2$, and let $p : G \to X$ be a continuous monomorphism such that $\alpha_j(p(G)) = \beta_j(p(G)) = p(G)$, $j = 1, 2, \dots n$. Let ξ_j be independent random variables with values in the group X and distributions μ_j. Assume that there exist elements $x_j \in X$ such that all distributions $\mu_j * E_{x_j}$ are concentrated in the subgroup $p(G)$. Assume that the linear forms $L_1 = \alpha_1\xi_1 + \cdots + \alpha_n\xi_n$ and $L_2 = \beta_1\xi_1 + \cdots + \beta_n\xi_n$ are independent. Then $\hat{\xi}_j = p^{-1}(\xi_j + x_j)$ are independent random variables with values in the group G, $\hat{\alpha}_j = p^{-1}\alpha_j p$, $\hat{\beta}_j = p^{-1}\beta_j p$ are topological automorphisms of the group G and the linear forms $\hat{L}_1 = \hat{\alpha}_1\hat{\xi}_1 + \cdots + \hat{\alpha}_n\hat{\xi}_n$ and $\hat{L}_2 = \hat{\beta}_1\hat{\xi}_1 + \cdots + \hat{\beta}_n\hat{\xi}_n$ are independent.*

PROOF. Since p is a continuous monomorphism and G is a complete separable metric group, by Suslin's theorem, images of Borel sets under the mapping p are also Borel sets. This implies that $\hat{\xi}_j$ are independent random variables with values in the group G and $\hat{\alpha}_j$, $\hat{\beta}_j$ are Borel automorphisms of the group G. Hence $\hat{\alpha}_j$, $\hat{\beta}_j \in \mathrm{Aut}(G)$ (see e.g. [**95**, §4.3.9]). It is obvious that the random variables $\hat{\xi}_j$ are independent. To prove the independence of the linear forms $\hat{L}_1$ and $\hat{L}_2$ it suffices to note that for any Borel subsets A_1 and A_2 of the group G we have

$$\{\omega : \hat{L}_i(\omega) \in A_i\} = \{\omega : L_i(\omega) \in p(A_i) - (\alpha_1 x_1 + \cdots + \alpha_n x_n)\}, \quad i = 1, 2.$$

□

REMARK 6.14. Let $X = \mathbb{R}^a \times K \times \mathbb{T}^{\mathfrak{m}}$, where $a \geq 0$, K is a compact connected Abelian group containing no subgroups topologically isomorphic to the circle group $\mathbb{T}$, $\mathfrak{m} \leq \aleph_0$. Let ξ_j, $j = 1, 2, \dots, n$, $n \geq 2$, be independent random variables with values in the group X and distributions μ_j with nonvanishing characteristic functions. Let $\alpha_j, \beta_j \in \mathrm{Aut}(X)$. Assume that the linear forms $L_1 = \alpha_1\xi_1 + \cdots + \alpha_n\xi_n$ and $L_2 = \beta_1\xi_1 + \cdots + \beta_n\xi_n$ are independent.

Assume K has finite dimension and $\dim(K) = l$. By Theorem 6.8, there exist a continuous monomorphism $p : G \to X$, where $G = \mathbb{R}^b \times \mathbb{T}^{\mathfrak{m}}$, $b = a + l$, and $x_j \in X$, $j = 1, 2, \dots, n$, such that all distributions $\mu_j * E_{x_j}$, $j = 1, 2, \dots, n$, are concentrated on the subgroup $p(G)$ and $p(G)$ is a characteristic subgroup. It follows from Proposition 6.13 that the study of possible distributions of independent random variables ξ_j is reduced to the study of possible distributions of independent random variables $\hat{\xi}_j$ with values in a group of the form $\mathbb{R}^b \times \mathbb{T}^{\mathfrak{m}}$, where $b \geq 0$, $\mathfrak{m} \leq \aleph_0$.

If K has infinite dimension, applying Theorem 6.11 instead of Theorem 6.8, we come to the conclusion that the study of possible distributions of independent random variables ξ_j is reduced to the study of possible distributions of independent random variables $\hat{\xi}_j$ with values in a group of the form $\mathbb{R}^{\aleph_0} \times \mathbb{T}^{\mathfrak{m}}$, where $\mathfrak{m} \leq \aleph_0$.

Let X be a locally compact Abelian group containing no subgroups topologically isomorphic to the circle group $\mathbb{T}$. Let α_j, β_j, $j = 1, 2, \dots, n$, $n \geq 2$, be topological automorphisms of X. Let ξ_j be independent random variables with values in the group X and distributions μ_j with nonvanishing characteristic functions. By Theorem 6.3, the independence of the linear forms $L_1 = \alpha_1\xi_1 + \cdots + \alpha_n\xi_n$ and $L_2 = \beta_1\xi_1 + \cdots + \beta_n\xi_n$ implies that all μ_j are Gaussian. Let us now study the case when a locally compact Abelian group X contains a subgroup topologically isomorphic to the circle group $\mathbb{T}$, but has no subgroups topologically isomorphic to the 2-dimensional torus $\mathbb{T}^2$. We need some lemmas.

LEMMA 6.15. *Let Y be an Abelian group and let β be an endomorphism of the group Y. Assume that the functions $f_j(y)$ do not vanish, satisfy the equation*

$$f_1(u+v)f_2(u+\beta v) = f_1(u)f_1(v)f_2(u)f_2(\beta v), \quad u, v \in Y, \tag{6.17}$$

and the conditions $f_1(0) = f_2(0) = 1$. Then each of the functions $f_j(y)$ satisfies the equation

$$f_j(u+v)f_j(u-v) = f_j^2(u)f_j(v)f_j(-v), \quad u \in (\beta - I)(Y), \ v \in Y, \ j = 1, 2. \tag{6.18}$$

PROOF. Let t and s be arbitrary elements of the group Y. Putting $u = t - s$, $v = s$ in equation (6.17) we obtain

$$f_1(t)f_2(t + (\beta - I)s) = f_1(t-s)f_1(s)f_2(t-s)f_2(\beta s), \quad t, \ s \in Y. \tag{6.19}$$

Let h be an arbitrary element of the group Y. Substituting $t + h$ for t and $s + h$ for s into equation (6.19) we infer that

$$\begin{aligned} f_1(t+h)&f_2(t + (\beta - I)s + \beta h) \\ &= f_1(t-s)f_1(s+h)f_2(t-s)f_2(\beta s + \beta h), \quad t, \ s, \ h \in Y. \end{aligned} \tag{6.20}$$

Dividing equation (6.20) by equation (6.19) we find

$$\frac{f_1(t+h)f_2(t + (\beta - I)s + \beta h)}{f_1(t)f_2(t + (\beta - I)s)} = \frac{f_1(s+h)f_2(\beta s + \beta h)}{f_1(s)f_2(\beta s)}, \quad t, s, h \in Y. \tag{6.21}$$

Let k be an arbitrary element of the group Y. Substituting $t + (\beta - I)k$ for t and $s - k$ for s into equation (6.21) we have

$$\begin{aligned} &\frac{f_1(t + h + (\beta - I)k)f_2(t + (\beta - I)s + \beta h)}{f_1(t + (\beta - I)k)f_2(t + (\beta - I)s)} \\ &\qquad = \frac{f_1(s + h - k)f_2(\beta s + \beta h - \beta k)}{f_1(s-k)f_2(\beta s - \beta k)}, \quad t, s, h, k \in Y. \end{aligned} \tag{6.22}$$

Dividing equation (6.22) by equation (6.21) we find

$$\begin{aligned} &\frac{f_1(p)f_1(p + h + (\varepsilon - I)k)}{f_1(t+h)f_1(p + (\beta - I)k)} \\ &\qquad = \frac{f_1(s+h-k)f_1(s)f_2(\beta s)f_2(\beta s + \beta h - \beta k)}{f_1(q-k)f_1(q+h)f_2(\varepsilon q + \beta h)f_2(\beta s - \beta k)}, \quad t, s, h, k \in Y. \end{aligned} \tag{6.23}$$

The right-hand side of equation (6.23) does not depend on p. Hence the left-hand side of equation (6.23) takes the same value for $p = -h$ and for $p = 0$. We obtain

$$\frac{f_1(-h)f_1((\beta - I)k)}{f_1(-h + (\beta - I)k)} = \frac{f_1(h + (\beta - I)k)}{f_1(h)f_1((\beta - I)k)}, \quad h, k \in Y.$$

It follows from this that

$$(6.24)\quad f_1((\beta - I)k + h)f_1((\beta - I)k - h) = f_1^2((\beta - I)k)f_1(h)f_1(-h), \quad h,\ k \in Y.$$

Putting $u = (\beta - I)k$, $v = h$ in equation (6.24) we get that the function $f_1(y)$ satisfies equation (6.18).

Substituting $w = \beta v$ in equation (6.17) and arguing similarly we get that the function $f_2(y)$ also satisfies equation (6.18). □

LEMMA 6.16. *Let Y be an Abelian group and let β be an endomorphism of the group Y. Assume that the functions $\psi_j(y)$ satisfy the equation*

$$(6.25)\qquad \psi_1(u + v) + \psi_2(u + \beta v) = P(u) + Q(v), \quad u, v \in Y,$$

where $P(y)$ and $Q(y)$ are some functions on Y. Then each of the functions $\psi_j(y)$ satisfies the equation

$$(6.26)\qquad \Delta_{(I-\beta)k}\Delta_h^2\psi_j(y) = 0, \quad k, h, y \in Y,\ j = 1, 2.$$

PROOF. For the proof we use the finite difference method. Let k be an arbitrary element of the group Y. Put $h = -\beta k$. Substitute $u + h$ for u and $v + k$ for v in equation (6.25). Subtracting equation (6.25) from the resulting equation we obtain

$$(6.27)\qquad \Delta_{(I-\beta)k}\psi_1(u + v) = \Delta_h P(u) + \Delta_k Q(v), \quad u, v \in Y.$$

Substituting $v = 0$ into equation (6.27) and subtracting the resulting equation from (6.27) we obtain

$$(6.28)\qquad \Delta_{(I-\beta)k}\Delta_v\psi_1(u) = \Delta_k Q(v) - \Delta_k Q(0), \quad u, v \in Y.$$

Let t be an arbitrary element of the group Y. Substitute $u + t$ for u in equation (6.28) and subtract equation (6.28) from the resulting equation. We get that the function $\psi_1(y)$ satisfies the equation

$$\Delta_{(I-\beta)k}\Delta_v\Delta_t\psi_1(u) = 0, \quad k, v, t, u \in Y.$$

Substituting here $v = t = h$, $u = y$, we get (6.26).

Substituting $w = \beta v$ in equation (6.25) and arguing similarly we get that the function $\psi_2(y)$ also satisfies equation (6.26). □

LEMMA 6.17. *Let α_j, β_j, $j = 1, 2, \dots, n$, $n \geq 2$, be topological automorphisms of the group $\mathbb{R}^{\aleph_0}$. Let ξ_j be independent random variables with values in $\mathbb{R}^{\aleph_0}$ and distributions μ_j. If the linear forms $L_1 = \alpha_1\xi_1 + \cdots + \alpha_n\xi_n$ and $L_2 = \beta_1\xi_1 + \cdots + \beta_n\xi_n$ are independent, then $\mu_j \in \Gamma(\mathbb{R}^{\aleph_0})$, $j = 1, 2, \dots, n$.*

PROOF. Each element $s \in \mathbb{R}^{\aleph_0 *}$ defines a linear continuous functional in the space $\mathbb{R}^{\aleph_0}$. Then for any distribution μ on the group $\mathbb{R}^{\aleph_0}$ we can consider its image $s(\mu) \in \mathrm{M}^1(\mathbb{R})$. We say that a distribution γ on the group $\mathbb{R}^{\aleph_0}$ is weak Gaussian if $s(\gamma) \in \Gamma(\mathbb{R})$ for all $s \in \mathbb{R}^{\aleph_0 *}$. It is not difficult to check that a distribution μ on the group $\mathbb{R}^{\aleph_0}$ is Gaussian if and only if μ is weak Gaussian. As has been noted in [**77**], the Skitovich–Darmois theorem holds true for weak Gaussian distributions on an arbitrary Fréchet space, in particular on the group $\mathbb{R}^{\aleph_0}$. □

Consider the group $\mathbb{R}^{\aleph_0} \times \mathbb{T}$. The Gaussian distribution on the group $\mathbb{R}^{\aleph_0} \times \mathbb{T}$ is defined in the same way as for the group $\mathbb{R}^{\aleph_0}$. We need the following lemma.

LEMMA 6.18. *Let $X = \mathbb{R}^{\aleph_0} \times \mathbb{T}$ with character group $Y = \mathbb{R}^{\aleph_0 *} \times \mathbb{Z}$. Let ξ_1 and ξ_2 be independent identically distributed random variables with values in X and distribution μ with nonvanishing characteristic function. If the linear forms $L_1 = \xi_1 + \xi_2$ and $L_2 = \xi_1 - \xi_2$ are independent, then μ is a Gaussian distribution.*

PROOF. Denote by (s, n), where $s \in \mathbb{R}^{\aleph_0 *}$, $n \in \mathbb{Z}$, elements of the group Y. It is easily seen that Lemma 6.1 holds for the group X. This implies that Lemma 4.7 is also true for the group X. Hence the characteristic function $\hat{\mu}(y)$ can be represented in the form (4.10). The lemma will be proved if we verify that the function $l(y)$ in (4.10) is a character of the group Y, i.e., $l(y)$ satisfies equation (4.25).

Swap u and v in equation (4.11). Multiplying the resulting equation by (4.10) we find

$$l^2(u+v) = l^2(u)l^2(v), \quad u, v \in Y. \tag{6.29}$$

Substituting $u = v = y$ into equation (4.11), we obtain $l(2y) = l^2(y)$, $y \in Y$. It follows from this and (6.29) that

$$l(2u+2v) = l(2u)l(2v), \quad u, v \in Y. \tag{6.30}$$

Equality (6.30) yields that the function $l(y)$ is a character of the subgroup $Y^{(2)}$.

By Lemma 6.10, we have $l(y) = (x_0, y)$, $x_0 \in X$, $y \in Y^{(2)}$. Put

$$m(y) = l(y)/(x_0, y).$$

The function $m(y)$ also satisfies equation (4.11) and $m(y) = 1$ for all $y \in Y^{(2)}$. Since $m(2y) = m^2(y)$ for all $y \in Y$, we get

$$m(y) = \pm 1, \quad y \in Y. \tag{6.31}$$

Equation (4.11) for the function $m(y)$ becomes the equation $m(u+v)m(u-v) = 1$ for all $u, v \in Y$. Substituting here $u+v$ for u we infer that $m(u+2v)m(u) = 1$ for all $u, v \in Y$. In view of (6.31), we have

$$m(u+2v) = m(u), \quad u, v \in Y,$$

i.e., the function $m(y)$ is invariant with respect to shifts by elements of $Y^{(2)}$. It is obvious that $Y = Y^{(2)} \cup ((0,1) + Y^{(2)})$. The function $m(y)$ takes a constant value on each coset. It follows from (6.31) that we have two possibilities: either $m(y) = 1$ for all $y \in Y$ or

$$m(y) = \begin{cases} 1 & \text{if} \quad y \in Y^{(2)}, \\ -1 & \text{if} \quad y \in (0,1) + Y^{(2)}. \end{cases}$$

In both cases the function $m(y)$ satisfies equation (4.25) on the group Y. Therefore, $m(y)$ is a character of the group Y. Hence $l(y)$ is also a character. □

Let X be a locally compact Abelian group. Assume that X contains a subgroup topologically isomorphic to the circle group $\mathbb{T}$, but does not contain subgroups topologically isomorphic to the 2-dimensional torus $\mathbb{T}^2$. Let α_j, β_j, $j = 1, 2$, be topological automorphisms of the group X. Let ξ_j be independent random variables with values in the group X and distributions μ_j with nonvanishing characteristic functions. We describe distributions which are characterized by the independence of the linear forms $L_1 = \alpha_1\xi_1 + \alpha_2\xi_2$ and $L_2 = \beta_1\xi_1 + \beta_2\xi_2$. It is easy to see that we can assume without loss of generality that $L_1 = \xi_1 + \xi_2$ and $L_2 = \xi_1 + \alpha\xi_2$, where $\alpha \in \mathrm{Aut}(X)$.

Taking into account Corollary 6.2 and Proposition 6.6, we can assume that X is a connected group. By Lemma 6.7, X is topologically isomorphic to a group of the form $\mathbb{R}^a \times K \times \mathbb{T}$, where $a \geq 0$ and K is a compact connected Abelian group containing no subgroups topologically isomorphic to the circle group $\mathbb{T}$. Let $X = \mathbb{R}^a \times K \times \mathbb{T}$. As has been noted in the proof of Theorem 6.8, $\mathbb{T}$ is a characteristic subgroup of the group X. Since $\mathrm{Aut}(\mathbb{T}) = \{\pm I\}$, we have two possibilities: either $\alpha|_{\mathbb{T}} = I$ or $\alpha|_{\mathbb{T}} = -I$.

THEOREM 6.19. *Let $X = \mathbb{R}^a \times K \times \mathbb{T}$, where $a \geq 0$ and K is a compact connected Abelian group containing no subgroups topologically isomorphic to the circle group $\mathbb{T}$. Let $\alpha \in \mathrm{Aut}(X)$. Let ξ_1 and ξ_2 be independent random variables with values in the group X and distributions μ_1 and μ_2 with nonvanishing characteristic functions. Assume that the linear forms $L_1 = \xi_1 + \xi_2$ and $L_2 = \xi_1 + \alpha\xi_2$ are independent. If $\alpha|_{\mathbb{T}} = I$, then $\mu_j \in \Gamma(X)$, $j = 1, 2$. If $\alpha|_{\mathbb{T}} = -I$, then $\mu_j = \gamma_j * \pi_j$, where $\gamma_j \in \Gamma(X)$ and π_j are signed measures on the subgroup $\mathbb{T}_{(2)}$, $j = 1, 2$.*

PROOF. There are two possibilities: either the dimension of the group K is finite or infinite.

1. Assume K has finite dimension and $\dim(K) = l$. Put $D = K^*$. By Theorem 1.4, $r_0(D) = l$. Taking into account Proposition 6.13 and Remark 6.14, it suffices to prove the theorem for a group of the form $X = \mathbb{R}^b \times \mathbb{T}$, where $b \geq 0$, because in the notation of Proposition 6.13, $\alpha|_{\mathbb{T}} = \hat{\alpha}|_{\mathbb{T}}$. Denote by Y the character group of the group X. Then $Y \cong \mathbb{R}^b \times \mathbb{Z}$. In order not to complicate the notation, suppose $Y = \mathbb{R}^b \times \mathbb{Z}$. Denote by (t, z), where $t \in \mathbb{R}^b$, $z \in \mathbb{T}$, elements of the group X and by (s, n), where $s \in \mathbb{R}^b$, $n \in \mathbb{Z}$, elements of the group Y. It is easy to see that each automorphism $\alpha \in \mathrm{Aut}(X)$ is determined by the matrix

$$\begin{pmatrix} A & v \\ 0 & \pm 1 \end{pmatrix}, \quad A \in \mathrm{Aut}(\mathbb{R}^b), \ v \in \mathbb{R}^b, \tag{6.32}$$

and the automorphisms α and $\widetilde{\alpha}$ act on the groups X and Y by the formulas

$$\alpha(t, z) = (At, e^{i\langle v, t\rangle} z^{\pm 1}), \ (t, z) \in X, \quad \widetilde{\alpha}(s, n) = (As + nv, \pm n), \ (s, n) \in Y,$$

respectively. Inasmuch as $\mathbb{R}^b$ is the connected component of the zero of the group Y, $\mathbb{R}^b$ is a characteristic subgroup of Y. Put $L = \mathrm{Ker}(I - \widetilde{\alpha}) \cap \mathbb{R}^b$ and first verify that the proof of the theorem is reduced to the case when $L = \{0\}$.

By Lemma 6.1, the characteristic functions $\hat{\mu}_j(y)$ satisfy equation (6.1) which takes the form

$$\hat{\mu}_1(u + v)\hat{\mu}_2(u + \widetilde{\alpha}v) = \hat{\mu}_1(u)\hat{\mu}_1(v)\hat{\mu}_2(u)\hat{\mu}_2(\widetilde{\alpha}v), \quad u, v \in Y. \tag{6.33}$$

Put $\nu_j = \mu_j * \bar{\mu}_j$, $j = 1, 2$. This implies that $\hat{\nu}_j(y) = |\hat{\mu}_j(y)|^2 > 0$ for all $y \in Y$. The characteristic functions $\hat{\nu}_j(y)$ also satisfy equation (6.33). Consider the restriction of equation (6.33) for the functions $\hat{\nu}_j(y)$ to the subgroup L. We have

$$\hat{\nu}_1(u + v)\hat{\nu}_2(u + v) = \hat{\nu}_1(u)\hat{\nu}_1(v)\hat{\nu}_2(u)\hat{\nu}_2(v), \quad u, v \in L. \tag{6.34}$$

Set $h(y) = \hat{\nu}_1(y)\hat{\nu}_2(y)$. It follows from (6.34) that the function $h(y)$ on the group L satisfies the equation

$$h(u + v) = h(u)h(v), \quad u, v \in Y,$$

i.e., $h(y)$ is a character of the group L. This implies that the restrictions of the characteristic functions $\hat{\nu}_j(y)$ to L are also characters of the subgroup L. Since $\hat{\nu}_j(y) > 0$ for all $y \in Y$, we have $\hat{\nu}_1(y) = \hat{\nu}_2(y) = 1$ for all $y \in L$. Applying Proposition 2.10, we get $\sigma(\nu_j) \subset G = A(X, L)$, $j = 1, 2$. Inasmuch as L is the kernel of a continuous linear operator in the space $\mathbb{R}^b$, L is a closed subspace in $\mathbb{R}^b$. It follows from this that $G = W \times \mathbb{T}$, where W is a closed subspace in $\mathbb{R}^b$. It is obvious that $\widetilde{\alpha}(L) = L$. Hence $\alpha(G) = G$. It is clear that if $L \neq \{0\}$, then G is a proper subgroup of X.

Consider a family of subgroups B of the group X having the properties:

(i) $B = V \times \mathbb{T}$, where V is a closed subspace in $\mathbb{R}^b$;
(ii) $\alpha(B) = B$;
(iii) $\sigma(\nu_j) \subset B$, $j = 1, 2$.

Let N be an intersection of all subgroups B of the group X having properties (i)–(iii). The subgroup N also possesses properties (i)–(iii) and N is the smallest closed subgroup having these properties. In order not to complicate the notation, suppose $N = \mathbb{R}^c \times \mathbb{T}$, where $c \leq b$, and $N^* = \mathbb{R}^c \times \mathbb{Z}$. Put $\beta = \alpha|_N$. If $\text{Ker}(I - \widetilde{\beta}) \cap \mathbb{R}^c \neq \{0\}$, then the above reasoning shows that the subgroup $A(N, \text{Ker}(I - \widetilde{\beta}) \cap \mathbb{R}^c)$ possesses properties (i)–(iii) and is a proper subgroup of N, contrary to the assumption. Since $\nu_j = \mu_j * \bar{\mu}_j$, it follows from Proposition 2.1 that the distributions μ_j can be replaced by their shifts λ_j in such a manner that $\sigma(\lambda_j) \subset N$, $j = 1, 2$. It follows from what has been said that we can assume, without loss of generality, that $L = \{0\}$. Let us note that the condition $L = \{0\}$ means that $(I - A) \in \text{Aut}(\mathbb{R}^b)$ in (6.32).

1a. Assume $\alpha|_\mathbb{T} = I$. This means that the matrix which corresponds to the automorphism α is of the form $\begin{pmatrix} A & v \\ 0 & 1 \end{pmatrix}$. Put $M = \text{Ker}(I - \widetilde{\alpha})$ and consider the restriction of equation (6.33) for the functions $\hat{\nu}_j(y)$ to the subgroup M. We come to equation (6.34), but for $u, v \in M$. Reasoning as above, we conclude that $\sigma(\nu_j) \subset F = A(X, M)$. We have $F = A(X, M) = A(X, \text{Ker}(I - \widetilde{\alpha})) = (I - \alpha)(X)$. In view of $X = \mathbb{R}^b \times \mathbb{T}$ and $L = \{0\}$, we have $F = (I - \alpha)(X) \cong \mathbb{R}^b$. Inasmuch as $\widetilde{\alpha}(M) = M$, the restriction of the automorphism α to the subgroup F is a topological automorphism of the group F. Let η_j be independent random variables with values in F and distributions ν_j. Since the characteristic functions $\hat{\nu}_j(y)$ satisfy equation (6.33), by Lemma 6.1, the linear forms $M_1 = \eta_1 + \eta_2$ and $M_2 = \eta_1 + \alpha\eta_2$ are independent. By the Ghurye–Olkin theorem, this implies that $\nu_j \in \Gamma(F)$. Next, applying Cramér's decomposition theorem for the Gaussian distribution on the group $\mathbb{R}^b$ and Proposition 2.1, we get $\mu_j \in \Gamma(X)$.

1b. Assume $\alpha|_\mathbb{T} = -I$. This means that the matrix corresponding to the automorphism α is of the form $\begin{pmatrix} A & v \\ 0 & -1 \end{pmatrix}$. Put $H = (I - \widetilde{\alpha})(Y)$. Inasmuch as $L = \text{Ker}(I - \widetilde{\alpha}) \cap \mathbb{R}^b = \{0\}$, we have $H = Y^{(2)} = \mathbb{R}^b \times \mathbb{Z}^{(2)} \cong \mathbb{R}^b \times \mathbb{Z}$. It follows from Lemma 6.15 that each of the characteristic functions $\hat{\mu}_j(y)$ satisfies equation (6.18) when $u, v \in Y^{(2)}$. Since $H \cong \mathbb{R}^b \times \mathbb{Z}$, we have $H^* \cong \mathbb{R}^b \times \mathbb{T}$. Taking into account that the group H^* has no subgroups topologically isomorphic to the 2-dimensional torus $\mathbb{T}^2$ and applying Lemma 6.1 and Proposition 4.16, we conclude that the restrictions of the characteristic functions $\hat{\mu}_j(y)$ to the subgroup H are the characteristic functions of Gaussian distributions. Thus, we have the representation

$$\hat{\mu}_j(y) = m_j(y)\exp\{-\varphi_j(y)\}, \quad y \in H, \ j = 1, 2,$$

where $m_j(y)$ are characters of the subgroup H and $\varphi_j(y)$ are continuous nonnegative functions on H satisfying equation (3.2). Replacing if necessary the distributions μ_j by their shifts we can assume that

$$\hat{\mu}_j(y) = \exp\{-\varphi_j(y)\}, \quad y \in H, \ j = 1, 2. \tag{6.35}$$

Put

$$l_j(y) = \frac{\hat{\mu}_j(y)}{|\hat{\mu}_j(y)|}, \quad y \in Y, \ j = 1, 2.$$

and verify that $l_j(y)$ are characters of the group Y. It follows from the equality $\hat{\mu}_j(y) = |\hat{\mu}_j(y)|$ for all $y \in H$ that $l_j(y) = 1$ for all $y \in H$. Each of the functions $l_j(y)$ satisfies equation (6.18) that takes the form

$$l_j(u+v)l_j(u-v) = 1, \quad u \in H, \ v \in Y. \tag{6.36}$$

Substituting $u = (s/2, 0)$, $v = (s/2, n)$ in equation (6.36), we obtain

$$l_j(s,n)l_j(0,-n) = 1, \quad s \in \mathbb{R}^b, \ n \in \mathbb{Z}.$$

Multiplying both sides of this equation by $l_j(0,n)$ and taking into account that $|l_j(y)| = 1$, $l_j(-y) = \overline{l_j(y)}$, we find

$$l_j(s,n) = l_j(0,n), \quad s \in \mathbb{R}^b, \ n \in \mathbb{Z}. \tag{6.37}$$

Obviously, the functions $l_j(y)$ satisfy equation (6.33). Taking into account (6.37), we can write equation (6.33) for the functions $l_j(y)$ in the form

$$l_1(0,m+n)l_2(0,m-n) = l_1(0,m)l_1(0,n)l_2(0,m)l_2(0,-n), \quad m,n \in \mathbb{Z}. \tag{6.38}$$

We get by induction from equation (6.38) that $l_j(0,n)$ are characters of the group $\mathbb{Z}$. Hence $l_j(y)$ are characters of the group Y. By Pontryagin's duality theorem, there exist elements $x_j \in X$ such that

$$l_j(y) = (x_j, y), \quad y \in Y, \ j = 1, 2. \tag{6.39}$$

Now we find the representations for $|\hat{\mu}_j(y)|$. Put $\psi_j(y) = -\ln|\hat{\mu}_j(y)|$. Then (6.33) implies that the functions $\psi_j(y)$ satisfy the equation

$$\psi_1(u+v) + \psi_2(u+\widetilde{\alpha}v) = P(u) + Q(v), \quad u, v \in Y, \tag{6.40}$$

where $P(y) = \psi_1(y) + \psi_2(y)$, $Q(y) = \psi_1(y) + \psi_2(\widetilde{\alpha}y)$.

By Lemma 6.16, each of the functions $\psi_j(y)$ satisfies the equation

$$\Delta_{(I-\widetilde{\alpha})k}\Delta_h^2\psi_j(y) = 0, \quad k, h, y \in Y. \tag{6.41}$$

Since $(I-\widetilde{\alpha})(Y) = Y^{(2)}$, we conclude from (6.41) that each of the functions $\psi_j(y)$ satisfies the equation

$$\Delta_{2k}\Delta_h^2\psi_j(y) = 0, \quad k, h, y \in Y.$$

Inasmuch as $H = Y^{(2)} = \overline{Y^{(2)}}$, a $\overline{Y^{(2)}}$-coset decomposition of the group Y can be written in the form $Y = H \cup ((0,1) + H)$. Lemma 5.1 and (6.35) imply that there exist real constants c_1, c_2 such that

$$\psi_j(y) = \begin{cases} \varphi_j(y), & \text{if } \ y \in H, \\ \varphi_j(y) + c_j, & \text{if } \ y \in (0,1) + H, \quad j = 1, 2. \end{cases} \tag{6.42}$$

It follows from (6.33) and (6.35) that the equality

$$\varphi_1(u+v)+\varphi_2(u+\widetilde{\alpha}v)=\varphi_1(u)+\varphi_1(v)+\varphi_2(u)+\varphi_2(\widetilde{\alpha}v) \tag{6.43}$$

holds true for all $u, v \in H$. In view of $H = Y^{(2)}$ and that each of the functions $\varphi_j(y)$ satisfies equation (3.2), equation (6.43) is valid for all $u, v \in Y$. Substituting $u, v \in (0,1)+H$ in equation (6.40) and taking into account (6.42) and (6.43), we find that $c_1 = -c_2$. Let $c_1 = -c_2 = -2\kappa$.

It follows from (6.42) that the functions $|\hat{\mu}_j(y)|$ are of the form

$$|\hat{\mu}_1(y)| = \exp\{-\varphi_1(y)+\kappa(1-(-1))^n\}, \quad y=(s,n)\in Y,$$

$$|\hat{\mu}_2(y)| = \exp\{-\varphi_2(y)-\kappa(1-(-1))^n\}, \quad y=(s,n)\in Y.$$

Taking into account (6.39), we finally obtain

$$\hat{\mu}_1(y) = (x_1,y)\exp\{-\varphi_1(y)+\kappa(1-(-1))^n\}, \quad y=(s,n)\in Y, \tag{6.44}$$

$$\hat{\mu}_2(y) = (x_2,y)\exp\{-\varphi_2(y)-\kappa(1-(-1))^n\}, \quad y=(s,n)\in Y. \tag{6.45}$$

Consider the signed measures

$$\pi_1 = \frac{(1+e^{2\kappa})}{2}E_{(0,1)}+\frac{(1-e^{2\kappa})}{2}E_{(0,-1)}, \ \pi_2 = \frac{(1+e^{-2\kappa})}{2}E_{(0,1)}+\frac{(1-e^{-2\kappa})}{2}E_{(0,-1)}$$

supported in $\mathbb{T}_{(2)}$ (it is clear that one of π_j is a distribution). The characteristic functions $\hat{\pi}_j(y)$ are of the form

$$\hat{\pi}_1(y) = \exp\{\kappa(1-(-1)^n)\}, \tag{6.46}$$

$$\hat{\pi}_2(y) = \exp\{-\kappa(1-(-1)^n)\}, \quad y=(s,n)\in Y.$$

Observe that $\pi_1 * \pi_2 = E_{(0,1)}$.

Denote by γ_j the Gaussian distribution on the group X with the characteristic function

$$\hat{\gamma}_j(y) = (x_j,y)\exp\{-\varphi_j(y)\}, \quad y\in Y, \ j=1,2. \tag{6.47}$$

It follows from (6.44)–(6.47) that $\hat{\mu}_j(y) = \hat{\gamma}_j(y)\hat{\pi}_j(y)$, $y \in Y$. This implies that $\mu_j = \gamma_j * \pi_j$, $j = 1, 2$.

2. Assume K has infinite dimension. It follows from Proposition 6.13 and Remark 6.14 that it suffices to prove the theorem for the group of the form $X = \mathbb{R}^{\aleph_0} \times \mathbb{T}$. Then $Y \cong \mathbb{R}^{\aleph_0 *} \times \mathbb{Z}$. In order not to complicate the notation, suppose $Y = \mathbb{R}^{\aleph_0 *} \times \mathbb{Z}$. We follow the scheme of the proof of the theorem in case 1. First note that using Lemma 6.10 we prove that Proposition 2.10 and Lemma 6.1 hold true for the group X. As in case 1 we consider the subgroup $L = \operatorname{Ker}(I-\widetilde{\alpha}) \cap \mathbb{R}^{\aleph_0 *}$ and reduce the proof of the theorem to the case when $L = \{0\}$. In doing so we use the fact that any closed linear subspace in $\mathbb{R}^{\aleph_0}$ is either finite-dimensional or topologically isomorphic to $\mathbb{R}^{\aleph_0}$ [**15**] and that Proposition 2.10 holds true for the group X.

2a. Let $\alpha|_{\mathbb{T}} = I$. We argue as in case 1a. We use the fact that Lemma 6.1 is valid for the group $\mathbb{R}^{\aleph_0}$ and Proposition 2.10 holds true for the group X. Instead of the Ghurye–Olkin theorem we apply Lemma 6.17. We also use the fact that Cramér's decomposition theorem for the Gaussian distribution holds true on the group $\mathbb{R}^{\aleph_0}$.

2b. Let $\alpha|_{\mathbb{T}} = -I$. We argue as in case 1b. We use the fact that Lemma 6.1 is valid for the group X. Instead of Proposition 4.16 we apply Lemma 6.18. In all other respects the proof repeats the proof in the case 1b. □

7. Shifts of idempotent distributions on discrete and compact totally disconnected Abelian groups

Let X be a second countable locally compact Abelian group. If X is discrete, this means that X is countable. In this section we characterise shifts of idempotent distributions in the class of countable discrete Abelian groups by the independence of two linear forms of independent random variables ξ_1 and ξ_2 with values in X. We consider the linear forms $L_1 = \alpha_1\xi_1 + \alpha_2\xi_2$ and $L_2 = \beta_1\xi_1 + \beta_2\xi_2$, where $\alpha_j, \beta_j \in \mathrm{Aut}(X)$. We prove that if X is a countable discrete Abelian group, then the independence of L_1 and L_2 implies that μ_j are shifts of idempotent distributions. Next we give a complete description of compact totally disconnected Abelian groups X for which the independence of L_1 and L_2 implies that μ_j are shifts of idempotent distributions. We also prove that for an arbitrary compact connected Abelian group X there exist topological automorphisms $\alpha_j, \beta_j \in \mathrm{Aut}(X)$ and independent random variables ξ_1 and ξ_2 with values in X and distributions μ_1 and μ_2 such that the linear forms L_1 and L_2 are independent, whereas μ_j are not represented as convolutions of Gaussian distributions and idempotent distributions. In other words, on such groups there is not even a weak analogue of the Skitovich–Darmois theorem. At the end of this section we prove a theorem where convolutions of Gaussian distributions and idempotent distributions are characterized.

THEOREM 7.1. *Let X be a finite Abelian group and let α_j, β_j, $j = 1, 2$, be automorphisms of X. Let ξ_1 and ξ_2 be independent random variables with values in the group X and distributions μ_1 and μ_2. If the linear forms $L_1 = \alpha_1\xi_1 + \alpha_2\xi_2$ and $L_2 = \beta_1\xi_1 + \beta_2\xi_2$ are independent, then $\mu_j \in I(X)$, $j = 1, 2$.*

PROOF. Denote by Y the character group of the group X. It is easy to see that we can assume without loss of generality that $L_1 = \xi_1 + \xi_2$ and $L_2 = \xi_1 + \alpha\xi_2$, where $\alpha \in \mathrm{Aut}(X)$. Put $a = \widetilde{\alpha}$. By Lemma 6.1, the characteristic functions $\hat{\mu}_j(y)$ satisfy equation (6.1) which takes the form (6.33). Put $\nu_j = \mu_j * \bar{\mu}_j$, $j = 1, 2$. Then $\hat{\nu}_j(y) = |\hat{\mu}_j(y)|^2 \geq 0$ for all $y \in Y$. The characteristic functions $\hat{\nu}_j(y)$ also satisfy equation (6.33). Set $f(y) = \hat{\nu}_1(y)$, $g(y) = \hat{\nu}_2(y)$ and rewrite equation (6.33) for the characteristic functions $\hat{\nu}_j(y)$ in the form

$$f(u+v)g(u+av) = f(u)g(u)f(v)g(av), \quad u, v \in Y. \tag{7.1}$$

We will prove that in this case $f(y) = g(y) = \widehat{m}_K(y)$ for all $y \in Y$, where K is a subgroup of X. Then statement (v) of Theorem 2.5 and (2.3) imply that $\mu_j \in I(X)$. Thus, we solve equation (7.1) assuming that $f(y) \geq 0$, $g(y) \geq 0$, $f(-y) = f(y)$, $g(-y) = g(y)$ for all $y \in Y$.

Set $b = I - a$. Two cases are possible: either $\mathrm{Ker}\, b = \{0\}$ or $\mathrm{Ker}\, b \neq \{0\}$.

1. $\mathrm{Ker}\, b = \{0\}$. Since Y is a finite group, we have $b \in \mathrm{Aut}(Y)$. Substituting $v = -u$ into equation (7.1), we get

$$g(bu) = f^2(u)g(u)g(au), \quad u \in Y. \tag{7.2}$$

In view of $0 \leq f(y) \leq 1$ and $0 \leq g(y) \leq 1$, it follows from equation (7.2) that

$$g(bu) \leq g(u), \quad u \in Y. \tag{7.3}$$

Inasmuch as Y is a finite group, the automorphism group $\mathrm{Aut}(Y)$ is also finite. Hence

$$b^n = I \tag{7.4}$$

for some natural n. We assume that n in (7.4) is the smallest one. We deduce from (7.3) and (7.4) that

$$g(u) = g(b^n u) \le \cdots \le g(bu) \le g(u), \quad u \in Y.$$

Therefore,

$$g(u) = g(bu) = \cdots = g(b^{n-1}u), \quad u \in Y.$$

Thus, the function $g(y)$ takes a constant value on each of the orbits

$$O_y = \{y, by, \dots, b^{n-1}y\}.$$

This value generally depends on $y \in Y$.

Substitute $u = -av$ into equation (7.1). We get

$$f(bv) = f(av)g^2(av)f(v), \quad v \in Y. \tag{7.5}$$

Arguing as above, we find from (7.5) that the function $f(y)$ also takes a constant value on each of the orbits O_y.

It is obvious that the group Y can be represented as a union of disjoint orbits. Denote by N the union of orbits where $g(y) > 0$ and represent Y as the union $Y = N \cup N'$, where $N' = \{y \in Y : g(y) = 0\}$. It follows from (7.2) that

$$1 = f^2(u)g(au), \quad u \in N. \tag{7.6}$$

This implies that $g(ay) = 1$ for all $y \in N$ and hence $a(N) \subset N$. Since a is a one-to-one mapping and N is a finite set, we have $a(N) = N$. Hence $g(y) = 1$ for all $y \in N$. Taking into account that $g(y) = 0$ for all $y \in N'$, we get

$$g(y) = \begin{cases} 1 & \text{if } \ y \in N, \\ 0 & \text{if } \ y \in N'. \end{cases} \tag{7.7}$$

By Proposition 2.10, N is a subgroup of Y. Set $K = A(X, N)$. By Theorem 1.8, we have $N = A(Y, K)$. In view of (2.3), it follows from (7.7) that $g(y) = \widehat{m}_K(y)$. Hence $\mu_2 = m_K$.

It also follows from (7.6) that $f(y) = 1$ for all $y \in N$. We will verify that if $y \in N'$, then $f(y) = 0$. Assume that $v \in N'$. It follows from $a(N') = N'$ that $av \in N'$, and hence $g(av) = 0$. Then (7.5) implies that $f(bv) = 0$. Since $f(v) = f(bv)$, we have $f(v) = 0$. Thus, we proved that $f(y) = \widehat{m}_K(y)$. Hence $\mu_1 = m_K$. In case 1 the theorem is proved.

2. $\operatorname{Ker} b \neq \{0\}$, i.e., there exists an element $y_0 \in Y$, $y_0 \neq 0$ such that $ay_0 = y_0$. Put $L = \operatorname{Ker} b = \{y \in Y : ay = y\}$. It is obvious that $a(L) = L$.

Consider the restriction of equation (7.1) to L. We get

$$f(u+v)g(u+v) = f(u)g(u)f(v)g(v), \quad u, v \in L. \tag{7.8}$$

Substituting $v = -u$ into equation (7.8), we find

$$1 = f^2(u)g^2(u), \quad u \in L.$$

It follows from this that $f(y) = g(y) = 1$ for all $y \in L$. Put $K = A(X, L)$. By Proposition 2.10, the functions $f(y)$ and $g(y)$ are L-invariant and the inclusions $\sigma(\mu_j) \subset K$, $j = 1, 2$, hold. We note that by Theorems 1.8 and 1.9, we have $K^* \cong Y/L$. Since the functions $f(y)$ and $g(y)$ are L-invariant, they induce some functions $\widetilde{f}([y])$ and $\widetilde{g}([y])$ on the factor-group Y/L, namely $\widetilde{f}([y]) = f(y)$, $\widetilde{g}([y]) = g(y)$, $y \in [y]$. In view of $a(L) = L$, the automorphism a also induces some automorphism $\hat{a}$ on the factor-group Y/L by the formula $\hat{a}[y] = [ay]$, $y \in [y]$. Therefore we can consider equation (7.1) on the factor-group Y/L. The induced homomorphism $\hat{b}$

can also satisfy the condition $\operatorname{Ker}\hat{b} \neq \{0\}$. Repeating this procedure in a finite number of steps we get the induced homomorphism $\hat{b}$ which is an automorphism. Then item 1 yields that $\widetilde{f}([y])$ and $\widetilde{g}([y])$ are the characteristic functions of some Haar distribution $m_{\widehat{K}}$, where $\widehat{K}$ is a subgroup of X. Returning to the original characteristic functions $f(y)$ and $g(y)$, we obtain the required statement. It should be noted that if $b^m y = 0$ for some m and for all $y \in Y$, then this means that μ_1 and μ_2 are degenerate distributions. □

The following assertion was proved in the course of the proof of Theorem 7.1.

COROLLARY 7.2. *Let X be a finite Abelian group and let α be an automorphism of X. Let ξ_1 and ξ_2 be independent random variables with values in the group X and distributions μ_1 and μ_2. If the linear forms $L_1 = \xi_1 + \xi_2$ and $L_2 = \xi_1 + \alpha\xi_2$ are independent, then $\mu_j = m_K * E_{x_j}$, where K is a subgroup of X and $x_j \in X$, $j = 1, 2$. Moreover, $\alpha(K) = K$.*

Now consider the case when X is a discrete Abelian group. First, we prove a statement which can be regarded as a group analogue of the Skitovich–Darmois theorem for discrete torsion-free Abelian groups.

PROPOSITION 7.3. *Let X be a discrete torsion-free Abelian group. Let α_j, β_j, $j = 1, 2, \dots, n$, $n \geq 2$, be automorphisms of X. Let ξ_j be independent random variables with values in the group X and distributions μ_j. If the linear forms $L_1 = \alpha_1\xi_1 + \cdots + \alpha_n\xi_n$ and $L_2 = \beta_1\xi_1 + \cdots + \beta_n\xi_n$ are independent, then all μ_j are degenerate distributions.*

PROOF. Denote by Y the character group of the group X. Taking into consideration the new independent random variables $\zeta_j = \alpha_j\xi_j$ we reduce the proof of the theorem to the case when $L_1 = \xi_1 + \cdots + \xi_n$ and $L_2 = \delta_1\xi_1 + \cdots + \delta_n\xi_n$, where $\delta_j \in \operatorname{Aut}(X)$. By Lemma 6.1, if L_1 and L_2 are independent, then the characteristic functions $\hat{\mu}_j(y)$ satisfy equation (6.1) which takes the form (6.3).

First we prove that the characteristic functions $\hat{\mu}_j(y)$ do not vanish. Put $\nu_j = \mu_j * \bar{\mu}_j$. This implies that $\hat{\nu}_j(y) = |\hat{\mu}_j(y)|^2 \geq 0$ for all $y \in Y$. The characteristic functions $\hat{\nu}_j(y)$ also satisfy equation (6.3). Set

$$A_j = \{y \in Y : \hat{\nu}_j(y) > 0\}, \quad A = \bigcap_{j=1}^{n} A_j,$$

$$B = \left\{y \in Y : \prod_{j=1}^{n} \hat{\nu}_j(\widetilde{\delta}_j y) > 0\right\}, \quad C = \bigcap_{j=1}^{n} \widetilde{\delta}_j(B).$$

It is easy to see that $A + C \subset A$. Indeed, let $u \in A$, $v \in C$. Then $v = \widetilde{\delta}_j w_j$, where $w_j \in B$, $j = 1, 2, \dots, n$. It results from equation (6.3) that $u + v = u + \widetilde{\delta}_j w_j \in A_j$, $j = 1, 2, \dots, n$. Therefore $u + v \in \bigcap_{j=1}^{n} A_j = A$. Put

$$(m)C = \{y \in Y : y = y_1 + \cdots + y_m,\ y_j \in C\}.$$

Then the inclusion $A + C \subset A$ implies that $A + \bigcup_{m=1}^{\infty} (m)C \subset A$. In view of $0 \in C$, we have $A + \bigcup_{m=1}^{\infty} (m)C = A$.

Since X is a discrete torsion-free Abelian group, by Theorems 1.2 and 1.3, Y is a compact connected Abelian group. Taking into account that $0 \in C$, C is an open set, and Y is a connected group, we have

$$Y = \bigcup_{m=1}^{\infty} (m)C.$$

This implies that $A = Y$, i.e., the characteristic functions $\hat{\nu}_j(y) > 0$ for all $y \in Y$. Hence the characteristic functions $\hat{\mu}_j(y)$, $j = 1, 2, \dots, n$, do not vanish.

Put $\varphi_j(y) = -\ln \hat{\nu}_j(y)$, $y \in Y$. It results from (6.3) that the functions $\varphi_j(y)$ satisfy the equation

$$\sum_{j=1}^{n} \varphi_j(u + \widetilde{\delta}_j v) = \sum_{j=1}^{n} \varphi_j(u) + \sum_{j=1}^{n} \varphi_j(\widetilde{\delta}_j v), \quad u, v \in Y. \tag{7.9}$$

Integrating equation (7.9) over the group Y with respect to the Haar distribution m_Y in the variable u and using that the Haar distribution m_Y is Y-invariant, we find

$$\sum_{j=1}^{n} \varphi_j(\widetilde{\delta}_j v) = 0, \quad v \in Y. \tag{7.10}$$

It follows from this that $\varphi_j(y) = 0$ for all $y \in Y$, $j = 1, 2, \dots, n$. This implies that $\hat{\nu}_j(y) = 1$ for all $y \in Y$, $j = 1, 2, \dots, n$. Thus, we proved that $\nu_j = E_0$, $j = 1, 2, \dots, n$. Hence all μ_j are degenerate distributions. □

COROLLARY 7.4. *Let Y be a compact connected Abelian group and let $\widetilde{\alpha}_j, \widetilde{\beta}_j \in \mathrm{Aut}(Y)$, $j = 1, 2, \dots, n$, $n \geq 2$. If characteristic functions $\hat{\mu}_j(y)$ on the group Y satisfy equation* (6.1), *then $\hat{\mu}_j(y) = (x_j, y)$, $x_j \in X$, $j = 1, 2, \dots, n$.*

REMARK 7.5. In the course of the proof of Proposition 7.3 we proved that the characteristic functions $\hat{\mu}_j(y)$ do not vanish. In so doing we used only the connectedness of the group Y. This implies in particular that if $Y = \mathbb{R}^m$ and characteristic functions $\hat{\mu}_j(y)$ on the group Y satisfy equation (6.1), then $\hat{\mu}_j(y)$ do not vanish. Taking into account this remark we see that the Ghurye–Olkin theorem [**60**] results from Lemma 6.1 and Theorem 6.3.

We recall that by Theorem 1.14, every locally compact Abelian group X is topologically isomorphic to a group of the form $\mathbb{R}^m \times G$, where $m \geq 0$ and G is a locally compact Abelian group containing a compact open subgroup. Corollary 7.4 allows us to prove the following statement.

PROPOSITION 7.6. *Let $X = \mathbb{R}^m \times G$, where $m \geq 0$ and the group G contains a compact open subgroup. Let δ_j, $j = 1, 2, \dots, n$, $n \geq 2$, be topological automorphisms of X. Let ξ_j be independent random variables with values in the group X and distributions μ_j. If the linear forms $L_1 = \xi_1 + \cdots + \xi_n$ and $L_2 = \delta_1\xi_1 + \cdots + \delta_n\xi_n$ are independent, then the distributions μ_j are supported in a coset of a subgroup $\mathbb{R}^m \times K$, where K is a compact subgroup of the group X.*

PROOF. Denote by Y the character group of the group X. By Theorem 1.15, we have $c_Y = L \times M$, where $L \cong \mathbb{R}^m$ and M is a compact connected Abelian group. Since by Theorem 1.10, $A(G, M) = b_G$, we conclude that $A(X, M) = \mathbb{R}^m \times b_G = \mathbb{R}^m \times b_X$.

Put $\nu_j = \mu_j * \bar{\mu}_j$, $j = 1, 2, \ldots, n$. Then $\hat{\nu}_j(y) = |\hat{\mu}_j(y)|^2 \geq 0$ for all $y \in Y$. By Lemma 6.1, the characteristic functions $\hat{\mu}_j(y)$ satisfy equation (6.1) which takes the form (6.3). The characteristic functions $\hat{\nu}_j(y)$ also satisfy equation (6.3). Note that M is a characteristic subgroup of the group Y. Hence we can consider the restriction of equation (6.3) to the subgroup M. By Corollary 7.4, $\hat{\nu}_j(y) = 1$ for all $y \in M$. By Proposition 2.10, we have $\sigma(\nu_j) \subset A(X, M) = \mathbb{R}^m \times b_X$. It follows from Proposition 2.1 that we can replace the distributions μ_j by their shifts λ_j such that the distributions λ_j are supported in the subgroup $\mathbb{R}^m \times b_X$. It should be noted that $\mathbb{R}^m \times b_X$ is a characteristic subgroup of the group X. Moreover, if η_j are independent random variables with values in the group $\mathbb{R}^m \times b_X$ and distributions λ_j, then by Corollary 6.2, the linear forms $M_1 = \alpha_1\eta_1 + \cdots + \alpha_n\eta_n$ and $M_2 = \beta_1\eta_1 + \cdots + \beta_n\eta_n$ are independent.

Thus, we may prove the proposition assuming that the group G consists of compact elements. Then by Theorem 1.10, the group $H = G^*$ is totally disconnected. By Theorem 1.17, every neighborhood of the zero in the group H contains a compact open subgroup. Denote by N this subgroup and choose it in such a way that all characteristic functions $\hat{\nu}_j(y) > 0$ for all $y \in N$. Applying Theorem 1.17 again and taking into account the continuity of topological automorphisms $\widetilde{\delta}_j$, we get that there exists a compact open subgroup $F \subset N$ such that $\widetilde{\delta}_j(F) \subset N$, $j = 1, 2, \ldots, n$. Set $\varphi_j(y) = -\ln \hat{\nu}_j(y)$, $y \in N$. Inasmuch as the characteristic functions $\hat{\nu}_j(y)$ satisfy equation (6.3), the functions $\varphi_j(y)$ satisfy equation (7.9) for $u \in N$, $v \in F$, i.e.,

$$\sum_{j=1}^{n} \varphi_j(u + \widetilde{\delta}_j v) = \sum_{j=1}^{n} \varphi_j(u) + \sum_{j=1}^{n} \varphi_j(\widetilde{\delta}_j v), \quad u \in N,\ v \in F. \tag{7.11}$$

Integrating equation (7.11) over the group N with respect to the Haar distribution m_N and using that the Haar distribution m_N is N-invariant, we find that the functions $\varphi_j(y)$ satisfy equation (7.10) on the subgroup F. This implies that $\hat{\nu}_j(\widetilde{\delta}_j v) = 1$ for all $v \in F$. Put $B = \bigcap_{j=1}^{s} \widetilde{\delta}_j(F)$. Then B is an open subgroup of H, and $\hat{\nu}_j(y) = 1$ for all $y \in B$, $j = 1, 2, \ldots, n$.

Put $K = A(G, B)$ and note that $A(X, B) = \mathbb{R}^m \times K$, where, by Theorem 1.11, K is a compact group. By Proposition 2.10, $\sigma(\nu_j) \subset \mathbb{R}^m \times K$, $j = 1, 2, \ldots, n$. It follows from Proposition 2.1 that each of the distributions μ_j is supported in a coset of the subgroup $\mathbb{R}^m \times K$.

It should be noted that generally speaking the subgroup $\mathbb{R}^m \times K$ need not be characteristic. □

COROLLARY 7.7. *Let $X = \mathbb{R}^m \times G$, where $m \geq 0$ and the group G contains a compact open subgroup. Let α_j, β_j, $j = 1, 2, \ldots, n$, $n \geq 2$, be topological automorphisms of X. Let ξ_j be independent random variables with values in the group X and distributions μ_j. If the linear forms $L_1 = \alpha_1\xi_1 + \cdots + \alpha_n\xi_n$ and $L_2 = \beta_1\xi_1 + \cdots + \beta_n\xi_n$ are independent, then each of the distributions μ_j is supported in a coset of the subgroup $\mathbb{R}^m \times b_X$.*

REMARK 7.8. Let $X = \mathbb{R}^m \times G$, where $m \geq 0$ and G is a locally compact Abelian group containing a compact open subgroup. Taking into account that $\mathbb{R}^m \times b_X$ is a characteristic subgroup of the group X, it follows from Corollaries 6.2 and 7.7 that the description of distributions which are characterized by the independence of the linear forms L_1 and L_2 is reduced to the case when the group X is of the form $\mathbb{R}^m \times G$, where G is a locally compact Abelian group such that all its elements are compact.

The following statement follows directly from the proof of Proposition 7.6.

LEMMA 7.9. *Let X be a discrete torsion Abelian group with character group Y and let δ_j, $j = 1, 2, \dots, n$, $n \geq 2$, be automorphisms of X. Let ξ_j be independent random variables with values in the group X and distributions μ_j with nonnegative characteristic functions. If the linear forms $L_1 = \xi_1 + \cdots + \xi_n$ and $L_2 = \delta_1\xi_1 + \cdots + \delta_n\xi_n$ are independent, then there is an open subgroup $B \subset Y$ such that $\hat{\mu}_j(y) = 1$, $j = 1, 2, \dots, n$, for all $y \in B$.*

LEMMA 7.10. *Let X be a locally compact Abelian group with character group Y. Let K be a compact subgroup of X and let α be a continuous endomorphism of the group X. Then the following statements are equivalent:*

(i) *for each $y \in Y$ if $\widetilde{\alpha} y \in A(Y, K)$, then $y \in A(Y, K)$;*
(ii) *$\alpha(K) \supset K$.*

PROOF. Let us assume that (i) is true. Suppose that $y \in A(Y, \alpha(K))$. Then $(\alpha x, y) = 1$ for all $x \in K$. This implies that $(x, \widetilde{\alpha} y) = 1$ for all $x \in K$, i.e., $\widetilde{\alpha} y \in A(Y, K)$ and in view of (i), $y \in A(Y, K)$. Therefore $A(Y, \alpha(K)) \subset A(Y, K)$. By Theorem 1.8, it follows from this that (ii) holds.

Let us assume that (ii) is true. We conclude from (ii) that

$$A(Y, K) \supset A(Y, \alpha(K)). \tag{7.12}$$

Suppose that $\widetilde{\alpha} y \in A(Y, K)$. Then $(x, \widetilde{\alpha} y) = 1$ for all $x \in K$. Since $\alpha = \widetilde{\widetilde{\alpha}}$, we have $(\alpha x, y) = 1$ for all $x \in K$, i.e., $y \in A(Y, \alpha(K))$. Hence (7.12) implies that $y \in A(Y, K)$. □

LEMMA 7.11. *Let X be a locally compact Abelian group, let K be a compact subgroup of X, and let $\alpha \in \mathrm{Aut}(X)$. Let ξ_1 and ξ_2 be independent identically distributed random variables with values in the group X and distribution m_K. Then the following statements are equivalent:*

(i) *the linear forms $L_1 = \xi_1 + \xi_2$ and $L_2 = \xi_1 + \alpha\xi_2$ are independent;*
(ii) *$(I - \alpha)(K) \supset K$.*

PROOF. Denote by Y the character group of the group X. Put $a = \widetilde{\alpha}$, $b = I - a$, $H = A(Y, K)$, $f(y) = \widehat{m}_K(y)$. Let us assume that (i) is true. By Lemma 6.1, the characteristic functions of the random variables ξ_j satisfy equation (6.1) which takes the form

$$f(u + v) f(u + av) = f^2(u) f(v) f(av), \quad u, v \in Y. \tag{7.13}$$

Substituting $v = -u$ into equation (7.13), we obtain

$$f(bu) = f^3(u) f(au), \quad u \in Y. \tag{7.14}$$

It follows from (7.14) that if $by \in H$, then $y \in H$. Therefore, by Lemma 7.10, (ii) holds.

Let us assume that (ii) is true. We verify that the function $f(y)$ satisfies equation (7.13). If $u, v \in H$ and $av \in H$, then both sides of equation (7.13) are equal to 1. If $u, v \in H$ and $av \notin H$, then both sides of equation (7.13) are equal to zero. If either $u \in H$, $v \notin H$ or $u \notin H$, $v \in H$, then both sides of equation

(7.13) are equal to zero. If $u, v \notin H$, then the right-hand side of equation (7.13) is equal to zero. If the left-hand side of equation (7.13) is not equal to zero, then $u+v, u+av \in H$. It follows from this that $bv \in H$. In view of Lemma 7.10, (ii) implies that $v \in H$. The obtained contradiction shows that the left-hand side of equation (7.13) is also equal to zero. Thus, the function $f(y)$ satisfies equation (7.13). By Lemma 6.1, the linear forms L_1 and L_2 are independent. □

REMARK 7.12. We note that when $\delta = -I$ Lemma 7.11 yields the description of distributions m_K which are Gaussian distributions in the sense of Bernstein on the group X. Namely, K must be a Corwin group (see Remark 4.4). We also note that if K is a finite group, then statement (ii) of Lemma 7.11 and the condition $(I - \alpha)(K) = K$ are equivalent.

LEMMA 7.13. *Let X be a locally compact Abelian group, let K_1 and K_2 be finite subgroups of X, and let $\alpha \in \mathrm{Aut}(X)$. Let ξ_1 and ξ_2 be independent random variables with values in the group X and distributions $\mu_1 = m_{K_1}$ and $\mu_2 = m_{K_2}$. If the linear forms $L_1 = \xi_1 + \xi_2$ and $L_2 = \xi_1 + \alpha\xi_2$ are independent, then $K_1 = K_2 = K$ and $\alpha(K) = K$.*

PROOF. Denote by Y the character group of the group X. Put $a = \widetilde{\alpha}$, $b = I - a$, $f(y) = \widehat{m}_{K_1}(y)$, $g(y) = \widehat{m}_{K_2}(y)$, $H_j = A(Y, K_j)$, $j = 1, 2$. We use representation (2.3) for the characteristic functions $\widehat{m}_{K_j}(y)$. By Lemma 6.1, the characteristic functions of the random variables ξ_j satisfy equation (6.1) which takes the form (7.1). Substituting $u = -av$ into equation (7.1), we get (7.5). We conclude from equation (7.5) that if $bv \in H_1$, then $v \in H_1$. Hence by Lemma 7.10, $(I - \alpha)(K_1) \supset K_1$. Taking into account that K_1 is a finite subgroup, we get $(I - \alpha)(K_1) = K_1$. This implies that $\alpha(K_1) \subset K_1$. Since $\alpha \in \mathrm{Aut}(X)$ and K_1 is a finite group, we have $\alpha(K_1) = K_1$. It follows from this that $a(H_1) = H_1$. Consider the restriction of equation (7.1) to the subgroup H_1. We have

$$g(u + av) = g(u)g(av), \quad u, v \in H_1.$$

This implies that $g(y) = 1$ for all $y \in H_1$, i.e., $H_1 \subset H_2$.

Substituting $v = -u$ into equation(7.1), we obtain (7.2). Arguing as above and considering equation (7.2) instead of equation (7.5), we prove that $H_2 \subset H_1$. Hence $H_1 = H_2$. By Theorem 1.8, it follows that $K_1 = K_2 = K$. □

EXAMPLE 7.14. Generally speaking, Lemma 7.13 is false if K_1 and K_2 are compact but not finite Abelian groups. Here is an example.

Let G be a compact Abelian group. Consider the direct product

$$X = \mathop{\mathbf{P}}_{j=-\infty}^{\infty} G_j,$$

where all $G_j = G$. Denote by H and Y the character group of the groups G and X respectively. By Theorem 1.7,

$$Y \cong \mathop{\mathbf{P}^*}_{j=-\infty}^{\infty} H_j,$$

where $H_j = H$. In order not to complicate the notation, suppose

$$Y = \mathop{\mathbf{P}^*}_{j=-\infty}^{\infty} H_j.$$

Denote by $(g_j)_{j=-\infty}^{\infty}$, $g_j \in G$ and by $(h_j)_{j=-\infty}^{\infty}$, $h_j \in H$ elements of the group X and Y respectively. Let $\alpha \in \mathrm{Aut}(X)$ be an automorphism of the form

$$\alpha(g_j)_{j=-\infty}^{\infty} = (g_{j-2})_{j=-\infty}^{\infty}, \quad (g_j)_{j=-\infty}^{\infty} \in X.$$

Put $a = \widetilde{\alpha}$. Then

$$a(h_j)_{j=-\infty}^{\infty} = (h_{j+2})_{j=-\infty}^{\infty}, \quad (h_j)_{j=-\infty}^{\infty} \in Y.$$

It is easy to see that

$$\mathrm{Ker}(I - a) = \{0\}. \tag{7.15}$$

Consider the subgroups of X of the form

$$K_1 = \mathop{\mathbf{P}}_{j \neq 1} G_j, \quad K_2 = \mathop{\mathbf{P}}_{j \neq 2} G_j.$$

It is obvious that $K_1 \neq K_2$ and $H_j = A(Y, K_j)$, $j = 1, 2$. We will check that the characteristic functions $f(y) = \widehat{m}_{K_1}(y)$ and $g(y) = \widehat{m}_{K_2}(y)$ satisfy equation (7.1). It suffices to verify that equation (7.1) is satisfied for all $u \neq 0$, $v \neq 0$. Inasmuch as $H_1 \cap H_2 = \{0\}$, the right-hand side of equation (7.1) is equal to zero for all $u \neq 0$, $v \neq 0$. If the left-hand side of equation (7.1) is not equal to zero, we have $u + v \in H_1, u + av \in H_2$. It follows from this that $(I - a)v \in H_1 \times H_2$. But it is possible only if $(I - a)v = 0$. In view of (7.15) this implies that $v = 0$, contrary to the assumption. Thus, the characteristic functions $f(y)$ and $g(y)$ satisfy equation (7.1). By Lemma 6.1, if ξ_1 and ξ_2 are independent random variables with values in X and distributions m_{K_1} and m_{K_2}, then the linear forms $L_1 = \xi_1 + \xi_2$ and $L_2 = \xi_1 + \alpha\xi_2$ are independent.

The following theorem can be considered as an analogue of the Skitovich–Darmois theorem for two independent random variables for discrete Abelian groups.

THEOREM 7.15. *Let X be a discrete Abelian group and let α_j, β_j, $j = 1, 2$, be automorphisms of X. Let ξ_1 and ξ_2 be independent random variables with values in the group X and distributions μ_1 and μ_2. If the linear forms $L_1 = \alpha_1\xi_1 + \alpha_2\xi_2$ and $L_2 = \beta_1\xi_1 + \beta_2\xi_2$ are independent, then $\mu_j \in I(X)$, $j = 1, 2$.*

PROOF. We keep the same notation as in the proof of Theorem 7.1. Thus, we can solve equation (7.1), assuming that $f(y) \geq 0$, $g(y) \geq 0$, $f(-y) = f(y)$, $g(-y) = g(y)$ for all $y \in Y$. We will prove that in this case $f(y) = g(y) = \widehat{m}_K(y)$, where K is a finite subgroup of the group X. Statement (v) of Theorem 2.5 and (2.3) imply that $\mu_j \in I(X)$.

Since X is a discrete Abelian group, b_X is a subgroup of X consisting of all elements of X of finite order. Taking into account Remark 7.8, we can assume that X is a torsion group.

Put $E_f = \{y \in Y : f(y) = 1\}$, $E_g = \{y \in Y : g(y) = 1\}$. Then by Proposition 2.10, we have $\sigma(\mu_1) \subset A(X, E_f)$, $\sigma(\mu_2) \subset A(X, E_g)$. Put $F = A(X, E_f)$, $G = A(X, E_g)$. In view of Lemma 7.9, there exists an open subgroup B of the group Y such that $B \subset E_f \cap E_g$. Set $S = A(X, B)$. Then F and G are subgroups of S. Since B is an open subgroup, by Theorem 1.11, S is a compact group. Taking into account that X is a discrete group, S is a finite group. Hence F and G are also finite groups.

Observe now that for all natural n the characteristic functions $f^n(y)$ and $g^n(y)$ also satisfy equation (7.1), i.e.,

$$f^n(u + v)g^n(u + av) = f^n(u)g^n(u)f^n(v)g^n(av), \quad u, v \in Y. \tag{7.16}$$

Obviously, there exist the limits

$$\bar{f}(y) = \lim_{n\to\infty} f^n(y) = \begin{cases} 1 & \text{if } \ y \in E_f, \\ 0 & \text{if } \ y \notin E_f, \end{cases} \quad \bar{g}(y) = \lim_{n\to\infty} g^n(y) = \begin{cases} 1 & \text{if } \ y \in E_g, \\ 0 & \text{if } \ y \notin E_g. \end{cases}$$

Since by Theorem 1.8, $E_f = A(Y, F)$ and $E_g = A(Y, G)$, it follows from (2.3) that

$$\widehat{m}_F(y) = \begin{cases} 1 & \text{if } \ y \in E_f, \\ 0 & \text{if } \ y \notin E_f, \end{cases} \qquad \widehat{m}_G(y) = \begin{cases} 1 & \text{if } \ y \in E_g, \\ 0 & \text{if } \ y \notin E_g. \end{cases}$$

Hence

$$\widehat{m}_F(y) = \bar{f}(y), \quad \widehat{m}_G(y) = \bar{g}(y).$$

Let ζ_1 and ζ_2 be independent random variables with values in the group X and distributions $\lambda_1 = m_F$ and $\lambda_2 = m_G$. We deduce from (7.16) that the characteristic functions $\bar{f}(y)$ and $\bar{g}(y)$ also satisfy equation (7.1). By Lemma 6.1, this implies that the linear forms $L_1 = \zeta_1 + \zeta_2$ and $L_2 = \zeta_1 + \alpha\zeta_2$ are independent. Note that the conditions of Lemma 7.13 are fulfilled. It follows from Lemma 7.13 that $F = G$ and $\alpha(G) = G$.

Let us return to the original random variables ξ_1 and ξ_2 and to the linear forms $L_1 = \xi_1 + \xi_2$ and $L_2 = \xi_1 + \alpha\xi_2$. Inasmuch as $\sigma(\mu_j) \subset G$, $j = 1, 2$, the random variables ξ_j take values in the finite group G. Since $\alpha(G) = G$, the conditions of Corollary 7.2 are fulfilled. By Corollary 7.2, $\mu_j = m_K * E_{g_j}$, where K is a subgroup of the group G, and $g_j \in G$, $j = 1, 2$. □

The following assertion was proved in the course of the proof of Theorem 7.15.

COROLLARY 7.16. *Let X be a discrete Abelian group and let α be an automorphism of X. Let ξ_1 and ξ_2 be independent random variables with values in the group X and distributions μ_1 and μ_2. If the linear forms $L_1 = \xi_1 + \xi_2$ and $L_2 = \xi_1 + \alpha\xi_2$ are independent, then $\mu_j = m_K * E_{x_j}$, where K is a subgroup of X and $x_j \in X$, $j = 1, 2$. Moreover, $\alpha(K) = K$.*

Now we describe compact totally disconnected Abelian groups X for which for two independent random variables the Skitovich–Darmois theorem holds. We need some lemmas.

LEMMA 7.17. *Let X be a compact Abelian group with character group Y. Assume that there exist an automorphism $\alpha \in \mathrm{Aut}(X)$ and an element $\widetilde{y} \in Y$ such that the following conditions are satisfied:*

(i) $\mathrm{Ker}(I - \widetilde{\alpha}) = \{0\}$;
(ii) $(I - \widetilde{\alpha})(Y) \cap \{0, \pm\widetilde{y}, \pm 2\widetilde{y}\} = \{0\}$;
(iii) $\widetilde{\alpha}\widetilde{y} \neq -\widetilde{y}$.

Then there exist independent identically distributed random variables ξ_1 and ξ_2 with values in the group X and distribution μ such that the linear forms $L_1 = \xi_1 + \xi_2$ and $L_2 = \xi_1 + \alpha\xi_2$ are independent, whereas $\mu \notin \Gamma(X) * I(X)$.

PROOF. Consider on the group X the function

$$r(x) = 1 + (1/2)\,\mathrm{Re}(x, \widetilde{y}).$$

Then $r(x) > 0$ for all $x \in X$ and

$$\int_X r(x) dm_X(x) = 1.$$

Denote by μ the distribution on the group X with the density $r(x)$ with respect to the Haar distribution m_X. It is obvious that $\mu \notin \Gamma(X) * I(X)$. Let ξ_1 and ξ_2 be independent identically distributed random variables with values in X and distribution μ. Put $f(y) = \hat{\mu}(y)$. We will verify that the linear forms $L_1 = \xi_1 + \xi_2$ and $L_2 = \xi_1 + \alpha\xi_2$ are independent. By Lemma 6.1, it suffices to show that the characteristic functions of the random variables ξ_j satisfy equation (6.1) which takes the form

$$f(u+v)f(u+\widetilde{\alpha}v) = f^2(u)f(v)f(\widetilde{\alpha}v), \quad u, v \in Y. \tag{7.17}$$

It is easily seen that

$$f(y) = \begin{cases} 1 & \text{if } \ y = 0, \\ \frac{1}{4} & \text{if } \ y = \pm\widetilde{y}, \\ 0 & \text{if } \ y \notin \{0, \pm\widetilde{y}\}, \end{cases} \tag{7.18}$$

if $2\widetilde{y} \neq 0$ and

$$f(y) = \begin{cases} 1 & \text{if } \ y = 0, \\ \frac{1}{2} & \text{if } \ y = \widetilde{y}, \\ 0 & \text{if } \ y \notin \{0, \widetilde{y}\}, \end{cases} \tag{7.19}$$

if $2\widetilde{y} = 0$. It suffices to show that equation (7.17) is satisfied for all $u \neq 0$, $v \neq 0$. Assume $2\widetilde{y} \neq 0$. We conclude from (i) and (iii) that $\widetilde{\alpha}\widetilde{y} \neq \pm\widetilde{y}$. Hence (7.18) implies that the right-hand side of equation (7.17) is equal to zero for all $v \neq 0$. If the left-hand side of equation (7.17) is not equal to zero, this implies that $u+v, u+\widetilde{\alpha}v \in \{0, \pm\widetilde{y}\}$. It follows from this that $(I - \widetilde{\alpha})v \in \{0, \pm\widetilde{y}, \pm 2\widetilde{y}\}$. Taking into account (ii), we obtain that $(I - \widetilde{\alpha})v = 0$. Then (i) implies that $v = 0$. The obtained contradiction proves that the left-hand side of equation (7.17) is also equal to zero. If $2\widetilde{y} = 0$, then we use (7.19) and argue similarly. □

LEMMA 7.18. *Let $X = \Delta_p^{\aleph_0}$. Then there exist a topological automorphism $\alpha \in \mathrm{Aut}(X)$ and independent identically distributed random variables ξ_1 and ξ_2 with values in X and distribution μ such that the linear forms $L_1 = \xi_1 + \xi_2$ and $L_2 = \xi_1 + \alpha\xi_2$ are independent, whereas $\mu \notin I(X)$.*

PROOF. Denote by Y the character group of the group X. We assume that elements of the group X are bilateral sequences $x = (x_k)_{k=-\infty}^{\infty}$, where $x_k \in \Delta_p$. Consider $\alpha \in \mathrm{Aut}(X)$ of the form

$$\alpha(x_k)_{k=-\infty}^{\infty} = (x_{k-1})_{k=-\infty}^{\infty}.$$

We have $\Delta_p^* \cong \mathbb{Z}(p^\infty)$. Then by Theorem 1.7, $Y \cong (\mathbb{Z}(p^\infty))^{\aleph_0 *}$. Denote by $y = (y_k)_{k=-\infty}^{\infty}$, where $y_k \in \mathbb{Z}(p^\infty)$, elements of the group Y. Then

$$\widetilde{\alpha}(y_k)_{k=-\infty}^{\infty} = (y_{k+1})_{k=-\infty}^{\infty}.$$

Put $\widetilde{y} = (y_k)_{k=-\infty}^{\infty}$, where $y_k = 0$ for all $k \neq 0$, and y_0 is an arbitrary nonzero element of the group $\mathbb{Z}(p^\infty)$. We will show that the topological automorphism α and the element $\widetilde{y}$ satisfy the conditions of Lemma 7.17.

Take $y = (y_k)_{k=-\infty}^{\infty} \in \mathrm{Ker}(I - \widetilde{\alpha})$. Then $\widetilde{\alpha}y = y$. It follows from this that $y_{k+1} = y_k$ for all $k \in \mathbb{Z}$. Since $y_k \neq 0$ only for a finite number of indices k, we have $y_k = 0$ for all $k \in \mathbb{Z}$. Thus, condition (i) of Lemma 7.17 is fulfilled. Observe that if $y = (y_k)_{k=-\infty}^{\infty} \in (I - \widetilde{\alpha})(Y)$, where $y \neq 0$, then at least two elements y_{k_1} and y_{k_2} are different from zero. This implies that condition condition (ii) of Lemma

7.17 holds. Obviously, condition (iii) of Lemma 7.17 is also fulfilled. Since the group X is totally disconnected, by Proposition 3.5, Gaussian distributions on X are degenerated. The statement of the lemma follows from Lemma 7.17. □

LEMMA 7.19. *Let p be a prime number. Consider the group*

$$X = \mathop{\mathbf{P}}_{m=1}^{\infty} \mathbb{Z}(p^{k_m}), \quad k_m \leq k_{m+1}, \ m = 1, 2, \ldots. \tag{7.20}$$

Then there exist a topological automorphism $\alpha \in \mathrm{Aut}(X)$ and independent identically distributed random variables ξ_1 and ξ_2 with values in X and distribution μ such that the linear forms $L_1 = \xi_1 + \xi_2$ and $L_2 = \xi_1 + \alpha\xi_2$ are independent, whereas $\mu \notin I(X)$.

PROOF. Denote by $t = (t_m)_{m=1}^{\infty}$, where $t_m \in \mathbb{Z}(p^{k_m})$, elements of the group X. Let $j \geq i$ and let π_{ij} be the epimorphism $\pi_{ij} : \mathbb{Z}(p^{k_j}) \to \mathbb{Z}(p^{k_i})$ defined as follows:

$$\pi_{ij}(t_j) = t_j (\mathrm{mod}\ p^{k_i}), \quad t_j \in \mathbb{Z}(p^{k_j}).$$

Define the homomorphism $\alpha : X \to X$ by the formula $\alpha(t_m)_{m=1}^{\infty} = (s_m)_{m=1}^{\infty}$, where

$$s_m = \begin{cases} t_m + \pi_{m,m+1}t_{m+1} + \pi_{m,m+2}t_{m+2} & \text{if} \quad m \text{ is odd,} \\ t_m + \pi_{m,m+1}t_{m+1} & \text{if} \quad m \text{ is even.} \end{cases} \tag{7.21}$$

It is obvious that the homomorphism α is continuous. We will show that α is a monomorphism. Assume that $\alpha(t_m)_{m=1}^{\infty} = 0$. Take two sequential natural numbers m and $m+1$ such that m is odd. We have

$$t_m + \pi_{m,m+1}t_{m+1} + \pi_{m,m+2}t_{m+2} = 0, \tag{7.22}$$

$$t_{m+1} + \pi_{m+1,m+2}t_{m+2} = 0. \tag{7.23}$$

Apply $\pi_{m,m+1}$ to both sides of equality (7.23). Inasmuch as $\pi_{m,m+1}\pi_{m+1,m+2} = \pi_{m,m+2}$, we obtain

$$\pi_{m,m+1}t_{m+1} + \pi_{m,m+2}t_{m+2} = 0. \tag{7.24}$$

Subtracting (7.24) from (7.22), we get $t_m = 0$ for all $m = 1, 3, 5, \ldots$. Then it follows from (7.23) that $t_{m+1} = 0$ for all $m = 1, 3, 5, \ldots$. Thus, all $t_m = 0$, i.e., α is a monomorphism.

We will verify that α is an epimorphism. For this purpose we will prove that for any $s = (s_m)_{m=1}^{\infty} \in X$ the equation $\alpha t = s$ has a solution. It suffices to show the existence of the solution of the system of equations

$$\begin{cases} t_m + \pi_{m,m+1}t_{m+1} + \pi_{m,m+2}t_{m+2} = s_m, & (7.25) \\ t_{m+1} + \pi_{m+1,m+2}t_{m+2} = s_{m+1} & (7.26) \end{cases}$$

for each odd m.

Apply $\pi_{m,m+1}$ to both sides of equation (7.26). We obtain

$$\pi_{m,m+1}t_{m+1} + \pi_{m,m+2}t_{m+2} = \pi_{m,m+1}s_{m+1}. \tag{7.27}$$

Subtracting (7.27) from (7.25) we get that $t_m = s_m - \pi_{m,m+1}s_{m+1}$ and find in such a way all t_m, where m is odd. Then we find from (7.26) all t_m, where m is even. Thus, we proved that $\alpha \in \mathrm{Aut}(X)$.

By Theorem 1.7,

$$Y \cong \mathop{\mathbf{P}^*}_{m=1}^{\infty} \mathbb{Z}(p^{k_m}), \quad k_m \leq k_{m+1}, \ m = 1, 2, \ldots.$$

Denote by $l = (l_m)_{m=1}^{\infty}$, where $l_m \in \mathbb{Z}(p^{k_m})$ and $l_m = 0$ for all but a finite set of indices, elements of the group Y. Let $\widetilde{y} = (y_m)_{m=1}^{\infty} \in Y$ be an element such that $y_1 \neq 0$ and $y_m = 0$ for all $m > 1$. We will prove that the topological automorphism α and the element $\widetilde{y}$ satisfy the conditions of Lemma 7.17.

Let $j \geq i$. It is easy to see that the homomorphism $\widetilde{\pi}_{ij} : \mathbb{Z}(p^{k_i}) \to \mathbb{Z}(p^{k_j})$ is defined as follows

$$\widetilde{\pi}_{ij} t_i = p^{k_j - k_i} t_i, \quad t_i \in \mathbb{Z}(p^{k_i}).$$

A direct verification shows that the homomorphism $\widetilde{\alpha}$ is of the form $\widetilde{\alpha}(l_m)_{m=1}^{\infty} = (h_m)_{m=1}^{\infty}$, where

$$h_m = \begin{cases} l_1 & \text{if } \ m = 1, \\ \widetilde{\pi}_{m-1,m} l_{m-1} + l_m & \text{if } \ m \text{ is even}, \\ \widetilde{\pi}_{m-2,m} l_{m-2} + \widetilde{\pi}_{m-1,m} l_{m-1} + l_m & \text{if } \ m \text{ is odd}, \ m \neq 1. \end{cases} \tag{7.28}$$

We conclude from (7.21) that $I - \alpha$ is an epimorphism. By statement (b) of Theorem 1.20, $I - \widetilde{\alpha}$ is a monomorphism. Thus, condition (i) of Lemma 7.17 is fulfilled. It follows from (7.28) that condition (ii) of Lemma 7.17 is true. Obviously, condition (iii) of Lemma 7.17 also holds. Since the group X is totally disconnected, by Proposition 3.5, Gaussian distributions on X are degenerated. The statement of the lemma follows from Lemma 7.17. □

REMARK 7.20. Let X be a locally compact Abelian group and let G be a closed subgroup of X. Let $\alpha_j, \beta_j \in \mathrm{Aut}(G)$, $j = 1, 2, \dots, n$, and assume that α_j, β_j can be extended to some topological automorphisms $\bar{\alpha}_j$, $\bar{\beta}_j$ of the group X. Assume that there exist independent random variables ξ_j with values in G and distributions μ_j such that the linear forms $L_1 = \alpha_1\xi_1 + \cdots + \alpha_n\xi_n$ and $L_2 = \beta_1\xi_1 + \cdots + \beta_n\xi_n$ are independent, whereas $\mu_j \notin \Gamma(G) * I(G)$. We can consider ξ_j as independent random variables taking values in the group X. Furthermore, the linear forms $L_1 = \bar{\alpha}_1\xi_1 + \cdots + \bar{\alpha}_n\xi_n$ and $L_2 = \bar{\beta}_1\xi_1 + \cdots + \bar{\beta}_n\xi_n$ are independent and $\mu_j \notin \Gamma(X) * I(X)$.

We also note that if G is a topological direct factor of X, then any topological automorphism $\alpha \in \mathrm{Aut}(G)$ can be extended to a topological automorphism of the group X.

LEMMA 7.21. *Let X be a compact totally disconnected Abelian group. Then either the group X is topologically isomorphic to a group of the form*

$$\mathop{\mathbf{P}}_{p \in \mathcal{P}} (\Delta_p^{n_p} \times G_p), \tag{7.29}$$

where n_p is a nonnegative integer and G_p is a finite p-group or for some prime number p there exists a topological direct factor K of the group X such that K is topologically isomorphic to either the group $\Delta_p^{\aleph_0}$ or a group of the form (7.20).

PROOF. Denote by Y the character group of the group X. By Theorems 1.2 and 1.5, Y is a discrete torsion Abelian group. We deduce from Theorem 1.24 that Y is a weak direct product of its p-components Y_p

$$Y = \mathop{\mathbf{P}^*}_{p \in \mathcal{P}} Y_p. \tag{7.30}$$

By Theorem 1.25, each subgroup Y_p can be represented as a direct product $Y_p = D_p \times N_p$, where D_p is the maximal divisible subgroup of Y_p and N_p is a countable reduced p-group. By Theorem 1.28, the group D_p can be represented in its turn as a weak direct product of groups each of which is isomorphic to the group $Z(p^\infty)$. Taking into account that $(Z(p^\infty))^* \cong \Delta_p$, we conclude from Theorem 1.7 that

$$X = \mathop{\mathbf{P}}_{p\in\mathcal{P}} X_p,$$

where

$$X_p \cong \Delta_p^{\mathfrak{n}} \times G_p, \quad \mathfrak{n} \leq \aleph_0, \ G_p \cong (N_p)^*. \tag{7.31}$$

Assume that the group X is not topologically isomorphic to a group of the form (7.29). This means that in (7.31) for some p either $\mathfrak{n} = \aleph_0$ or G_p is an infinite group. If $\mathfrak{n} = \aleph_0$, then the group X has a topological direct factor G topologically isomorphic to the group $\Delta_p^{\aleph_0}$. If G_p is an infinite group, then the group N_p is also infinite. As is well known every countable infinite p-group has a direct factor isomorphic to the group

$$\mathop{\mathbf{P}^*}_{n=1}^{\infty} \mathbb{Z}(p^{k_m}), \quad k_m \leq k_{m+1}, \quad m = 1, 2, \ldots$$

[**59**, Proposition 77.5]. It follows from this that the group N_p also has such a direct factor. In this case by Theorem 1.7, the group G_p has a topological direct factor G, topologically isomorphic to a group of the form (7.20). □

THEOREM 7.22. *Let X be a compact totally disconnected Abelian group and let α_j, β_j, $j = 1, 2$, be topological automorphisms of X. Let ξ_1 and ξ_2 be independent random variables with values in the group X and distributions μ_1 and μ_2. The independence of the linear forms $L_1 = \alpha_1\xi_1 + \alpha_2\xi_2$ and $L_2 = \beta_1\xi_1 + \beta_2\xi_2$ implies that $\mu_j \in I(X)$, $j = 1, 2$, if and only if X is topologically isomorphic to a group of the form* (7.29).

PROOF. Necessity. Assume that the group X is not topologically isomorphic to a group of the form (7.29). By Lemma 7.21, for some prime number p the group X has a topological direct factor G such that G is topologically isomorphic either to the group $\Delta_p^{\aleph_0}$ or to a group of the form (7.20). If G is topologically isomorphic to the group $\Delta_p^{\aleph_0}$, then by Lemma 7.18, there exist an automorphism $\alpha \in \mathrm{Aut}(G)$ and independent random variables ξ_1 and ξ_2 with values in G and distributions μ_j such that the linear forms $L_1 = \xi_1 + \xi_2$ and $L_2 = \xi_1 + \alpha\xi_2$ are independent, whereas $\mu_j \notin I(G)$, $j = 1, 2$. If G is topologically isomorphic to a group of the form (7.20), then by Lemma 7.19, the analogous statement is also true. Since the subgroup G is a topological direct factor of X, the necessity results from Remark 7.20.

Sufficiency. Assume that the group X is topologically isomorphic to a group of the form (7.29). Denote by Y the character group of the group X. By Lemma 6.1, the characteristic functions $\hat\mu_j(y)$ satisfy equation (6.1) which takes the form

$$\begin{aligned}\hat\mu_1(\widetilde\alpha_1 u + \widetilde\beta_1 v)&\hat\mu_2(\widetilde\alpha_2 u + \widetilde\beta_2 v) \\ &= \hat\mu_1(\widetilde\alpha_1 u)\hat\mu_1(\widetilde\beta_1 v)\hat\mu_2(\widetilde\alpha_2 u)\hat\mu_2(\widetilde\beta_2 v), \quad u, v \in Y.\end{aligned}\tag{7.32}$$

Set $\nu_j = \mu_j * \bar\mu_j$, $j = 1, 2$. This implies that $\hat\nu_j(y) = |\hat\mu_j(y)|^2 \geq 0$ for all $y \in Y$. The characteristic functions $\hat\nu_j(y)$ also satisfy equation (7.32). If we prove that $\nu_j \in I(X)$, then statement (v) of Theorem 2.5 and (2.3) implies that $\mu_j \in I(X)$. Thus, we can solve equation (7.32) assuming that $\hat\mu_j(y) \geq 0$, $j = 1, 2$ for all $y \in Y$. The sufficiency will be proved if we show that the functions $\hat\mu_j(y)$ take only the values 0 and 1. Indeed, in this case $\hat\mu_j^2(y) = \hat\mu_j(y)$ for all $y \in Y$, $j = 1, 2$. This implies that $\mu_j^{*2} = \mu_j$, i.e., $\mu_j \in I(X)$, $j = 1, 2$.

Let $y_0 \in Y$. Taking into account (7.30), we have

$$y_0 = \sum_{j=1}^{n} y_j,$$

where $y_j \in Y_{p_j}$. It is obvious that $p_j^{k_j} y_j = 0$, $j = 1, 2, \ldots, n$ for some natural k_j. Consider the subgroups $B_j = \{y \in Y_{p_j} : p_j^{k_j} y = 0\}$. Then $y_0 \in B = B_1 \times \cdots \times B_n$. Since the group X is topologically isomorphic to a group of the form (7.29), each of the subgroups B_j is finite. Hence the subgroup B is also finite. Obviously, B is a characteristic subgroup of the group Y. It is clear that for any $u, v \in Y$ there exists a subgroup B of the form above such that $u, v \in B$. Consider the restriction of equation (7.32) to the subgroup B. By Corollary 2.8, the restriction of the characteristic functions $\hat\mu_j(y)$ to B are the characteristic functions of some distributions on the factor-group $X/A(X, B)$. Note that by Theorems 1.8 and 1.9, we have $(X/A(X, B))^* \cong B$. Taking into account that the group B is finite, this implies that the factor-group $X/A(X, B)$ is also finite. By Theorem 7.1, the characteristic functions $\hat\mu_j(y)$ take only values 0 and 1. The sufficiency is also proved. □

It is easy to see that if $L_1 = \xi_1 + \xi_2$ and $L_2 = \xi_1 + \alpha\xi_2$, where $\alpha \in \mathrm{Aut}(X)$, we obtain from the proof of the sufficiency in Theorem 7.22 the following assertion.

COROLLARY 7.23. *Let X be a compact totally disconnected Abelian group of the form* (7.29) *and let α be a topological automorphism of X. Let ξ_1 and ξ_2 be independent random variables with values in the group X and distributions μ_1 and μ_2. If the linear forms $L_1 = \xi_1 + \xi_2$ and $L_2 = \xi_1 + \alpha\xi_2$ are independent then $\mu_j = m_K * E_{x_j}$, where K is a subgroup of X and $x_j \in X$, $j = 1, 2$. Moreover, $\alpha(K) = K$.*

Consider now the case when X is a compact connected Abelian group. We prove that on these groups there is not even a weak analogue of the Skitovich–Darmois theorem. The following theorem is valid.

THEOREM 7.24. *Let X be a compact connected Abelian group. Then there exist topological automorphisms α_j, β_j, $j = 1, 2$, of X and independent random variables ξ_1 and ξ_2 with values in the group X and distributions μ_j such that the linear forms $L_1 = \alpha_1\xi_1 + \alpha_2\xi_2$ and $L_2 = \beta_1\xi_1 + \beta_2\xi_2$ are independent, whereas $\mu_j \notin \Gamma(X) * I(X)$, $j = 1, 2$.*

PROOF. Denote by Y the character group of the group X. By Theorems 1.2 and 1.3, Y is a discrete torsion-free Abelian group. Two cases are possible: either $f_p \notin \mathrm{Aut}(X)$ for some prime number p or $f_p \in \mathrm{Aut}(X)$ for all prime numbers p.

1. Suppose that

$$f_p \notin \mathrm{Aut}(X) \tag{7.33}$$

for some prime number p. We assume that p in (7.33) is the smallest one. Inasmuch as X is a connected Abelian group, it follows from Theorem 1.13 that $X^{(n)} = X$ for all natural n. Therefore, if $f_p \notin \mathrm{Aut}(X)$, then $\mathrm{Ker}\, f_p \neq \{0\}$.

Let $p = 2$. Then $\mathrm{Ker}\, f_2 = \{x \in X : 2x = 0\} \neq \{0\}$, i.e., the group X contains an element of order 2. Since $c_X = X$, it follows from Proposition 5.6 that there exist independent random variables ξ_1 and ξ_2 with values in the group X and distributions μ_j such that the linear forms $L_1 = \xi_1 + \xi_2$ and $L_2 = \xi_1 - \xi_2$ are independent, whereas $\mu_j \notin \Gamma(X) * I(X)$, $j = 1, 2$.

Assume $p \geq 3$. Put $q = 1 - p$. Since p is the smallest natural number with property (7.33), we have $f_q \in \mathrm{Aut}(X)$. By Theorem 1.12, $\mathrm{Ker}\, f_p = A(X, Y^{(p)})$. In view of $\mathrm{Ker}\, f_p \neq \{0\}$, this implies that $Y^{(p)} \neq Y$. Take $\widetilde{y} \notin Y^{(p)}$ and verify that the automorphism $\alpha = f_q$ and the element $\widetilde{y}$ satisfy the conditions of Lemma 7.17. Observe that $\widetilde{f_n} = f_n$ for all integers n and $I - f_q = f_p$. Inasmuch as Y is a torsion-free group, $\mathrm{Ker}(I - \widetilde{f_q}) = \{0\}$, i.e., condition (i) of Lemma 7.17 is fulfilled. We have $(I - \widetilde{f_q})(Y) = \widetilde{f_p}(Y) = Y^{(p)}$. Since $p \geq 3$, the numbers 2 and p are relatively prime. This implies that there exist some integers m and n such that $2m + pn = 1$. It follows from this that $y = 2my + pny$. Therefore, if $\widetilde{y} \notin Y^{(p)}$, then $2\widetilde{y} \notin Y^{(p)}$. This implies that condition (ii) of Lemma 7.17 is true. Since $\widetilde{f_q} + I = \widetilde{f_{2-p}}$, $p \geq 3$, and Y is a torsion-free group, condition (iii) of Lemma 7.17 also holds. Thus, in the case when (7.33) is fulfilled, the theorem follows from Lemma 7.17.

2. $f_p \in \mathrm{Aut}(X)$ for all prime numbers p. This means that X is a torsion-free group. Since X is a compact connected Abelian group, it follows from Theorem 1.16 that the group X is topologically isomorphic to a group of the form $(\Sigma_{\boldsymbol{a}})^{\mathfrak{n}}$, where $\boldsymbol{a} = (2, 3, 4, \dots)$, $\mathfrak{n} \leq \aleph_0$. Taking into account Remark 7.20, it suffices to prove the theorem for the group $\Sigma_{\boldsymbol{a}}$. Inasmuch as $(\Sigma_{\boldsymbol{a}})^* \cong \mathbb{Q}$, in order not to complicate the notation, suppose $(\Sigma_{\boldsymbol{a}})^* = \mathbb{Q}$.

Let H be a subgroup of $\mathbb{Q}$ of the form

$$H = \left\{\frac{m}{5^n} : n = 0, 1, \dots; \ m \in \mathbb{Z}\right\}.$$

Set $G = H^*$, $K = A(G, H^{(2)})$. It follows from Theorems 1.8 and 1.9 that $K \cong \mathbb{Z}(2)$. Let λ be an arbitrary nonidempotent distribution on the group G such that λ is supported in K. It is easy to see the characteristic function $\hat{\lambda}(h)$ is of the form

$$\hat{\lambda}(h) = \begin{cases} 1 & \text{if} \quad h \in H^{(2)}, \\ c & \text{if} \quad h \notin H^{(2)}, \end{cases} \tag{7.34}$$

where $-1 < c < 1$, $c \neq 0$. Consider on the group $\mathbb{Q}$ the function

$$g(y) = \begin{cases} \hat{\lambda}(y) & \text{if} \quad y \in H, \\ 0 & \text{if} \quad y \notin H. \end{cases} \tag{7.35}$$

By Proposition 2.9, $g(y)$ is a positive definite function. By Bochner's theorem, there exists a distribution $\mu \in \mathrm{M}^1(\Sigma_{\boldsymbol{a}})$ such that $\hat{\mu}(y) = g(y)$. It is obvious that $\mu \notin \Gamma(\Sigma_{\boldsymbol{a}}) * I(\Sigma_{\boldsymbol{a}})$.

Let ξ_1 and ξ_2 be independent identically distributed random variables with values in the group X and distribution μ. We verify that the linear forms $L_1 = \xi_1 + 2\xi_2$ and $L_2 = 2\xi_1 - \xi_2$ are independent. In view of Lemma 6.1, it suffices to show that the characteristic functions of the random variables ξ_j satisfy equation (6.1) which takes the form

$$\hat{\mu}(u + 2v)\hat{\mu}(2u - v) = \hat{\mu}(u)\hat{\mu}(2u)\hat{\mu}(2v)\hat{\mu}(-v), \quad u, v \in \mathbb{Q}. \tag{7.36}$$

It follows from (7.34) that if $u, v \in H$, then equation (7.36) becomes an equality. It follows from (7.35) that both sides of equation (7.36) are equal to zero if either $u \in H$, $v \notin H$ or $v \in H$, $u \notin H$. Assume that $u, v \notin H$. Then the right-hand side of equation (7.36) is equal to zero. If the left-hand side of equation (7.36) is not equal to zero, then $u+2v, 2u-v \in H$. This implies that $5u \in H$ and hence $u \in H$. The obtained contradiction shows that equation (7.36) becomes an equality for all $u, v \in \mathbb{Q}$. □

We note that if $\mu \in \Gamma(X) * I(X)$, then μ is invariant with respect to a compact subgroup K of the group X and, under the natural homomorphism $X \to X/K$, induces a Gaussian distribution on the factor-group X/K. Thus, Theorem 7.24 shows that even a weak analogue of the Skitovich–Darmois theorem is not valid for compact connected Abelian groups.

To conclude this section, we prove a theorem where convolutions of Gaussian distributions and idempotent distributions are characterized. Unlike the previous theorems proved in this chapter, here some conditions are imposed on the densities of distributions of random variables. It is convenient for us to formulate as a lemma the following standard statement.

LEMMA 7.25. *Let K be a locally compact Abelian group, $T \in \mathrm{Aut}(K)$, $\mu \in \mathrm{M}^1(K)$ and $\nu = T(\mu)$. Assume that the distribution μ is absolutely continuous with respect to m_K and $\rho_\mu(k)$ is the density of μ with respect to m_K. Then the distribution ν is also absolutely continuous with respect to m_K and $\rho_\nu(Tk) = c\rho_\mu(k)$ a.e. on K, where $\rho_\nu(k)$ is the density of ν with respect to m_K, c is a constant.*

We recall that by Theorem 1.14, each locally compact Abelian group X is topologically isomorphic to a group of the form $\mathbb{R}^m \times G$, where $m \geq 0$ and G is a locally compact Abelian group containing a compact open subgroup.

THEOREM 7.26. *Let $X = \mathbb{R}^m \times G$, where $m \geq 0$, and G is a locally compact Abelian group containing a compact open subgroup. Let α_j, β_j, $j = 1, 2$, be topological automorphisms of the group X. Let ξ_1 and ξ_2 be independent random variables with values in the group X and distributions μ_1 and μ_2. Assume that the following conditions are satisfied:*

(i) *$\alpha_1^{-1}\alpha_2 - \beta_1^{-1}\beta_2 \in \mathrm{Aut}(X)$;*
(ii) *μ_j has a continuous density $r_j(x)$ with respect to m_X such that $r_j(x) > 0$ for all $x \in X$, $j = 1, 2$.*

If the linear forms $L_1 = \alpha_1\xi_1 + \alpha_2\xi_2$ and $L_2 = \beta_1\xi_1 + \beta_2\xi_2$ are independent, then G is a compact group and $\mu_j = \gamma_j \times m_G = \gamma_j * m_G$, where $\gamma_j \in \Gamma(\mathbb{R}^m)$, $j = 1, 2$.

PROOF. It follows from condition (ii) that $\sigma(\mu_j) = X$, $j = 1, 2$. Applying Proposition 7.6, we obtain that G is a compact group.

Let $T : X^2 \to X^2$ be the mapping

$$T(x_1, x_2) = (\alpha_1 x_1 + \alpha_2 x_2, \beta_1 x_1 + \beta_2 x_2).$$

It follows from (i) that $T \in \mathrm{Aut}(X^2)$. Let E be a Borel subset of X^2. We have

$$\begin{aligned} T(\mu_{(\xi_1,\xi_2)})(E) = \mu_{(\xi_1,\xi_2)}(T^{-1}(E)) &= \{\omega \in \Omega : (\xi_1(\omega), \xi_2(\omega)) \in T^{-1}(E)\} \\ &= \{\omega \in \Omega : (L_1(\omega), L_2(\omega)) \in E\} = \mu_{(L_1,L_2)}(E). \end{aligned}$$

It means that

$$\mu_{(L_1,L_2)} = T(\mu_{(\xi_1,\xi_2)}). \tag{7.37}$$

The independence of ξ_1 and ξ_2 and condition (ii) imply that the distribution $\mu_{(\xi_1,\xi_2)}$ is absolutely continuous with respect to m_{X^2} and has a continuous density. Taking into account (7.37) and applying Lemma 7.25 to the group $K = X^2$, we conclude that the distribution $\mu_{(L_1,L_2)}$ is absolutely continuous with respect to m_{X^2} and also has a continuous density. Denote by $\rho(x_1, x_2)$ this density. The absolute continuity of $\mu_{(L_1,L_2)}$ with respect to m_{X^2} implies that the distributions μ_{L_1} and μ_{L_2} are absolutely continuous with respect to m_X. Assuming that the Haar measures are scaled so that $m_{X^2} = m_X \times m_X$, the independence of the linear forms L_1 and L_2 then implies that the densities $\rho(x_1, x_2)$, $\rho_1(x)$, and $\rho_2(x)$ satisfy the equation

$$\rho(x_1, x_2) = \rho_1(x_1)\rho_2(x_2) \quad \text{for a.e.} \quad (x_1, x_2) \in X^2. \tag{7.38}$$

It follows from (7.38) that there exist continuous functions $\widetilde{\rho}_j(x)$, $j = 1, 2$, such that $\rho_j(x) = \widetilde{\rho}_j(x)$ a.e. on X. Thus, we can assume that the densities $\rho_j(x)$, $j = 1, 2$, are continuous.

Taking into account (7.37), the independence of the random variables ξ_1 and ξ_2, the independence of the linear forms L_1 and L_2 and applying Lemma 7.25 to the group $K = X^2$, we conclude that the densities $\rho_j(x)$ and $r_j(x)$ satisfy the equation

$$\rho_1(\alpha_1 x_1 + \alpha_2 x_2)\rho_2(\beta_1 x_1 + \beta_2 x_2) = c r_1(x_1) r_2(x_2) \text{ for a.e. } (x_1, x_2) \in X^2. \tag{7.39}$$

The continuity of the functions $r_j(x)$ and $\rho_j(x)$ and (7.39) implies that

$$\rho_1(\alpha_1 x_1 + \alpha_2 x_2)\rho_2(\beta_1 x_1 + \beta_2 x_2) = c r_1(x_1) r_2(x_2), \quad x_1, x_2 \in X. \tag{7.40}$$

It follows from conditions (i), (ii), and (7.40) that $\rho_j(x) > 0$ for all $x \in X$, $j = 1, 2$.

Denote by $x = (t, g)$, where $t \in \mathbb{R}^m$, $g \in G$, elements of the group X. Let $\alpha \in \mathrm{Aut}(X)$. Since G is a compact Abelian group, it is easy to see that there exist $A \in \mathrm{Aut}(\mathbb{R}^m)$, $\kappa \in \mathrm{Aut}(G)$, and a continuous homomorphism $a : \mathbb{R}^m \to G$ such that α acts in X by the formula $\alpha(t, g) = (At, at + \kappa g)$, $t \in \mathbb{R}^m$, $g \in G$. Let $\alpha_j, \beta_j \in \mathrm{Aut}(X)$ such that $\alpha_j(t, g) = (A_j t, a_j t + \kappa_j g)$, $\beta_j(t, g) = (B_j t, b_j t + \iota_j g)$, where $A_j, B_j \in \mathrm{Aut}(\mathbb{R}^m)$, $\kappa_j, \iota_j \in \mathrm{Aut}(G)$, and a_j, b_j are continuous homomorphisms from $\mathbb{R}^m$ to G, $j = 1, 2$.

Put $\psi_j(x) = \ln \rho_j(x)$, $\varphi_j(x) = \ln r_j(x)$, $j = 1, 2$. It follows from (7.40) that the equation

$$\begin{aligned} \psi_1(A_1 t_1 + A_2 t_2, a_1 t_1 + a_2 t_2 + \kappa_1 g_1 + \kappa_2 g_2) \\ + \psi_2(B_1 t_1 + B_2 t_2, b_1 t_1 + b_2 t_2 + \iota_1 g_1 + \iota_2 g_2) \\ = \ln c + \varphi_1(t_1, g_1) + \varphi_2(t_2, g_2), \quad (t_1, g_1,), (t_2, g_2) \in X \end{aligned} \tag{7.41}$$

holds. Taking into account that the group G is compact and the functions $\varphi_j(x)$ and $\psi_j(x)$ are continuous, we can integrate equation (7.41) over the subgroup G with respect to the Haar distribution m_G in the variable g_1. Using that the Haar distribution m_G is G-invariant, we conclude that the function $\varphi_2(t, g)$ and hence the function $r_2(t, g)$, does not depend on g. Analogically, integrating equation (7.41) over the subgroup G with respect to the Haar distribution m_G in the variable g_2, we conclude that the function $\varphi_1(t, g)$ and hence the function $r_1(t, g)$, also does not depend on g. Thus, $r_j(t, g) = r_j(t)$ for all $t \in \mathbb{R}^m$, $g \in G$, $j = 1, 2$. Denote by γ_j the distribution on $\mathbb{R}^m$ with the density $r_j(t)$ with respect to $m_{\mathbb{R}^m}$. Let p be the homomorphism $p : X \to \mathbb{R}^m$ defined as follows: $p(t, g) = t$ for all $t \in \mathbb{R}^m$, $g \in G$.

Let E be a Borel subset in $\mathbb{R}^m$. We have

$$p(\mu_j)(E) = \mu_j(p^{-1}(E)) = \mu_j(E \times G) = \int\limits_{E\times G} r_j(t,g)dm_X(t,g)$$
$$= \int\limits_{E} r_j(t)dm_{\mathbb{R}^m}(t) = \gamma_j(E), \quad j = 1,2.$$

Thus, $\gamma_j = p(\mu_j)$. It is easy to see that $\mu_j = \gamma_j \times m_G = \gamma_j * m_G$, $j = 1,2$.

Let us verify that γ_j are Gaussian distributions in $\mathbb{R}^m$. Set $\eta_j = p(\xi_j)$, $j = 1,2$. Then η_j are independent random variables with values in $\mathbb{R}^m$ and distributions γ_j, $j = 1,2$. Consider the linear forms $M_1 = A_1\eta_1 + A_2\eta_2$ and $M_2 = B_1\eta_1 + B_2\eta_2$. Since $p\alpha_j = A_j$ and $p\beta_j = B_j$, we have $M_j = p(L_j)$, $j = 1,2$. The linear forms M_1 and M_2 are also independent. By the Ghurye–Olkin theorem, γ_j are Gaussian distributions. □

As a corollary of Theorem 7.26 we get the following characterization of the Haar distribution on a compact Abelian group.

COROLLARY 7.27. *Let X be a compact Abelian group and let topological automorphisms α_j, β_j, $j = 1,2$, of X satisfy condition* (i) *of Theorem* 7.26. *Let ξ_1 and ξ_2 be independent random variables with values in the group X and distributions μ_1 and μ_2 satisfying condition* (ii) *of Theorem* 7.26. *If the linear forms $L_1 = \alpha_1\xi_1 + \alpha_2\xi_2$ and $L_2 = \beta_1\xi_1 + \beta_2\xi_2$ are independent, then $\mu_1 = \mu_2 = m_X$.*

REMARK 7.28. Theorem 7.26 is not true if either condition (i) or condition (ii) of Theorem 7.26 is not fulfilled. We construct the corresponding examples on $\boldsymbol{a}$-adic solenoids.

First we prove that Theorem 7.26 fails if condition (i) of Theorem 7.26 is not satisfied. Let $\boldsymbol{a} = (2,2,2,\dots)$. Consider the $\boldsymbol{a}$-adic solenoid $\Sigma_{\boldsymbol{a}}$ with character group

$$H_{\boldsymbol{a}} = \left\{\frac{m}{2^n} : n = 0,1,\dots;\ m \in \mathbb{Z}\right\}.$$

Put

$$r(x) = 1 + \mathrm{Re}(x,1), \quad x \in \Sigma_{\boldsymbol{a}}.$$

It is obvious that $r(x) \geq 0$ for all $x \in \Sigma_{\boldsymbol{a}}$ and

$$\int\limits_{\Sigma_{\boldsymbol{a}}} r(x)dm_{\Sigma_{\boldsymbol{a}}}(x) = 1.$$

Let μ be the distribution on $\Sigma_{\boldsymbol{a}}$ with the density $r(x)$ with respect to $m_{\Sigma_{\boldsymbol{a}}}$. Then the characteristic function of the distribution μ is of the form

$$\hat{\mu}(y) = \begin{cases} 1 & \text{if} \quad y = 0, \\ \frac{1}{2} & \text{if} \quad y = \pm 1, \\ 0 & \text{if} \quad y \notin \{0,\pm 1\}. \end{cases} \tag{7.42}$$

Let ξ_1 and ξ_2 be independent identically distributed random variables with values in the group $\Sigma_{\boldsymbol{a}}$ and distribution μ. Let α be a topological automorphism of the group $\Sigma_{\boldsymbol{a}}$ of the form $\alpha x = -2x$. Inasmuch as $I - \alpha \notin \mathrm{Aut}(\Sigma_{\boldsymbol{a}})$, condition (i) of Theorem 7.26 is not fulfilled. Obviously, condition (ii) of Theorem 7.26 holds.

Using (7.42) it is not difficult to verify that the characteristic function $\hat{\mu}(y)$ satisfies the equation

$$\hat{\mu}(u+v)\hat{\mu}(u-2v)=\hat{\mu}^2(u)\hat{\mu}(v)\hat{\mu}(-2v),\quad u,v\in H_{\boldsymbol{a}}. \tag{7.43}$$

By Lemma 6.1, it follows from (7.43) that the linear forms $L_1=\xi_1+\xi_2$ and $L_2=\xi_1-2\xi_2$ are independent.

Prove now that Theorem 7.26 fails if condition (ii) of Theorem 7.26 is not satisfied. Let $\boldsymbol{a}=(2,3,4,\dots)$. Consider the $\boldsymbol{a}$-adic solenoid $\Sigma_{\boldsymbol{a}}$ with character group $\mathbb{Q}$. Consider on the group $\mathbb{Q}$ the functions

$$f_1(y)=\begin{cases}1 & \text{if}\quad y\in\mathbb{Z}^{(3)},\\ \frac{1}{2} & \text{if}\quad y\in\mathbb{Z}\setminus\mathbb{Z}^{(3)},\\ 0 & \text{if}\quad y\notin\mathbb{Z},\end{cases}\qquad f_2(y)=\begin{cases}1 & \text{if}\quad y\in\mathbb{Z}^{(2)},\\ \frac{1}{2} & \text{if}\quad y\in\mathbb{Z}\setminus\mathbb{Z}^{(2)},\\ 0 & \text{if}\quad y\notin\mathbb{Z}.\end{cases} \tag{7.44}$$

It is easy to verify that the restrictions of the functions $f_j(y)$ to the subgroup $\mathbb{Z}$ are positive definite functions. By Proposition 2.9, $f_j(y)$ are positive definite functions on the group $\mathbb{Q}$. By Bochner's theorem, there exist the distributions $\mu_j\in\mathrm{M}^1(\Sigma_{\boldsymbol{a}})$ such that $\hat{\mu}_j(y)=f_j(y)$, $j=1,2$. Let ξ_1 and ξ_2 be independent random variables with values in the group $\Sigma_{\boldsymbol{a}}$ and distributions μ_j. Let α_j be topological automorphisms of the group $\Sigma_{\boldsymbol{a}}$ of the form $\alpha_1x=3x$, $\alpha_2x=2x$, $x\in\Sigma_{\boldsymbol{a}}$. Then $\alpha_1-\alpha_2\in\mathrm{Aut}(\Sigma_{\boldsymbol{a}})$, i.e., condition (i) of Theorem 7.26 holds. By Proposition 2.10, it follows from (7.42) that the supports of μ_j are contained in proper closed subgroups of $\Sigma_{\boldsymbol{a}}$ and hence condition (ii) of Theorem 7.26 is not fulfilled. Using (7.44) it is not difficult to verify that the characteristic functions $\hat{\mu}_j(y)$ satisfy the equation

$$\hat{\mu}_1(u+3v)\hat{\mu}_2(u+2v)=\hat{\mu}_1(u)\hat{\mu}_2(u)\hat{\mu}_1(3v)\hat{\mu}_2(2v),\quad u,v\in\mathbb{Q}. \tag{7.45}$$

By Lemma 6.1, it follows from (7.45) that the linear forms $L_1=\xi_1+\xi_2$ and $L_2=3\xi_1+2\xi_2$ are independent.

8. Gaussian distributions on the cylinder $\mathbb{R}\times\mathbb{T}$

Consider the cylinder $\mathbb{R}\times\mathbb{T}$. Let α_j, β_j, $j=1,2$, be topological automorphisms of the group $\mathbb{R}\times\mathbb{T}$. Let ξ_j be independent random variables with values in $\mathbb{R}\times\mathbb{T}$ and distributions μ_j with nonvanishing characteristic functions. Taking into account Theorem 6.19, it is not difficult to prove that if the linear forms $L_1=\alpha_1\xi_1+\alpha_2\xi_2$ and $L_2=\beta_1\xi_1+\beta_2\xi_2$ are independent, then μ_j need not be Gaussian. They can be both either Gaussian distributions or convolutions of Gaussian distributions with signed measures. Surprising enough is the fact that if we consider three linear forms of three independent random variables with values in the group $\mathbb{R}\times\mathbb{T}$, then only Gaussian distributions are characterised by the independence of these linear forms. This is the main result of the section.

Although we will not use this, we note that the group $\mathbb{R}\times\mathbb{T}$ is topologically isomorphic to the multiplicative group of nonzero complex numbers.

Some of the lemmas in this section are proved for arbitrary locally compact Abelian groups. We always assume that these groups are second countable.

LEMMA 8.1. *Let X be a locally compact Abelian group with character group Y. Let α_i, β_i, $i=1,2$, be continuous endomorphisms of X. Let ξ_j, $j=1,2,3$, be independent random variables with values in the group X and distributions μ_j. The* *linear forms* $L_1 = \xi_1+\xi_2+\xi_3$, $L_2 = \alpha_1\xi_1+\alpha_2\xi_2+\xi_3$, *and* $L_3 = \beta_1\xi_1+\beta_2\xi_2+\xi_3$ *are independent if and only if the characteristic functions* $\hat{\mu}_j(y)$ *satisfy the equation*

$$(8.1)\quad \hat{\mu}_1(u+\widetilde{\alpha}_1 v+\widetilde{\beta}_1 w)\hat{\mu}_2(u+\widetilde{\alpha}_2 v+\widetilde{\beta}_2 w)\hat{\mu}_3(u+v+w)$$
$$= \hat{\mu}_1(u)\hat{\mu}_2(u)\hat{\mu}_3(u)\hat{\mu}_1(\widetilde{\alpha}_1 v)\hat{\mu}_2(\widetilde{\alpha}_2 v)\hat{\mu}_3(v)\hat{\mu}_1(\widetilde{\beta}_1 w)\hat{\mu}_2(\widetilde{\beta}_2 w)\hat{\mu}_3(w), \quad u,v,w\in Y.$$

The proof of the lemma is the same as the proof of Lemma 6.1 and we omit it.

LEMMA 8.2. *Let* a_i, b_i, $i=1,2$, *be nonzero real numbers. Let* ξ_j, $j=1,2,3$, *be independent random variables with values in the group* $\mathbb{R}$ *and distributions* μ_j. *If the linear forms* $L_1=\xi_1+\xi_2+\xi_3$, $L_2=a_1\xi_1+a_2\xi_2+\xi_3$, *and* $L_3=b_1\xi_1+b_2\xi_2+\xi_3$ *are independent, then all* μ_j *are either degenerate distributions or nondegenerate Gaussian distributions and the following statements hold:*

(i) $a_1b_2-a_1a_2b_2-a_1b_1b_2-a_2b_1+a_1a_2b_1+a_2b_1b_2=0$;

(ii) *all possible combinations of signs of* a_i, b_i *are described in the table:*

a_1	a_2	b_1	b_2
+	−	−	+
+	−	−	−
−	+	+	−
−	+	−	−
−	−	+	−
−	−	−	+

(iii) $a_1\neq a_2$, $b_1\neq b_2$;

(iv) $a_2b_1-a_1b_2\neq 0$;

(v) $\det\begin{pmatrix} a_1-1 & a_2-1\\ b_1-1 & b_2-1\end{pmatrix}\neq 0.$

PROOF. By Lemma 8.1, the characteristic functions $\hat{\mu}_j(y)$ satisfy equation (8.1), which takes the form

$$(8.2)\quad \hat{\mu}_1(u+a_1v+b_1w)\hat{\mu}_2(u+a_2v+b_2w)\hat{\mu}_3(u+v+w)$$
$$= \hat{\mu}_1(u)\hat{\mu}_2(u)\hat{\mu}_3(u)\hat{\mu}_1(a_1v)\hat{\mu}_2(a_2v)\hat{\mu}_3(v)\hat{\mu}_1(b_1w)\hat{\mu}_2(b_2w)\hat{\mu}_3(w), \quad u,v,w\in\mathbb{R}.$$

By the Skitovich–Darmois theorem, the independence of the linear forms L_1 and L_2 implies that all μ_j are Gaussian distributions, i.e., the characteristic functions $\hat{\mu}_j(s)$ are represented in the form

$$(8.3)\quad \hat{\mu}_j(s)=\exp\{-\sigma_j s^2+i\tau_j s\},\quad s\in\mathbb{R},$$

where $\sigma_j\geq 0$ and $\tau_j\in\mathbb{R}$, $j=1,2,3$. Substituting (8.3) in equation (8.2) we get that the following equalities hold:

$$(8.4)\quad \sigma_1a_1+\sigma_2a_2+\sigma_3=0,$$

$$(8.5)\quad \sigma_1b_1+\sigma_2b_2+\sigma_3=0,$$

$$(8.6)\quad \sigma_1a_1b_1+\sigma_2a_2b_2+\sigma_3=0.$$

It follows easily from (8.4)–(8.6) that either all μ_j are degenerate distributions or all μ_j are nondegenerate distributions.

Assume that all μ_j are nondegenerate distributions. Then in (8.4)–(8.6) $\sigma_j>0$, $j=1,2,3$. Prove statements (i)–(v).

(i) Since all $\sigma_j \neq 0$ in (8.4)–(8.6), we have

$$\det\begin{pmatrix} a_1 & a_2 & 1 \\ b_1 & b_2 & 1 \\ a_1b_1 & a_2b_2 & 1 \end{pmatrix} = 0,$$

i.e., (i) holds.

(ii) Inasmuch as all $\sigma_j > 0$, statement (ii) follows directly from (8.4)–(8.6).

(iii) Suppose $a_1 = a_2$. Then it follows from (ii) that $a_i < 0$, $i = 1, 2$, and (8.5) and (8.6) imply that $(1 - a_1)\sigma_3 = 0$. Since $a_1 < 0$, it follows from this that $\sigma_3 = 0$. Then we get from (8.4) that $\sigma_1 = \sigma_2 = 0$. The obtained contradiction proves (iii) in the case when $a_1 = a_2$. In the case when $b_1 = b_2$ we argue in the same way.

(iv) Assume $a_2b_1 - a_1b_2 = 0$. Then

$$b_2 = a_2b_1/a_1. \tag{8.7}$$

Taking into account (8.7), we rewrite (8.5) in the form

$$\frac{b_1}{a_1}(\sigma_1a_1 + \sigma_2a_2) + \sigma_3 = 0.$$

It follows from this and from (8.4) that

$$\left(1 - \frac{b_1}{a_1}\right)\sigma_3 = 0. \tag{8.8}$$

Taking into account (ii), (8.7) implies that either $a_1 > 0$, $a_2 < 0$, $b_1 < 0$, and $b_2 > 0$ or $a_1 < 0$, $a_2 > 0$, $b_1 > 0$, and $b_2 < 0$. Since in both cases $a_1b_1 < 0$, it follows from (8.8) that $\sigma_3 = 0$. Then taking into account that $a_ib_i < 0$, we get from (8.6) that $\sigma_1 = \sigma_2 = 0$. The obtained contradiction proves (iv).

(v) Assume that $\det\begin{pmatrix} a_1 - 1 & a_2 - 1 \\ b_1 - 1 & b_2 - 1 \end{pmatrix} = 0$, i.e.,

$$(a_1 - 1)(b_2 - 1) = (a_2 - 1)(b_1 - 1). \tag{8.9}$$

Taking into account (8.9), we rewrite the left-hand side of equality (i) in the following way:

$$\begin{aligned} &a_1b_2 - a_1a_2b_2 - a_1b_1b_2 + a_1a_2b_1b_2 - a_2b_1 + a_1a_2b_1 + a_2b_1b_2 - a_1a_2b_1b_2 \\ &\qquad = a_1b_2(1 - a_2 - b_1 + a_2b_1) - a_2b_1(1 - a_1 - b_2 + a_1b_2) \\ &= a_1b_2(a_2 - 1)(b_1 - 1) - a_2b_1(a_1 - 1)(b_2 - 1) = (a_2 - 1)(b_1 - 1)(a_1b_2 - a_2b_1) = 0. \end{aligned}$$

Since, according to (iv), $a_2b_1 - a_1b_2 \neq 0$, we have either $a_2 = 1$ or $b_1 = 1$. It follows from (ii) that in these cases $a_1 < 0$, $b_2 < 0$, but this contradicts (8.9). ☐

The next statement follows directly from Example 5.9.

LEMMA 8.3. *Consider the circle group $\mathbb{T}$ with character group $\mathbb{Z}$. Let ξ_1 and ξ_2 be independent random variables with values in $\mathbb{T}$ and distributions μ_1 and μ_2 with nonvanishing characteristic functions. Assume that the sum $\xi_1 + \xi_2$ and difference $\xi_1 - \xi_2$ are independent. Then $\mu_j = \gamma * \pi_j * E_{x_j}$, where $\gamma \in \Gamma^s(\mathbb{T})$, π_j are signed measures on $\mathbb{T}_{(2)}$ such that $\pi_1 * \pi_2 = E_1$, $x_j \in \mathbb{T}$. To put it another way, the characteristic functions $\hat\mu_j(n)$ are represented in the form*

$$\begin{aligned} \hat\mu_1(n) &= \exp\{-\sigma n^2 + i\theta_1 n + \kappa(1 - (-1)^n)\}, \quad n \in \mathbb{Z}, \\ \hat\mu_2(n) &= \exp\{-\sigma n^2 + i\theta_2 n - \kappa(1 - (-1)^n)\}, \quad n \in \mathbb{Z}, \end{aligned} \tag{8.10}$$

where $\sigma \geq 0$, $\kappa \in \mathbb{R}$, $0 \leq \theta_j < 2\pi$, $j = 1, 2$.

LEMMA 8.4. *Consider the circle group* $\mathbb{T}$ *with character group* $\mathbb{Z}$. *Let* $\alpha_{ij} \in \mathrm{Aut}(\mathbb{T})$, $i, j = 1, 2, 3$. *Let* ξ_j, $j = 1, 2, 3$, *be independent random variables with values in* $\mathbb{T}$ *and distributions* μ_j *with nonvanishing characteristic functions. If the linear forms* $L_1 = \alpha_{11}\xi_1 + \alpha_{12}\xi_2 + \alpha_{13}\xi_3$, $L_2 = \alpha_{21}\xi_1 + \alpha_{22}\xi_2 + \alpha_{23}\xi_3$, *and* $L_3 = \alpha_{31}\xi_1 + \alpha_{32}\xi_2 + \alpha_{33}\xi_3$ *are independent, then all* μ_j *are degenerate distributions.*

PROOF. It is obvious that $\mathrm{Aut}(\mathbb{T}) = \{I, -I\}$. We note that if for some two linear forms, say for L_1 and L_2, we have $L_1 = \pm L_2$, then all μ_j are degenerate distributions. So, we can assume that $L_i \neq \pm L_j$, $i, j = 1, 2, 3$, and $i \neq j$. This easily implies that we can suppose without loss of generality that $L_1 = \xi_1 + \xi_2 + \xi_3$, $L_2 = \xi_1 - \xi_2 + \xi_3$, and $L_3 = -\xi_1 + \xi_2 + \xi_3$.

Put $\eta_1 = \xi_1 + \xi_3$, $\eta_2 = \xi_2$. Then η_1 and η_2 are independent random variables. The independence of L_1 and L_2 implies that the sum $\eta_1 + \eta_2$ and difference $\eta_1 - \eta_2$ are also independent. Applying Lemma 8.3 to the independent random variables η_1 and η_2, we obtain that there exists $\sigma_2 \geq 0$ such that

$$|\hat{\mu}_1(2n)||\hat{\mu}_3(2n)| = |\hat{\mu}_2(2n)| = \exp\{-4\sigma_2 n^2\}, \quad n \in \mathbb{Z}. \tag{8.11}$$

The same arguments as above show that the independence of the linear forms L_1 and L_3 and the linear forms $-L_2$ and L_3 imply that there exist $\sigma_1 \geq 0$ and $\sigma_3 \geq 0$ such that

$$|\hat{\mu}_2(2n)||\hat{\mu}_3(2n)| = |\hat{\mu}_1(2n)| = \exp\{-4\sigma_1 n^2\}, \quad n \in \mathbb{Z}, \tag{8.12}$$

$$|\hat{\mu}_1(2n)||\hat{\mu}_2(2n)| = |\hat{\mu}_3(2n)| = \exp\{-4\sigma_3 n^2\}, \quad n \in \mathbb{Z}. \tag{8.13}$$

It follows from (8.11)–(8.13) that

$$\sigma_1 + \sigma_3 = \sigma_2, \quad \sigma_2 + \sigma_3 = \sigma_1, \quad \sigma_1 + \sigma_2 = \sigma_3.$$

Hence

$$\sigma_1 = \sigma_2 = \sigma_3 = 0. \tag{8.14}$$

Since $\hat{\mu}_j(n)$ are the characteristic functions of some distributions, (8.14) implies that $\kappa = 0$ in (8.3). Hence all μ_j are degenerate distributions. □

Let X be a locally compact Abelian group with character group Y and let $\alpha_i, \beta_i \in \mathrm{Aut}(X)$, $i = 1, 2$. We introduce the notation

$$L = (\widetilde{\alpha}_1 - I)(Y) + (\widetilde{\beta}_1 - I)(Y),$$
$$M = (\widetilde{\alpha}_2 - I)(Y) + (\widetilde{\beta}_2 - I)(Y),$$
$$N = (\widetilde{\alpha}_2 - \widetilde{\alpha}_1)(Y) + (\widetilde{\beta}_2 - \widetilde{\beta}_1)(Y),$$

and retain them until the end of this section.

LEMMA 8.5. *Let* X *be a locally compact Abelian group with character group* Y *and let* $\alpha_i, \beta_i \in \mathrm{Aut}(X)$, $i = 1, 2$. *Let* ξ_j, $j = 1, 2, 3$, *be independent random variables with values in the group* X *and distributions* μ_j *with nonvanishing characteristic functions. Assume that the linear forms* $L_1 = \xi_1 + \xi_2 + \xi_3$, $L_2 = \alpha_1\xi_1 + \alpha_2\xi_2 + \xi_3$, *and* $L_3 = \beta_1\xi_1 + \beta_2\xi_2 + \xi_3$ *are independent. Put* $\nu_j = \mu_j * \bar{\mu}_j$, $j = 1, 2, 3$. *Then* $\nu = \nu_1 * \nu_2 * \nu_3 \in \Gamma(X)$ *and the functions* $\psi_j(y) = -\ln \hat{\nu}_j(y)$ *satisfy the equations*

$$\Delta_h \Delta_k \Delta_l \psi_1(y) = 0, \quad h, y \in Y, \ k \in N, \ l \in L, \tag{8.15}$$

$$\Delta_h \Delta_k \Delta_l \psi_2(y) = 0, \quad h, y \in Y, \ k \in N, \ l \in M, \tag{8.16}$$

$$\Delta_h\Delta_k\Delta_l\psi_3(y)=0,\quad h,y\in Y,\ k\in L,\ l\in M. \tag{8.17}$$

PROOF. By Lemma 8.1, the characteristic functions $\hat\mu_j(y)$ satisfy equation (8.1). We have $\hat\nu_j(y)=|\hat\mu_j(y)|^2>0$ for all $y\in Y$. The characteristic functions $\hat\nu_j(y)$ also satisfy equation (8.1). It follows from (8.1) that

$$\begin{aligned}\psi_1(u+\widetilde\alpha_1 v+\widetilde\beta_1 w)+\psi_2(u+\widetilde\alpha_2 v+\widetilde\beta_2 w)+\psi_3(u+v+w)\\ =A(u)+B(v)+C(w),\quad u,v,w\in Y,\end{aligned} \tag{8.18}$$

where

$$\begin{aligned}A(y)&=\psi_1(y)+\psi_2(y)+\psi_3(y),\\ B(y)&=\psi_1(\widetilde\alpha_1 y)+\psi_2(\widetilde\alpha_2 y)+\psi_3(y),\\ C(y)&=\psi_1(\widetilde\beta_1 y)+\psi_2(\widetilde\beta_2 y)+\psi_3(y).\end{aligned}$$

We use the finite difference method to solve equation (8.18). Let k_1 and m_1 be arbitrary elements of the group Y. Set $h_1=-k_1-m_1$, then $h_1+k_1+m_1=0$. Substitute $u+h_1$ for u, $v+k_1$ for v, and $w+m_1$ for w in equation (8.18). Subtracting equation (8.18) from the resulting equation, we obtain

$$\begin{aligned}\Delta_{l_{11}}\psi_1(u+\widetilde\alpha_1 v+\widetilde\beta_1 w)+\Delta_{l_{12}}\psi_2(u+\widetilde\alpha_2 v+\widetilde\beta_2 w)\\ =\Delta_{h_1}A(u)+\Delta_{k_1}B(v)+\Delta_{m_1}C(w),\quad u,v,w\in Y,\end{aligned} \tag{8.19}$$

where $l_{11}=(\widetilde\alpha_1-I)k_1+(\widetilde\beta_1-I)m_1$ and $l_{12}=(\widetilde\alpha_2-I)k_1+(\widetilde\beta_2-I)m_1$. Let k_2 and m_2 be arbitrary elements of the group Y. Set $h_2=-\widetilde\alpha_2 k_2-\widetilde\beta_2 m_2$. Then $h_2+\widetilde\alpha_2 k_2+\widetilde\beta_2 m_2=0$. Substitute $u+h_2$ for u, $v+k_2$ for v, and $w+m_2$ for w in equation (8.19). Subtracting equation (8.19) from the resulting equation, we find

$$\begin{aligned}\Delta_{l_{21}}\Delta_{l_{11}}\psi_1(u+\widetilde\alpha_1 v+\widetilde\beta_1 w)=\Delta_{h_2}\Delta_{h_1}A(u)\\ +\Delta_{k_2}\Delta_{k_1}B(v)+\Delta_{m_2}\Delta_{m_1}C(w),\quad u,v,w\in Y,\end{aligned} \tag{8.20}$$

where $l_{21}=(\widetilde\alpha_2-\widetilde\alpha_1)k_2+(\widetilde\beta_2-\widetilde\beta_1)m_2$. Let k_3 and m_3 be arbitrary elements of the group Y. Set $h_3=-\widetilde\alpha_1 k_3-\widetilde\beta_1 m_3$. Then $h_3+\widetilde\alpha_1 k_3+\widetilde\beta_1 m_3=0$. Substitute $u+h_3$ for u, $v+k_3$ for v, and $w+m_3$ for w in equation (8.20). Subtracting equation (8.20) from the resulting equation, we get

$$\begin{aligned}\Delta_{h_3}\Delta_{h_2}\Delta_{h_1}A(u)+\Delta_{k_3}\Delta_{k_2}\Delta_{k_1}B(v)\\ +\Delta_{m_3}\Delta_{m_2}\Delta_{m_1}C(w)=0,\quad u,v,w\in Y.\end{aligned} \tag{8.21}$$

Let h_4 be an arbitrary element of the group Y. Substitute $u+h_4$ for u in equation (8.21). Subtracting equation (8.21) from the resulting equation, we obtain

$$\Delta_{h_4}\Delta_{h_3}\Delta_{h_2}\Delta_{h_1}A(u)=0,\quad h_j,u\in Y. \tag{8.22}$$

Substituting $h_1=h_2=h_3=h_4=h$ in equation (8.22), we get

$$\Delta_h^4 A(u)=0,\quad h,u\in Y.$$

Thus, $A(y)$ is a polynomial of degree ≤ 3. By Theorem 1.29, there exist symmetric k-additive functions $g_k(y_1,\dots,y_k)$, $k=1,2,3$, such that

$$A(y)=g_3^*(y)+g_2^*(y)+g_1^*(y)+g_0,\quad y\in Y,$$

where $g_k^*(y)=g_k(y,\dots,y)$ and $g_0=\text{const}$. It is obvious that $g_k^*(-y)=(-1)^k g_k^*(y)$. Inasmuch as $A(-y)=A(y)$ and $A(0)=0$, we have $A(y)=g_2^*(y)$. It is not difficult to verify that the function $g_2^*(y)$ satisfies equation (3.2). Hence $\nu\in\Gamma(X)$.

Since $A(y) = g_2^*(y) = g_2(y, y)$, it is easily seen that the function $A(y)$ satisfies the equation

$$\Delta_{h_3}\Delta_{h_2}\Delta_{h_1}A(u) = 0, \quad h_j, u \in Y. \tag{8.23}$$

The same arguments as above show that each of the functions $B(y)$ and $C(y)$ also satisfies equation (8.23).

Let h be an arbitrary element of the group Y. Substitute $u+h$ for u in equation (8.20). Subtracting equation (8.20) from the resulting equation and taking into account (8.23), we find

$$\Delta_h\Delta_{l_{21}}\Delta_{l_{11}}\psi_1(u + \widetilde{\alpha}_1 v + \widetilde{\beta}_1 w) = 0. \tag{8.24}$$

Substituting $v = w = 0$ in equation (8.24) and taking into account the expressions for l_{11} and l_{21}, we obtain that the function $\psi_1(y)$ satisfies equation (8.15). Reasoning as above we get that each of the functions $\psi_2(y)$ and $\psi_3(y)$ satisfies equations (8.16) and (8.17) respectively. □

Consider the group $\mathbb{R}\times\mathbb{T}$ with character group $Y = \mathbb{R}\times\mathbb{Z}$. Denote by $x = (t, z)$, where $t \in \mathbb{R}$, $z \in \mathbb{T}$, elements of the group X and by $y = (s, n)$, where $s \in \mathbb{R}$, $n \in \mathbb{Z}$, elements of the group Y. It is easy to verify that every automorphism $\alpha \in \mathrm{Aut}(X)$ is defined by a matrix $\begin{pmatrix} a & c \\ 0 & p \end{pmatrix}$, where $a, c \in \mathbb{R}$, $a \neq 0$, $p = \pm 1$ and α operates on X as follows

$$\alpha(t, z) = (at, e^{ict}z^p), \quad t \in \mathbb{R},\ z \in \mathbb{T}.$$

Then the adjoint automorphism $\widetilde{\alpha} \in \mathrm{Aut}(Y)$ operates on Y as follows

$$\widetilde{\alpha}(s, n) = (as + cn, pn), \quad s \in \mathbb{R},\ n \in \mathbb{Z}.$$

We identify the automorphisms α and $\widetilde{\alpha}$ with the corresponding matrix $\begin{pmatrix} a & c \\ 0 & p \end{pmatrix}$.

LEMMA 8.6. *Let $X = \mathbb{R} \times \mathbb{T}$. Let $\alpha_i = \begin{pmatrix} a_i & c_i \\ 0 & p_i \end{pmatrix}$, $\beta_i = \begin{pmatrix} b_i & d_i \\ 0 & q_i \end{pmatrix}$, $i = 1, 2$, be topological automorphisms of X. Let ξ_j, $j = 1, 2, 3$, be independent random variables with values in the group X and distributions μ_j with nonvanishing characteristic functions. If the linear forms $L_1 = \xi_1+\xi_2+\xi_3$, $L_2 = \alpha_1\xi_1+\alpha_2\xi_2+\xi_3$, and $L_3 = \beta_1\xi_1+\beta_2\xi_2+\xi_3$ are independent, then either all μ_j are degenerate distributions or all μ_j are nondegenerate distributions and there are the following possibilities for the subgroups L, M, and N:*

1. *$L = \mathbb{R}$, $M = \mathbb{R}$, $N = \mathbb{R}$;*
2. *$L = \mathbb{R}$, $M = Y^{(2)}$, $N = Y^{(2)}$;*
3. *$L = Y^{(2)}$, $M = \mathbb{R}$, $N = Y^{(2)}$;*
4. *$L = Y^{(2)}$, $M = Y^{(2)}$, $N = \mathbb{R}$;*
5. *$L = Y^{(2)}$, $M = Y^{(2)}$, $N = Y^{(2)}$.*

PROOF. By Lemma 8.1, the characteristic functions $\hat{\mu}_j(y)$ satisfy equation (8.1). Substituting $v = (s_2, 0)$, $w = (s_3, 0)$ in equation (8.1), we obtain

$$\begin{aligned}\hat{\mu}_1(s_1 + a_1s_2 + b_1s_3, 0)\hat{\mu}_2(s_1 + a_2s_2 + b_2s_3, 0)\hat{\mu}_3(s_1 + s_2 + s_3, 0)\\ = \hat{\mu}_1(s_1, 0)\hat{\mu}_2(s_1, 0)\hat{\mu}_3(s_1, 0)\hat{\mu}_1(a_1s_2, 0)\hat{\mu}_2(a_2s_2, 0)\\ \times\hat{\mu}_3(s_2, 0)\hat{\mu}_1(b_1s_3, 0)\hat{\mu}_2(b_2s_3, 0)\hat{\mu}_3(s_3, 0), \quad s_1, s_2, s_3 \in \mathbb{R}.\end{aligned} \tag{8.25}$$

Set $\nu_j = \mu_j * \bar{\mu}_j$, $j = 1, 2, 3$. This implies that $\hat{\nu}_j(y) = |\hat{\mu}_j(y)|^2 > 0$ for all $y \in Y$. The characteristic functions $\hat{\nu}_j(y)$ also satisfy equation (8.1). Hence they satisfy equation (8.25). It follows from (8.25) and Lemma 8.2 for the group $\mathbb{R}$ that if at least one of the functions $\hat{\nu}_j(s, 0)$ is the characteristic function of a nondegenerate distribution, then all $\hat{\nu}_j(s, 0)$ are the characteristic functions of nondegenerate Gaussian distributions. To put it another way, either $\hat{\nu}_j(s, 0) = 1$ for all $s \in \mathbb{R}$ or $\hat{\nu}_j(s, 0) = e^{-\sigma_j s^2}$ for all $s \in \mathbb{R}$, where $\sigma_j > 0$, $j = 1, 2, 3$.

Assume that $\hat{\nu}_j(s, 0) = 1$ for all $s \in \mathbb{R}$, $j = 1, 2, 3$. By Proposition 2.10, this implies the inclusions $\sigma(\nu_j) \subset A(X, \mathbb{R}) = \mathbb{T}$, $j = 1, 2, 3$. Hence by Proposition 2.1, some shifts of the distributions μ_j are supported in $\mathbb{T}$. Since $\mathbb{T}$ is a characteristic subgroup of the group X, by Lemma 8.4, all μ_j are degenerate distributions.

Suppose now that $\hat{\nu}_j(s, 0) = e^{-\sigma_j s^2}$ for all $s \in \mathbb{R}$, $\sigma_j > 0$, $j = 1, 2, 3$. Then statement (v) of Lemma 8.2 implies that the following conditions hold:

(a) either $a_1 \neq 1$ or $b_1 \neq 1$;
(b) either $a_2 \neq 1$ or $b_2 \neq 1$.

It follows from (a) that there are only two possibilities for L: either $L = \mathbb{R}$ or $L = Y^{(2)}$. Reasoning as above we get from (b) that there are only two possibilities for M: either $M = \mathbb{R}$ or $M = Y^{(2)}$. Obviously, each of these possibilities can be realized.

Let $L = \mathbb{R}$ and $M = \mathbb{R}$. Then $p_i = q_i = 1$, where $i = 1, 2$. Hence $p_2 - p_1 = 0$ and $q_2 - q_1 = 0$. Taking into account statement (iii) of Lemma 8.2, this implies that for N there exists the only possibility, $N = \mathbb{R}$.

Let $L = \mathbb{R}$ and $M = Y^{(2)}$. Then $p_1 = q_1 = 1$ and either $p_2 = -1$ or $q_2 = -1$. Hence either $p_2 - p_1 = -2$ or $q_2 - q_1 = -2$. Taking into account statement (iii) of Lemma 8.2, this implies that for N there exists the only possibility, $N = Y^{(2)}$.

Let $L = Y^{(2)}$ and $M = \mathbb{R}$. Then either $p_1 = -1$ or $q_1 = -1$ and $p_2 = q_2 = 1$. Hence either $p_2 - p_1 = 2$ or $q_2 - q_1 = 2$. Taking into account statement (iii) of Lemma 8.2, this implies that for N there exists also the only possibility, $N = Y^{(2)}$.

Let $L = Y^{(2)}$ and $M = Y^{(2)}$. Then either $p_1 = -1$ or $q_1 = -1$ and either $p_2 = -1$ or $q_2 = -1$. If $p_1 = p_2$ and $q_1 = q_2$, then by statement (iii) of Lemma 8.2, $N = \mathbb{R}$. If either $p_1 \neq p_2$ or $q_1 \neq q_2$, then by statement (iii) of Lemma 8.2, $N = Y^{(2)}$. □

LEMMA 8.7. *Consider the group $\mathbb{R}\times\mathbb{Z}$ and denote by $y = (s, n)$, $s \in \mathbb{R}$, $n \in \mathbb{Z}$, its elements. Let $f(y)$ be a continuous function on $\mathbb{R}\times\mathbb{Z}$ satisfying the equation*

$$\Delta_k^2 \Delta_h f(y) = 0, \quad k \in \mathbb{R},\ h, y \in \mathbb{R}\times\mathbb{Z}, \tag{8.26}$$

and such that $f(-y) = f(y)$ for all $y \in \mathbb{R}\times\mathbb{Z}$. Then

$$f(s, n) = \sigma s^2 + \kappa(n)s + \lambda(n), \quad s \in \mathbb{R},\ n \in \mathbb{Z}, \tag{8.27}$$

where $\kappa(-n) = -\kappa(n)$ and $\lambda(-n) = \lambda(n)$ for all $n \in \mathbb{Z}$.

PROOF. Put $f_n(s) = f(s, n)$, where $s \in \mathbb{R}$, $n \in \mathbb{Z}$. Substituting $y = (s, n)$, $k = h = (w, 0)$ in equation (8.26), we get that the function $f_n(s)$ for any fixed n satisfies the equation

$$\Delta_w^3 f_n(s) = 0, \quad w, s \in \mathbb{R}. \tag{8.28}$$

It follows from (8.28) that the function $f_n(s)$ is of the form

$$f_n(s) = \sigma(n)s^2 + \kappa(n)s + \lambda(n), \quad s \in \mathbb{R}. \tag{8.29}$$

Substitute representation (8.29) for the function $f_n(s)$ into equation (8.26), supposing that $h = (0, m)$, $k = (w, 0)$, $m \in \mathbb{Z}$, $w \in \mathbb{R}$. We get

$$\Delta_k^2 \Delta_h f_n(s) = \Delta_h \Delta_w^2 f_n(s) = \Delta_h \sigma(n) \Delta_w^2 s^2 = 2w^2 \Delta_m \sigma(n) = 0, \quad w \in \mathbb{R},\ m \in \mathbb{Z}.$$

This implies that

$$\Delta_m \sigma(n) = 0, \quad m, n \in \mathbb{Z}.$$

Hence $\sigma(n) = \sigma = \text{const}$.

Since $f(-y) = f(y)$ for all $y \in \mathbb{R} \times \mathbb{Z}$, it is obvious that $\kappa(-n) = -\kappa(n)$ and $\lambda(-n) = \lambda(n)$ for all $n \in \mathbb{Z}$. It should be noted that any function $f(s, n)$ of the form (8.27) satisfies equation (8.26). □

Now we can prove the main result of this section.

THEOREM 8.8. *Let $X = \mathbb{R} \times \mathbb{T}$ and let $\alpha_{ij} \in \operatorname{Aut}(X)$, $i, j = 1, 2, 3$. Let ξ_j be independent random variables with values in the group X and distributions μ_j with nonvanishing characteristic functions. If the linear forms $L_1 = \alpha_{11}\xi_1 + \alpha_{12}\xi_2 + \alpha_{13}\xi_3$, $L_2 = \alpha_{21}\xi_1 + \alpha_{22}\xi_2 + \alpha_{23}\xi_3$, and $L_3 = \alpha_{31}\xi_1 + \alpha_{32}\xi_2 + \alpha_{33}\xi_3$ are independent, then all μ_j are either degenerate distributions or Gaussian distributions such that their supports are cosets of a subgroup of X topologically isomorphic to the group $\mathbb{R}$.*

PROOF. The proof of the theorem is long and it is based on Lemmas 8.1–8.7. So, we explain in some words the idea of the proof. By Lemma 8.5, under the conditions of the theorem on an arbitrary locally compact Abelian group X the convolution

$$\nu = \mu_1 * \bar{\mu}_1 * \mu_2 * \bar{\mu}_2 * \mu_3 * \bar{\mu}_3$$

is a Gaussian distribution. If X contains no subgroups topologically isomorphic to the circle group $\mathbb{T}$, then on the group X, Cramér's decomposition theorem for the Gaussian distribution holds (Theorem 3.14). Applying this theorem, we find that all distributions μ_j are Gaussian. In our case on the group $\mathbb{R} \times \mathbb{T}$ a Gaussian distribution may have non-Gaussian factors. Thus, generally speaking, if $\nu \in \mathrm{M}^1(\mathbb{R} \times \mathbb{T})$ and ν is a Gaussian distribution, it does not imply that factors of ν are also Gaussian. We prove that under the conditions of the theorem the Gaussian distribution ν is not arbitrary, but ν is concentrated on a subgroup of $\mathbb{R} \times \mathbb{T}$ topologically isomorphic to $\mathbb{R}$. By Cramér's decomposition theorem for the Gaussian distribution on the real line, it follows from this that all distributions μ_j are also Gaussian.

Let X be a locally compact Abelian group. First we note that if μ is the distribution of a random variable ξ with values in the group X and $\alpha \in \operatorname{Aut}(X)$, then the characteristic function of a random variable $\alpha\xi$ is $\hat{\mu}(\widetilde{\alpha} y)$. This implies that $\mu \in \Gamma(X)$ if and only if $\alpha(\mu) \in \Gamma(X)$. Thus, putting $\zeta_j = \alpha_{1j}\xi_j$, $j = 1, 2, 3$, we reduce the proof of the theorem to the case when L_1, L_2, and L_3 are of the form $L_1 = \xi_1 + \xi_2 + \xi_3$, $L_2 = \delta_{21}\xi_1 + \delta_{22}\xi_2 + \delta_{23}\xi_3$, and $L_3 = \delta_{31}\xi_1 + \delta_{32}\xi_2 + \delta_{33}\xi_3$, where $\delta_{ij} \in \operatorname{Aut}(X)$. We also note that if $\alpha, \beta \in \operatorname{Aut}(X)$, then the linear forms L_1, L_2, and L_3 are independent if and only if the linear forms L_1, αL_2, and βL_3 are independent. Thus, in proving the theorem we may assume that $L_1 = \xi_1 + \xi_2 + \xi_3$, $L_2 = \alpha_1\xi_1 + \alpha_2\xi_2 + \xi_3$, and $L_3 = \beta_1\xi_1 + \beta_2\xi_2 + \xi_3$, where

$$\alpha_i = \begin{pmatrix} a_i & c_i \\ 0 & p_i \end{pmatrix}, \quad \beta_i = \begin{pmatrix} b_i & d_i \\ 0 & q_i \end{pmatrix}, \quad i = 1, 2,$$

are topological automorphisms of the group $X = \mathbb{R} \times \mathbb{T}$.

Put $\nu_j = \mu_j * \bar{\mu}_j$, $j = 1, 2, 3$. Taking into consideration Lemma 8.6, we can prove the theorem assuming that all μ_j are nondegenerate distributions. Moreover, it follows from the proof of Lemma 8.6 that we have in this case $\hat{\nu}_j(s, 0) = e^{-\sigma_j s^2}$, $s \in \mathbb{R}$, where $\sigma_j > 0$, $j = 1, 2, 3$. We note that then σ_j satisfy equations (8.4)–(8.6) and statements (i)–(v) of Lemma 8.2 hold.

By Lemma 8.1, the characteristic functions $\hat{\mu}_j(y)$ satisfy equation (8.1). This implies that $\hat{\nu}_j(y) = |\hat{\mu}_j(y)|^2 > 0$ for all $y \in Y$. The characteristic functions $\hat{\nu}_j(y)$ also satisfy equation (8.1). Put $\psi_j(y) = -\ln \hat{\nu}_j(y)$. Then (8.1) implies that the functions $\psi_j(y)$ satisfy equation (8.18).

By Lemma 8.5, the function $\psi_1(s, n)$ satisfies equation (8.15). Apply now Lemma 8.6 and note that the inclusion

$$\mathbb{R} \subset N \cap L \tag{8.30}$$

is always valid. Then it follows from (8.15) and (8.30) that the function $\psi_1(s, n)$ satisfies equation (8.26). Reasoning as above, we get that each of the functions $\psi_2(s, n)$ and $\psi_3(s, n)$ also satisfies equation (8.26). Applying Lemma 8.7 to the functions $\psi_j(s, n)$ we obtain the representations

$$\psi_j(s, n) = \sigma_j s^2 + \kappa_j(n)s + \lambda_j(n), \quad s \in \mathbb{R},\ n \in \mathbb{Z}, \tag{8.31}$$

where $\kappa_j(-n) = -\kappa_j(n)$, $\lambda_j(-n) = \lambda_j(n)$ for all $n \in \mathbb{Z}$, and $\kappa_j(0) = \lambda_j(0) = 0$, $j = 1, 2, 3$.

By Lemma 8.5, $\nu = \nu_1 * \nu_2 * \nu_3 \in \Gamma(X)$. Then we have

$$\hat{\nu}(s, n) = \exp\{-(\sigma s^2 + \kappa s n + \lambda n^2)\}, \quad s \in \mathbb{R},\ n \in \mathbb{Z}, \tag{8.32}$$

where $\sigma > 0$, $\kappa \in \mathbb{R}$, $\lambda \geq 0$, and

$$\psi_1(s, n) + \psi_2(s, n) + \psi_3(s, n) = \sigma s^2 + \kappa s n + \lambda n^2, \quad s \in \mathbb{R},\ n \in \mathbb{Z}. \tag{8.33}$$

We will check that $\kappa_j(n)$ in (8.31) are linear functions. Substitute $u = (0, n)$, $v = (s_2, 0)$, $w = (s_3, 0)$ in equation (8.18). Note that $\widetilde{\alpha}_1 v = (a_1 s_2, 0)$, $\widetilde{\alpha}_2 v = (a_2 s_2, 0)$, $\widetilde{\beta}_1 w = (b_1 s_3, 0)$, and $\widetilde{\beta}_2 w = (b_2 s_3, 0)$. Taking into account representations (8.31), we find from (8.18) that

$$\begin{aligned} &\sigma_1(a_1 s_2 + b_1 s_3)^2 + \kappa_1(n)(a_1 s_2 + b_1 s_3) + \lambda_1(n) + \sigma_2(a_2 s_2 + b_2 s_3)^2 \\ &\quad + \kappa_2(n)(a_2 s_2 + b_2 s_3) + \lambda_2(n) + \sigma_3(s_2 + s_3)^2 + \kappa_3(n)(s_2 + s_3) + \lambda_3(n) \\ &\qquad = \lambda_1(n) + \lambda_2(n) + \lambda_3(n) + \sigma_1(a_1 s_2)^2 + \sigma_2(a_2 s_2)^2 + \sigma_3 s_2^2 \\ &\qquad\qquad + \sigma_1(b_1 s_3)^2 + \sigma_2(b_2 s_3)^2 + \sigma_3 s_3^2, \quad s_2, s_3 \in \mathbb{R},\ n \in \mathbb{Z}. \end{aligned} \tag{8.34}$$

Setting $s_2 = 1, s_3 = 0$ in (8.34), we get

$$a_1 \kappa_1(n) + a_2 \kappa_2(n) + \kappa_3(n) = 0, \quad n \in \mathbb{Z}. \tag{8.35}$$

Setting $s_2 = 0, s_3 = 1$ in (8.34), we find that

$$b_1 \kappa_1(n) + b_2 \kappa_2(n) + \kappa_3(n) = 0, \quad n \in \mathbb{Z}. \tag{8.36}$$

It follows from (8.31) and (8.33) that

$$\kappa_1(n) + \kappa_2(n) + \kappa_3(n) = \kappa n, \quad n \in \mathbb{Z}. \tag{8.37}$$

Subtracting from (8.35) and (8.36) equation (8.37), we obtain

$$(a_1 - 1)\kappa_1(n) + (a_2 - 1)\kappa_2(n) = -\kappa n, \quad n \in \mathbb{Z}. \tag{8.38}$$

$$(b_1 - 1)\kappa_1(n) + (b_2 - 1)\kappa_2(n) = -\kappa n, \quad n \in \mathbb{Z}. \tag{8.39}$$

Taking into account that by statement (v) of Lemma 8.2, the inequality

$$\det\begin{pmatrix} a_1-1 & a_2-1 \\ b_1-1 & b_2-1 \end{pmatrix} \neq 0$$

holds, it follows from (8.37)–(8.39) that $\kappa_j(n)$ are linear functions, i.e., $\kappa_j(n) = \kappa_j n$, $n \in \mathbb{Z}$, where $\kappa_j \in \mathbb{R}$. Thus, representations (8.31) take the form

$$\psi_j(s,n) = \sigma_j s^2 + \kappa_j sn + \lambda_j(n), \quad s \in \mathbb{R},\ n \in \mathbb{Z},\ j = 1,2,3. \tag{8.40}$$

Observe that it follows from (8.33) and (8.40) that

$$\lambda_1(n) + \lambda_2(n) + \lambda_3(n) = \lambda n^2, \quad n \in \mathbb{Z}.$$

Let $u = (s_1, n_1)$, $v = (s_2, n_2)$, $w = (s_3, n_3)$. Then

$$\widetilde{\alpha}_1 v = (a_1 s_2 + c_1 n_2, p_1 n_2), \quad \widetilde{\alpha}_2 v = (a_2 s_2 + c_2 n_2, p_2 n_2),$$

$$\widetilde{\beta}_1 w = (b_1 s_3 + d_1 n_3, q_1 n_3), \quad \widetilde{\beta}_2 w = (b_2 s_3 + d_2 n_3, q_2 n_3).$$

Substituting representations (8.40) into equation (8.18), we get that equalities (8.4)–(8.6) hold. Moreover, the following equalities also hold:

$$\kappa_1 a_1 + \kappa_2 a_2 + \kappa_3 = 0, \tag{8.41}$$

$$\kappa_1 b_1 + \kappa_2 b_2 + \kappa_3 = 0, \tag{8.42}$$

$$2\sigma_1 c_1 + 2\sigma_2 c_2 + \kappa_1 p_1 + \kappa_2 p_2 + \kappa_3 = 0, \tag{8.43}$$

$$2\sigma_1 d_1 + 2\sigma_2 d_2 + \kappa_1 q_1 + \kappa_2 q_2 + \kappa_3 = 0, \tag{8.44}$$

$$2\sigma_1 a_1 d_1 + 2\sigma_2 a_2 d_2 + \kappa_1 a_1 q_1 + \kappa_2 a_2 q_2 + \kappa_3 = 0, \tag{8.45}$$

$$2\sigma_1 b_1 c_1 + 2\sigma_2 b_2 c_2 + \kappa_1 b_1 p_1 + \kappa_2 b_2 p_2 + \kappa_3 = 0, \tag{8.46}$$

$$\begin{aligned} n_1 n_2(\kappa_1 c_1 + \kappa_2 c_2) &+ n_1 n_3(\kappa_1 d_1 + \kappa_2 d_2) \\ &+ n_2 n_3(2\sigma_1 c_1 d_1 + 2\sigma_2 c_2 d_2 + \kappa_1 c_1 q_1 + \kappa_1 d_1 p_1 + \kappa_2 c_2 q_2 + \kappa_2 d_2 p_2) \\ &\quad + \lambda_1(n_1 + p_1 n_2 + q_1 n_3) + \lambda_2(n_1 + p_2 n_2 + q_2 n_3) \\ &\qquad + \lambda_3(n_1 + n_2 + n_3) = \lambda(n_1^2 + n_2^2 + n_3^2), \quad n_j \in \mathbb{Z}. \end{aligned} \tag{8.47}$$

We find from (8.33) and (8.40) that

$$\sigma_1 + \sigma_2 + \sigma_3 = \sigma, \tag{8.48}$$

and

$$\kappa_1 + \kappa_2 + \kappa_3 = \kappa. \tag{8.49}$$

Prove that the support of ν is a subgroup of X topologically isomorphic to $\mathbb{R}$. It follows from (8.32) that this statement will be proved if we check that

$$4\sigma\lambda = \kappa^2, \tag{8.50}$$

i.e., taking into account (8.48) and (8.49), we should prove that

$$4(\sigma_1 + \sigma_2 + \sigma_3)\lambda = (\kappa_1 + \kappa_2 + \kappa_3)^2. \tag{8.51}$$

The proof of (8.51) is the series of elementary and boring computations. Note that by statement (iv) of Lemma 8.2, we have $a_2 b_1 - a_1 b_2 \neq 0$. Then it follows from (8.4) and (8.5) that

$$\sigma_1 = \frac{b_2 - a_2}{a_2 b_1 - a_1 b_2}\sigma_3, \quad \sigma_2 = \frac{a_1 - b_1}{a_2 b_1 - a_1 b_2}\sigma_3. \tag{8.52}$$

Similarly we find from (8.41) and (8.42) that

$$\kappa_1 = \frac{b_2 - a_2}{a_2 b_1 - a_1 b_2}\kappa_3, \quad \kappa_2 = \frac{a_1 - b_1}{a_2 b_1 - a_1 b_2}\kappa_3. \tag{8.53}$$

Each of the numbers p_j and q_j can take the values ± 1. We have here some cases and consider each of them separately.

I. Let $p_1 = p_2 = 1$. Then (8.43) implies that

$$\kappa_1 + \kappa_2 + \kappa_3 = -2\sigma_1 c_1 - 2\sigma_2 c_2. \tag{8.54}$$

Putting $n_1 = -n_2 = 1$, $n_3 = 0$ in (8.47), we get

$$-\kappa_1 c_1 - \kappa_2 c_2 = 2\lambda. \tag{8.55}$$

Substituting (8.54) and (8.55) into (8.51) and taking into account (8.52) and (8.53), we make sure that we obtain an equality. Thus, in case I equality (8.50) is proved.

II. Let $q_1 = q_2 = 1$. Then (8.44) implies that

$$\kappa_1 + \kappa_2 + \kappa_3 = -2\sigma_1 d_1 - 2\sigma_2 d_2. \tag{8.56}$$

Putting $n_1 = -n_3 = 1$, $n_2 = 0$ in (8.47), we obtain

$$-\kappa_1 d_1 - \kappa_2 d_2 = 2\lambda. \tag{8.57}$$

Substituting (8.56) and (8.57) into (8.51) and taking into account (8.52) and (8.53), we make sure that we obtain an equality. Thus, in case II equality (8.50) is proved.

III. Let either $p_1 = p_2 = q_1 = q_2 = -1$ or $p_1 = q_1 = -1$ and $p_2 = q_2 = 1$ or $p_1 = q_1 = 1$ and $p_2 = q_2 = -1$. Putting $n_1 = 0$, $n_2 = -n_3 = 1$ in (8.47), we get

$$-(2\sigma_1 c_1 d_1 + 2\sigma_2 c_2 d_2 + \kappa_1 c_1 p_1 + \kappa_1 d_1 p_1 + \kappa_2 c_2 p_2 + \kappa_2 d_2 p_2) = 2\lambda. \tag{8.58}$$

Note that, by statement (iii) of Lemma 8.2, we have $a_1 \neq a_2$ and $b_1 \neq b_2$. It follows from (8.43) and (8.46) that

$$c_1 = \frac{(1 - b_2)\kappa_3 - \kappa_1 p_1 (b_2 - b_1)}{2\sigma_1 (b_2 - b_1)}, \quad c_2 = \frac{(1 - b_1)\kappa_3 - \kappa_2 p_2 (b_1 - b_2)}{2\sigma_2 (b_1 - b_2)}. \tag{8.59}$$

We find from (8.44) and (8.45) that

$$d_1 = \frac{(1 - a_2)\kappa_3 - \kappa_1 q_1 (a_2 - a_1)}{2\sigma_1 (a_2 - a_1)}, \quad d_2 = \frac{(1 - a_1)\kappa_3 - \kappa_2 q_2 (a_1 - a_2)}{2\sigma_2 (a_1 - a_2)}. \tag{8.60}$$

Finding λ from (8.58) and substituting the obtained expression into (8.51), we get that we should check the equality

$$\begin{aligned} &- 2(2\sigma_1 c_1 d_1 + 2\sigma_2 c_2 d_2 + \kappa_1 c_1 p_1 + \kappa_1 d_1 p_1 + \kappa_2 c_2 p_2 + \kappa_2 d_2 p_2) \\ &\qquad\qquad \times (\sigma_1 + \sigma_2 + \sigma_3) = (\kappa_1 + \kappa_2 + \kappa_3)^2. \end{aligned} \tag{8.61}$$

Substitute (8.52), (8.53), (8.59), and (8.60) into (8.61). After elementary computations we see that the verification of (8.61) is reduced to the verification of the equality

$$\begin{aligned} (a_2 b_1 - a_1 b_2)((1 - b_2)(1 - a_2)(a_1 - b_1) + (1 - b_1)(1 - a_1)(b_2 - a_2)) \\ = -(b_2 - b_1)(a_2 - a_1)(a_1 - b_1)(b_2 - a_2). \end{aligned} \tag{8.62}$$

The validity of equality (8.62) follows from statement (i) of Lemma 8.2. Thus, in case III equality (8.50) is proved.

To consider the remaining cases, put $n_1 = n_2 = n_3 = 1$ into (8.47). Taking into account that $\lambda_j(-n) = \lambda_j(n)$ for all $n \in \mathbb{Z}$, $j = 1, 2$, we get

$$\kappa_1 c_1 + \kappa_2 c_2 + \kappa_1 d_1 + \kappa_2 d_2 + 2\sigma_1 c_1 d_1 + 2\sigma_2 c_2 d_2 + \kappa_1 c_1 q_1 + \kappa_1 d_1 p_1 + \kappa_2 c_2 q_2 + \kappa_2 d_2 p_2 + \lambda_1(1) + \lambda_2(1) + \lambda_3(3) = 3\lambda. \tag{8.63}$$

IV. Let $p_1 = p_2 = -1$ and $q_1 = -q_2$. Putting $n_1 = 2$, $n_2 = 1$, $n_3 = 0$ in (8.47), we obtain

$$2(\kappa_1 c_1 + \kappa_2 c_2) + \lambda_1(1) + \lambda_2(1) + \lambda_3(3) = 5\lambda.$$

It follows from this and from (8.63) that

$$\kappa_1 c_1 + \kappa_2 c_2 - (2\sigma_1 c_1 d_1 + 2\sigma_2 c_2 d_2 + \kappa_1 c_1 q_1 - \kappa_2 c_2 q_1) = 2\lambda.$$

V. Let $q_1 = q_2 = -1$ and $p_1 = -p_2$. Substituting $n_1 = 2$, $n_2 = 0$, $n_3 = 1$ in (8.47), we get

$$2(\kappa_1 d_1 + \kappa_2 d_2) + \lambda_1(1) + \lambda_2(1) + \lambda_3(3) = 5\lambda.$$

It follows from this and from (8.63) that

$$\kappa_1 d_1 + \kappa_2 d_2 - (2\sigma_1 c_1 d_1 + 2\sigma_2 c_2 d_2 + \kappa_1 d_1 p_1 - \kappa_2 d_2 p_1) = 2\lambda.$$

VI. Let either $p_1 = q_2 = -1$, $p_2 = q_1 = 1$ or $p_1 = q_2 = 1$, $p_2 = q_1 = -1$. Putting $n_1 = 0$, $n_2 = 2$, $n_3 = 1$ in (8.47), we find

$$2(2\sigma_1 c_1 d_1 + 2\sigma_2 c_2 d_2 + \kappa_1 c_1 q_1 + \kappa_1 d_1 p_1 + \kappa_2 c_2 q_2 + \kappa_2 d_2 p_2) + \lambda_1(1) + \lambda_2(1) + \lambda_3(3) = 5\lambda.$$

It follows from this and from (8.63) that

$$2\sigma_1 c_1 d_1 + 2\sigma_2 c_2 d_2 + \kappa_1 c_1 p_2 + \kappa_1 d_1 p_1 + \kappa_2 c_2 p_1 + \kappa_2 d_2 p_2 - (\kappa_1 d_1 + \kappa_2 d_2) - (\kappa_1 c_1 + \kappa_2 c_2) = 2\lambda.$$

Then in each of these cases we reason as in case III. Thus, we have proved that equality (8.50) holds. So, we have obtained the representation

$$\hat{\nu}(s, n) = \exp\{-(\sigma s^2 + \kappa s n + \lambda n^2)\} = \exp\left\{-\sigma\left(s + \frac{\kappa}{2\sigma}n\right)^2\right\}, \quad s \in \mathbb{R},\ n \in \mathbb{Z}.$$

Set $H = \{y \in Y : \hat{\nu}(y) = 1\}$, $G = A(X, H)$. It is obvious that

$$H = \left\{n\left(-\frac{\kappa}{2\sigma}, 1\right) : n \in \mathbb{Z}\right\}.$$

Then

$$G = \{(t, e^{it\frac{\kappa}{2\sigma}}) : t \in \mathbb{R}\}.$$

It is clear that $G \cong \mathbb{R}$. We obtain from Proposition 2.10 that $\sigma(\nu) \subset G$. Since $\nu = \nu_1 * \nu_2 * \nu_3 = \mu_1 * \bar{\mu}_1 * \mu_2 * \bar{\mu}_2 * \mu_3 * \bar{\mu}_3$, it follows from Proposition 2.1 that the distributions μ_j may be replaced by their shifts λ_j such that

$$\nu = \lambda_1 * \bar{\lambda}_1 * \lambda_2 * \bar{\lambda}_2 * \lambda_3 * \bar{\lambda}_3$$

and $\sigma(\lambda_j) \subset G$, $j = 1, 2, 3$. Since ν is a Gaussian distribution and $G \cong \mathbb{R}$, by Cramér's decomposition theorem for the Gaussian distribution on the real line, we get that all λ_j are Gaussian distributions. Hence all μ_j are also Gaussian distributions. $\square$

REMARK 8.9. As established in the proof of Theorem 8.8, if L_1, L_2, and L_3 are of the form $L_1 = \xi_1 + \xi_2 + \xi_3$, $L_2 = \alpha_1\xi_1 + \alpha_2\xi_2 + \xi_3$, $L_3 = \beta_1\xi_1 + \beta_2\xi_2 + \xi_3$ and μ_j are nondegenerate distributions with nonvanishing characteristic functions, then supports of μ_j are cosets of the one-parameter subgroup $G = \{(t, e^{it\frac{\kappa}{2\sigma}}) : t \in \mathbb{R}\}$ in the group $\mathbb{R} \times \mathbb{T}$.

We will check that the subgroup G is invariant with respect to all topological automorphisms α_i, β_i. Taking into account (8.48), (8.49), (8.52), and (8.53), we get that

$$G = \left\{ \left(t, e^{it\frac{\kappa_3}{2\sigma_3}} \right) : t \in \mathbb{R} \right\}.$$

Verify that the subgroup G is invariant with respect to α_1. Inasmuch as

$$\alpha_1 \left(t, e^{it\frac{\kappa_3}{2\sigma_3}} \right) = \left(a_1 t, e^{ic_1 t} e^{itp_1\frac{\kappa_3}{2\sigma_3}} \right) = \left(a_1 t, e^{ia_1 t\frac{\kappa_3}{2\sigma_3}\left(\frac{2c_1\sigma_3}{a_1\kappa_3} + \frac{p_1}{a_1}\right)} \right),$$

it suffices to show that

$$\frac{2c_1\sigma_3}{a_1\kappa_3} + \frac{p_1}{a_1} = 1. \tag{8.64}$$

To check equality (8.64) substitute first the representation for c_1 from (8.59) in (8.64). Then substitute the representation for σ_1 and κ_1 from (8.52) and (8.53) into the obtained expression. After simple transformations we see that (8.64) is reduced to equality (i) of Lemma 8.2.

Reasoning as above we get that the subgroup G is invariant with respect to α_2, β_1, β_2. Obviously, the restriction of each of the automorphisms α_i, β_i to the subgroup G is a topological automorphism of G.

It follows from what has been said that the statement of Theorem 8.8 that μ_j are Gaussian distributions, can be made without Cramér's decomposition theorem for the Gaussian distribution on the real line, but directly from the Skitovich–Darmois theorem on the real line.

We note that the property: G is invariant with respect to all topological automorphisms α_i, β_i imposes the strong restrictions on α_i, β_i. Indeed, let G be an arbitrary subgroup of the group $\mathbb{R} \times \mathbb{T}$ topologically isomorphic to $\mathbb{R}$. Then G is of the form $G = \{(t, e^{i\omega t}) : t \in \mathbb{R}\}$, where ω is a fixed real number. The subgroup G is invariant with respect to an automorphism $\alpha = \begin{pmatrix} a & c \\ 0 & p \end{pmatrix} \in \mathrm{Aut}(\mathbb{R} \times \mathbb{T})$ if and only if $c = (a - p)\omega$.

Let $\omega = 0$, i.e., $G = \{(t, 1) : t \in \mathbb{R}\}$. Then $c_1 = c_2 = d_1 = d_2 = 0$, i.e., α_i, β_i are diagonal matrices. Moreover, statements (i)–(v) of Lemma 8.2 hold for a_i and b_i.

Let $\omega \neq 0$. Assume also that no α_i, β_i is equal to $\pm I$. It follows from this that in addition to statements (i)–(v) of Lemma 8.2 the equalities

$$\frac{c_1}{a_1 - p_1} = \frac{c_2}{a_2 - p_2} = \frac{d_1}{b_1 - q_1} = \frac{d_2}{b_2 - q_2} \tag{8.65}$$

also hold. In particular, if at least one of these equalities does not hold, then all μ_j are degenerate distributions. If some of the automorphisms α_i, β_i are equal to $\pm I$, then in (8.65) the expressions corresponding to these automorphisms are omitted.

REMARK 8.10. Let G be an arbitrary one-parameter subgroup of the group $\mathbb{R} \times \mathbb{T}$ of the form $G = \{(t, e^{i\omega t}) : t \in \mathbb{R}\}$, where ω is a fixed real number. Let a_i, b_i, $i = 1, 2$, be nonzero real numbers such that the system of equations (8.4)–(8.6) has a solution $\sigma_1 > 0, \sigma_2 > 0, \sigma_3 > 0$. Consider the automorphisms $\alpha_i, \beta_i \in \mathrm{Aut}(\mathbb{R} \times \mathbb{T})$ of the form

$$\alpha_1 = \begin{pmatrix} a_1 & (a_1 - p_1)\omega \\ 0 & p_1 \end{pmatrix}, \quad \alpha_2 = \begin{pmatrix} a_2 & (a_2 - p_2)\omega \\ 0 & p_2 \end{pmatrix},$$

$$\beta_1 = \begin{pmatrix} b_1 & (b_1 - q_1)\omega \\ 0 & q_1 \end{pmatrix}, \quad \beta_2 = \begin{pmatrix} b_2 & (b_2 - q_2)\omega \\ 0 & q_2 \end{pmatrix}.$$

Let ξ_j, $j = 1, 2, 3$, be independent Gaussian random variables with values in the group $\mathbb{R} \times \mathbb{T}$ and distributions μ_j having the characteristic functions

$$\hat{\mu}_1(s, n) = \exp\{-\sigma_1(s + \omega n)^2\}, \quad \hat{\mu}_2(s, n) = \exp\{-\sigma_2(s + \omega n)^2\},$$

$$\hat{\mu}_3(s, n) = \exp\{-\sigma_3(s + \omega n)^2\}, \quad s \in \mathbb{R}, \ n \in \mathbb{Z}.$$

Applying Proposition 2.10, it is easy to verify that the support of each of the distributions μ_j coincides with the subgroup G. Taking into account (8.4)–(8.6), it is easy to verify directly that the characteristic functions $\hat{\mu}_j(s, n)$ satisfy equation (8.1). Hence by Lemma 8.1, the linear forms $L_1 = \xi_1 + \xi_2 + \xi_3$, $L_2 = \alpha_1\xi_1 + \alpha_2\xi_2 + \xi_3$, and $L_3 = \beta_1\xi_1 + \beta_2\xi_2 + \xi_3$ are independent.

It follows from what has been said that Theorem 8.8 cannot be strengthened by narrowing the class of distributions that are characterized by the independence of the linear forms L_1, L_2, and L_3 (compare with Lemma 8.4).

REMARK 8.11. Consider the circle group $\mathbb{T}$. Then $\mathrm{Aut}(\mathbb{T}) = \pm I$. Let $\alpha_{ij} \in \mathrm{Aut}(\mathbb{T})$, $i, j = 1, 2$. Consider independent random variables ξ_1 and ξ_2 with values in the circle group $\mathbb{T}$ and distributions μ_1 and μ_2 with nonvanishing characteristic functions. Assume that the linear forms $L_1 = \alpha_{11}\xi_1 + \alpha_{12}\xi_2$ and $L_2 = \alpha_{21}\xi_1 + \alpha_{22}\xi_2$ are independent. As follows from Theorem 5.8 and Example 5.9, the distributions μ_j are either Gaussian distributions or convolutions of Gaussian distributions and signed measures. By Lemma 8.4, if we consider three independent random variables ξ_j with values in the group $\mathbb{T}$ and distributions μ_j and three linear forms, then their independence implies that all μ_j are degenerate distributions. So, we could surmise that by increasing the number n of independent random variables, the class of distributions which are characterized by the independence of n linear forms of n independent random variables, decreases. It turns out that this statement is not true. Indeed, let ξ_j, $j = 1, 2, 3, 4$, be independent random variables with values in the group $\mathbb{T}$ and distributions μ_j. Consider the linear forms $L_1 = \xi_1 + \xi_2 + \xi_3 + \xi_4$, $L_2 = \xi_1 + \xi_2 - \xi_3 - \xi_4$, $L_3 = \xi_1 - \xi_2 + \xi_3 - \xi_4$, and $L_4 = \xi_1 - \xi_2 - \xi_3 + \xi_4$. Take $\mu_1 = \mu_2 = \gamma * \pi_1$, $\mu_3 = \mu_4 = \gamma * \pi_2$, where $\gamma \in \Gamma(\mathbb{T})$ and π_j are signed measures supported in $\mathbb{T}_{(2)}$ such that $\pi_1 * \pi_2 = E_1$. It is easy to verify that in this case the linear forms L_1, L_2, L_3, and L_4 are independent, whereas all $\mu_j \notin \Gamma(\mathbb{T})$.

Notes

The classical theorem, where the Gaussian distribution on the real line is characterised by the independence of two linear forms of n independent random variables was proved independently by V.P. Skitovich [**94**] and G. Darmois [**20**]. This theorem was generalized by S.G. Ghurye and I. Olkin [**60**] to the multivariable case when instead of random variables, random vectors ξ_j in the space $\mathbb{R}^m$ are considered and coefficients of the linear forms L_1 and L_2 are nonsingular matrices. In this case the independence of L_1 and L_2 also implies that all independent random vectors ξ_j are Gaussian.

First a generalization of the Skitovich–Darmois theorem on locally compact Abelian groups was considered by J.H. Stapleton in [**96**]. He proved that the Haar distribution on a compact connected Abelian group X is characterised by the independence of n linear forms of n independent random variables with values in X. He considered linear forms with integer coefficients and assumed that the distributions of random variables and the coefficients of linear forms satisfy some conditions. Probably, the paper [**96**] was the first article where a classical characterization theorem was studied on a locally compact Abelian group. In [**92**] (see also [**37**]), K. Schmidt proved that the Gaussian distribution on an arbitrary locally compact Abelian group X is characterised by the independence of two linear forms of two independent random variables with values in X. In doing so, coefficients of the forms were topological automorphisms of X satisfying some conditions and the characteristic functions of the random variables did not vanish.

A systematic study of the Skitovich–Darmois theorem on various classes of locally compact Abelian groups was started by G.M. Feldman in the papers [**26**] and [**27**] (see also [**36**, §10–15], where one can find additional references). Chapter III summarizes the main results obtained in this direction. Theorem 6.3 and Proposition 6.6 were proved by G.M. Feldman in [**32**]. Theorems 6.8, 6.11, 6.19 and Remark 6.9 belong to G.M. Feldman and M.V. Myronyuk [**56**]. Lemma 6.15 essentially belongs to Lajko [**72**], who proved it for the group $\mathbb{R}$. Theorem 7.1 belongs to G.M. Feldman [**31**]. Proposition 7.3 was proved by G.M. Feldman in [**30**]. Theorem 7.15 belongs to G.M. Feldman and P. Graczyk [**53**]. Lemmas 7.18, 7.19, 7.21 and Theorems 7.22 and 7.24 were proved by G.M. Feldman and P. Graczyk in [**52**]. Theorem 7.26 was proved by G.M. Feldman in [**40**]. We note that the Skitovich–Darmois theorem on the 2-dimensional torus $\mathbb{T}^2$ and on the groups $\mathbb{R}\times\mathbb{T}$ and $\mathbb{R}\times\Sigma_{\boldsymbol{a}}$ was studied by G.M. Feldman and M.V. Myronyuk in [**81**] and [**54**] respectively. We also observe that the main results of article [**54**] follows from Theorem 6.19. The results of Section 8 were obtained by G.M. Feldman and M.V. Myronyuk in [**82**].

In the papers [**27, 29, 30**] G.M. Feldman studied the following problem for various classes of locally compact Abelian groups: For which groups is an analogue of the Skitovich–Darmois theorem valid for an arbitrary number n of independent random variables and two linear forms. In so doing, the characteristic functions of random variables can vanish. Unlike the case when $n=2$, the set of groups where an analogue of the Skitovich–Darmois theorem is true for an arbitrary n turns out to be very poor. For example, in the class of compact totally disconnected Abelian groups these are just the groups of the form $G=\mathbb{Z}(2^{m_1})\times\cdots\times\mathbb{Z}(2^{m_l})$, where $0\le m_1<\cdots<m_l$, Δ_2, and $G\times\Delta_2$. Moreover, in this case, only degenerate distributions are characterized. See more about this in [**36**, §14].

I.P. Mazur proved that Theorems 7.15 and 7.22 are true if we consider n linear forms of n independent random variables [**74, 75**].

CHAPTER IV

Characterization of probability distributions through the symmetry of the conditional distribution of one linear form given another

9. Locally compact Abelian groups containing no elements of order 2

Heyde's theorem states: Let ξ_j, $j = 1, 2, \dots, n$, $n \geq 2$, be independent random variables and let α_j, β_j be nonzero real numbers. Assume that the conditions $\beta_i\alpha_i^{-1} + \beta_j\alpha_j^{-1} \neq 0$ for all $i \neq j$ are satisfied. If the conditional distribution of the linear form $L_2 = \beta_1\xi_1 + \cdots + \beta_n\xi_n$ given $L_1 = \alpha_1\xi_1 + \cdots + \alpha_n\xi_n$ is symmetric, then all random variables ξ_j are Gaussian.

In this section, we begin the study of group analogues of Heyde's theorem. Let X be a second countable locally compact Abelian group. We assume that the group X contains no elements of order 2 and consider independent random variables ξ_j with values in X and distributions μ_j with nonvanishing characteristic functions. All theorems in this section are proved under these assumptions. Consider the linear forms $L_1 = \alpha_1\xi_1 + \cdots + \alpha_n\xi_n$ and $L_2 = \beta_1\xi_1 + \cdots + \beta_n\xi_n$, where $\alpha_j, \beta_j \in \mathrm{Aut}(X)$. First, we prove that if $\beta_i\alpha_i^{-1} \pm \beta_j\alpha_j^{-1} \in \mathrm{Aut}(X)$ for all $i \neq j$ and the conditional distribution of the linear form L_2 given L_1 is symmetric, then all μ_j are Gaussian distributions. Then, for two independent random variables we prove that if the conditional distribution of the linear form $L_2 = \xi_1 + \alpha\xi_2$ given $L_1 = \xi_1 + \xi_2$ is symmetric and $\mathrm{Ker}(I + \alpha) = \{0\}$, then μ_j are Gaussian distributions. It turns out that this condition on the topological automorphism α cannot be weakened. Finally, for two independent random variables taking values either in a finite Abelian group or in a group of the form $\mathbb{R} \times G$, where G is a finite Abelian group, or in an $\boldsymbol{a}$-adic solenoid $\Sigma_{\boldsymbol{a}}$ or in a group of the form $\Sigma_{\boldsymbol{a}} \times G$, where $\boldsymbol{a} = (2, 3, 4, \dots)$ and G is a finite Abelian group, we give a complete description of distributions which are characterized by the symmetry of the conditional distribution of the linear form $L_2 = \xi_1 + \alpha\xi_2$ given $L_1 = \xi_1 + \xi_2$. In the latter cases we do not impose any conditions on the topological automorphism α.

LEMMA 9.1. *Let X be a locally compact Abelian group with character group Y. Let α_j, β_j, $j = 1, 2, \dots, n$, $n \geq 2$, be continuous endomorphisms of X. Let ξ_j be independent random variables with values in the group X and distributions μ_j. The conditional distribution of the linear form $L_2 = \beta_1\xi_1 + \cdots + \beta_n\xi_n$ given $L_1 = \alpha_1\xi_1 + \cdots + \alpha_n\xi_n$ is symmetric if and only if the characteristic functions $\hat\mu_j(y)$ satisfy the equation*

$$\prod_{j=1}^{n} \hat\mu_j(\widetilde{\alpha}_j u + \widetilde{\beta}_j v) = \prod_{j=1}^{n} \hat\mu_j(\widetilde{\alpha}_j u - \widetilde{\beta}_j v), \quad u, v \in Y. \tag{9.1}$$

PROOF. Taking into account that $\hat{\mu}_j(y) = \mathbf{E}[(\xi_j, y)]$ and the independence of the random variables ξ_j, we have

$$(9.1) \iff \prod_{j=1}^{n} \mathbf{E}[(\xi_j, \widetilde{\alpha}_j u + \widetilde{\beta}_j v)] = \prod_{j=1}^{n} \mathbf{E}[(\xi_j, \widetilde{\alpha}_j u - \widetilde{\beta}_j v)], \quad u, v \in Y.$$

$$\iff \mathbf{E}\Big[\prod_{j=1}^{n}(\xi_j, \widetilde{\alpha}_j u + \widetilde{\beta}_j v)\Big] = \mathbf{E}\Big[\prod_{j=1}^{n}(\xi_j, \widetilde{\alpha}_j u - \widetilde{\beta}_j v)\Big], \quad u, v \in Y.$$

$$\iff \mathbf{E}\Big[\prod_{j=1}^{n}(\xi_j, \widetilde{\alpha}_j u)\prod_{j=1}^{n}(\xi_j, \widetilde{\beta}_j v)\Big] = \mathbf{E}\Big[\prod_{j=1}^{n}(\xi_j, \widetilde{\alpha}_j u)\prod_{j=1}^{n}(\xi_j, -\widetilde{\beta}_j v)\Big], \quad u, v \in Y.$$

$$\iff \mathbf{E}\Big[\prod_{j=1}^{n}(\alpha_j\xi_j, u)\prod_{j=1}^{n}(\beta_j\xi_j, v)\Big] = \mathbf{E}\Big[\prod_{j=1}^{n}(\alpha_j\xi_j, u)\prod_{j=1}^{n}(-\beta_j\xi_j, v)\Big], \quad u, v \in Y.$$

$$(9.2) \qquad \iff \mathbf{E}[(L_1, u)(L_2, v)] = \mathbf{E}[(L_1, u)(-L_2, v)], \quad u, v \in Y.$$

Let $(\Omega, \mathfrak{A}, P)$ be a probabilistic space, where the random variables ξ_j are defined. Then (9.2) is equivalent to the statement that the random vectors (L_1, L_2) and $(L_1, -L_2)$ are identically distributed, i.e.,

$$P\{\omega \in \Omega : L_1(\omega) \in A,\ L_2(\omega) \in B\} = P\{\omega \in \Omega : L_1(\omega) \in A,\ L_2(\omega) \in -B\}$$

for all $A, B \in \mathfrak{B}(X)$. The equality obtained is equivalent to the symmetry of the conditional distribution of the linear form L_2 given L_1. □

Equation (9.1) is called the *Heyde functional equation.* In what follows, it is convenient for us to formulate such a special case of Lemma 9.1.

COROLLARY 9.2. *Let X be a locally compact Abelian group with character group Y. Let α be a continuous endomorphism of X. Let ξ_1 and ξ_2 be independent random variables with values in the group X and distributions μ_1 and μ_2. The conditional distribution of the linear form $L_2 = \xi_1 + \alpha\xi_2$ given $L_1 = \xi_1 + \xi_2$ is symmetric if and only if the characteristic functions $\hat{\mu}_j(y)$ satisfy the equation*

$$(9.3) \qquad \hat{\mu}_1(u + v)\hat{\mu}_2(u + \widetilde{\alpha}v) = \hat{\mu}_1(u - v)\hat{\mu}_2(u - \widetilde{\alpha}v), \quad u, v \in Y.$$

THEOREM 9.3. *Let X be a locally compact Abelian group containing no elements of order 2. Let α_j, β_j, $j = 1, 2, \dots, n$, $n \geq 2$, be topological automorphisms of X such that $\beta_i\alpha_i^{-1} \pm \beta_j\alpha_j^{-1} \in \mathrm{Aut}(X)$ for all $i \neq j$. Let ξ_j be independent random variables with values in the group X and distributions μ_j with nonvanishing characteristic functions. If the conditional distribution of the linear form $L_2 = \beta_1\xi_1 + \cdots + \beta_n\xi_n$ given $L_1 = \alpha_1\xi_1 + \cdots + \alpha_n\xi_n$ is symmetric, then $\mu_j \in \Gamma(X)$, $j = 1, 2, \dots, n$.*

PROOF. We can put $\zeta_j = \alpha_j\xi_j$ and reduce the proof of the theorem to the case when $L_1 = \xi_1 + \cdots + \xi_n$ and $L_2 = \delta_1\xi_1 + \cdots + \delta_n\xi_n$, $\delta_j \in \mathrm{Aut}(X)$. The conditions $\beta_i\alpha_i^{-1} \pm \beta_j\alpha_j^{-1} \in \mathrm{Aut}(X)$ for all $i \neq j$ are transformed into the conditions $\delta_i \pm \delta_j \in \mathrm{Aut}(X)$ for all $i \neq j$. Denote by Y the character group of the group X. It follows from statement (c) of Theorem 1.20 that $\widetilde{\delta}_i \pm \widetilde{\delta}_j \in \mathrm{Aut}(Y)$ for all $i \neq j$.

By Lemma 9.1, the characteristic functions $\hat\mu_j(y)$ satisfy equation (9.1) which takes the form

$$\prod_{j=1}^{n}\hat\mu_j(u+\widetilde\delta_j v)=\prod_{j=1}^{n}\hat\mu_j(u-\widetilde\delta_j v),\quad u,v\in Y. \tag{9.4}$$

Set $\nu_j=\mu_j*\bar\mu_j$. Then $\hat\nu_j(y)=|\hat\mu_j(y)|^2>0$ for all $y\in Y$. The characteristic functions $\hat\nu_j(y)$ also satisfy equation (9.4). Put $\varphi_j(y)=-\ln\hat\nu_j(y)$. We conclude from (9.4) that the functions $\varphi_j(y)$ satisfy the equation

$$\sum_{j=1}^{n}[\varphi_j(u+\widetilde\delta_j v)-\varphi_j(u-\widetilde\delta_j v)]=0,\quad u,v\in Y. \tag{9.5}$$

We use the finite difference method to solve equation (9.5). Let k_1 be an arbitrary element of the group Y. Set $h_1=\widetilde\delta_n k_1$, then $h_1-\widetilde\delta_n k_1=0$. Substitute $u+h_1$ for u and $v+k_1$ for v in equation (9.5). Subtracting equation (9.5) from the resulting equation we obtain

$$\sum_{j=1}^{n}\Delta_{l_{1,j}}\varphi_j(u+\widetilde\delta_j v)-\sum_{j=1}^{n-1}\Delta_{l_{1,j+n}}\varphi_j(u-\widetilde\delta_j v)=0,\quad u,v\in Y, \tag{9.6}$$

where $l_{1,j}=h_1+\widetilde\delta_j k_1=(\widetilde\delta_n+\widetilde\delta_j)k_1$, $j=1,2,\dots,n$, $l_{1,j+n}=h_1-\widetilde\delta_j k_1=(\widetilde\delta_n-\widetilde\delta_j)k_1$, $j=1,2,\dots,n-1$. Let k_2 be an arbitrary element of the group Y. Put $h_2=\widetilde\delta_{n-1}k_2$, then $h_2-\widetilde\delta_{n-1}k_2=0$. Substitute $u+h_2$ for u and $v+k_2$ for v in equation (9.6). Subtracting equation (9.6) from the resulting equation we obtain

$$\sum_{j=1}^{n}\Delta_{l_{2,j}}\Delta_{l_{1,j}}\varphi_j(u+\widetilde\delta_j v)-\sum_{j=1}^{n-2}\Delta_{l_{2,j+n}}\Delta_{l_{1,j+n}}\varphi_j(u-\widetilde\delta_j v)=0,\quad u,v\in Y,$$

where $l_{2,j}=h_2+\widetilde\delta_j k_2=(\widetilde\delta_n+\widetilde\delta_j)k_2$, $j=1,2,\dots,n$, $l_{2,j+n}=h_2-\widetilde\delta_j k_2=(\widetilde\delta_n-\widetilde\delta_j)k_2$, $j=1,2,\dots,n-2$. Arguing as above in n steps we get the equation

$$\sum_{j=1}^{n}\Delta_{l_{n,j}}\Delta_{l_{n-1,j}}\dots\Delta_{l_{1,j}}\varphi_j(u+\widetilde\delta_j v)=0,\quad u,v\in Y, \tag{9.7}$$

where $l_{p,j}=(\widetilde\delta_{n-p+1}+\widetilde\delta_j)k_p$, $p=1,2,\dots,n$, $j=1,2,\dots,n$. Let k_{n+1} be an arbitrary element of the group Y. Set $h_{n+1}=-\widetilde\delta_n k_{n+1}$, hence $h_{n+1}+\widetilde\delta_n k_{n+1}=0$. Substitute $u+h_n$ for u and $v+k_n$ for v in equation (9.7). Subtracting equation (9.7) from the resulting equation we obtain

$$\sum_{j=1}^{n-1}\Delta_{l_{n+1,j}}\Delta_{l_{n,j}}\Delta_{l_{n-1,j}}\dots\Delta_{l_{1,j}}\varphi_j(u+\widetilde\delta_j v)=0,\quad u,v\in Y, \tag{9.8}$$

where $l_{n+1,j}=h_{n+1}+\widetilde\delta_j k_{n+1}=(\widetilde\delta_j-\widetilde\delta_n)k_{n+1}$, $j=1,2,\dots,n-1$. Equation (9.8) does not contain the function φ_n. Arguing similarly we sequentially exclude the functions φ_{n-1}, φ_{n-2}, $\dots$, φ_2 from equation (9.8). Finally we obtain

$$\Delta_{l_{2n-1,1}}\Delta_{l_{2n-2,1}}\dots\Delta_{l_{1,1}}\varphi_1(u+\widetilde\delta_1 v)=0,\quad u,v\in Y, \tag{9.9}$$

where $l_{p,1}=(\widetilde\delta_1+\widetilde\delta_{n-p+1})k_p$, $p=1,2,\dots,n-1$, $l_{n,1}=2\widetilde\delta_1 k_n$, $l_{n+p,1}=(\widetilde\delta_1-\widetilde\delta_{n-p+1})k_{n+p}$, $p=1,2,\dots,n-1$.

Taking into account that k_p are arbitrary elements of the group Y and $\widetilde{\delta}_1 \pm \widetilde{\delta}_p \in \mathrm{Aut}(Y)$ for $p = 2, 3, \dots, n$, we can assume that in (9.9) $l_{n,1} = 2k$, $l_{p,1} = h$, $p = 1, 2, \dots, n-1, n+1, n+2, \dots, 2n-1$, where k and h are arbitrary elements of the group Y. Substituting $v = 0$ into the resulting equation, we obtain

$$\Delta_{2k}\Delta_h^{2n-2}\varphi_1(u) = 0, \tag{9.10}$$

where k, h, and u are arbitrary elements of the group Y. Since the group X has no elements of order 2, by Theorem 1.12, the subgroup $Y^{(2)}$ is dense in Y. We deduce from (9.10) that

$$\Delta_h^{2n-1}\varphi_1(u) = 0, \quad u, h \in Y,$$

i.e., the function $\varphi_1(y)$ is a continuous polynomial. Inasmuch as the group X contains no elements of order 2, X has no subgroups topologically isomorphic to the circle group $\mathbb{T}$. By Theorem 3.15, $\nu_1 \in \Gamma(X)$. Applying Theorem 3.14, we get $\mu_1 \in \Gamma(X)$. Arguing as above we prove that all μ_j are Gaussian distributions. □

REMARK 9.4. Theorem 9.3 is not valid if a locally compact Abelian group X contains an element of order 2. For the proof observe that if K is a closed subgroup of X, $\mu \in \mathrm{M}^1(X)$, and $\sigma(\mu) \subset K$, then $\hat\mu(y) = 1$ for all $y \in A(Y, K)$. Hence by Proposition 2.10, $\hat\mu(y + l) = \hat\mu(y)$ for all $y \in Y$, $l \in A(Y, K)$.

Put $G = X_{(2)}$. By Theorem 1.12, $A(Y, G) = \overline{Y^{(2)}}$. This implies that if $\sigma(\mu) \subset G$, then $\hat\mu(y + 2h) = \hat\mu(y)$. Hence $\hat\mu(y + h) = \hat\mu(y - h)$ for all $y, h \in Y$. It follows from this that if α_j, β_j, $j = 1, 2, \dots, n$, $n \geq 2$, are topological automorphisms of the group X and ξ_j are independent random variables with values in the subgroup G and distributions μ_j, then the characteristic functions $\hat\mu_j(y)$ satisfy equation (9.1). By Lemma 9.1, the conditional distribution of the linear form $L_2 = \beta_1\xi_1 + \cdots + \beta_n\xi_n$ given $L_1 = \alpha_1\xi_1 + \cdots + \alpha_n\xi_n$ is symmetric. If $G \neq \{0\}$, then $\mu_j \notin \Gamma(X)$, $j = 1, 2, \dots, n$.

Theorem 9.3 is proved under very strong restrictions on topological automorphisms α_j and β_j, namely $\beta_i\alpha_i^{-1} \pm \beta_j\alpha_j^{-1} \in \mathrm{Aut}(X)$ for all $i \neq j$. In the case of two independent random variables we can prove that Theorem 9.3 holds true under weaker conditions on α_j and β_j. First, prove the following lemma.

LEMMA 9.5. *Let X be a locally compact Abelian group. Let ξ_1 and ξ_2 be independent random variables with values in X and distributions μ_1 and μ_2. Then the following statements hold.*

1. *If $\mu_1 = \mu_2$, then the conditional distribution of the linear form $L_2 = \xi_1 - \xi_2$ given $L_1 = \xi_1 + \xi_2$ is symmetric.*
2. *If the group X contains no elements of order 2 and the conditional distribution of the linear form $L_2 = \xi_1 - \xi_2$ given $L_1 = \xi_1 + \xi_2$ is symmetric, then $\mu_1 = \mu_2$.*

PROOF. Denote by Y the character group of the group X.

Statement 1. If $\mu_1 = \mu_2$ and $\alpha = -I$, then the characteristic functions $\hat\mu_j(y)$ satisfy equation (9.3). By Corollary 9.2, this implies that the conditional distribution of the linear form $L_2 = \xi_1 - \xi_2$ given $L_1 = \xi_1 + \xi_2$ is symmetric.

Statement 2. By Corollary 9.2, the characteristic functions $\hat\mu_j(y)$ satisfy equation (9.3) which takes the form

$$\hat\mu_1(u + v)\hat\mu_2(u - v) = \hat\mu_1(u - v)\hat\mu_2(u + v), \quad u, v \in Y. \tag{9.11}$$

Substituting $u = v = y$ in equation (9.11), we get

$$\hat{\mu}_1(2y) = \hat{\mu}_2(2y), \quad y \in Y. \tag{9.12}$$

Since the group X has no elements of order 2, by Theorem 1.12, the subgroup $Y^{(2)}$ is dense in Y. It follows from equation (9.12) that $\hat{\mu}_1(y) = \hat{\mu}_2(y)$ for all $y \in Y$. Hence $\mu_1 = \mu_2$. □

LEMMA 9.6. *Let Y be an Abelian group and let β be an endomorphism of Y. Assume that the functions $f_j(y)$ satisfy the equation*

$$f_1(u+v)f_2(u+\beta v) = f_1(u-v)f_2(u-\beta v), \quad u, v \in Y. \tag{9.13}$$

Then the functions $f_j(y)$ satisfy the equation

$$\begin{aligned} f_1((I+\beta)u + 2v)&f_2(2\beta u + (I+\beta)v) \\ &= f_1((I+\beta)u)f_2(2\beta u)f_1(2v)f_2((I+\beta)v), \quad u, v \in Y. \end{aligned} \tag{9.14}$$

PROOF. Substituting $u = \beta y$, $v = -y$ into equation (9.13), we obtain

$$f_1((\beta - I)y) = f_1((I+\beta)y)f_2(2\beta y), \quad y \in Y. \tag{9.15}$$

Substituting $u = -y$, $v = y$ into equation (9.13), we get

$$f_2((\beta - I)y) = f_1(-2y)f_2(-(I+\beta)y), \quad y \in Y.$$

Rewrite this equation in the form

$$f_2(-(\beta - I)y) = f_1(2y)f_2((I+\beta)y), \quad y \in Y. \tag{9.16}$$

Let $s, t \in Y$. Substituting $u = s + \beta t$, $v = s + t$ into equation (9.13), we obtain

$$f_1((I+\beta)t + 2s)f_2(2\beta t + (I+\beta)s) = f_1((\beta - I)t)f_2(-(\beta - I)s), \quad s, t \in Y.$$

Taking into account (9.15) and (9.16), this equation can be written in the form (9.14). □

Let X be a locally compact Abelian group with character group Y and let α be a continuous endomorphism of X. Let ξ_1 and ξ_2 be independent random variables with values in the group X and distributions μ_1 and μ_2. Taking into account Lemma 6.1 and Corollary 9.2 and applying Lemma 9.6 to the functions $f_j(y) = \hat{\mu}_j(y)$, $j = 1, 2$, and $\beta = \widetilde{\alpha}$, we obtain the following statement.

COROLLARY 9.7. *Let X be a locally compact Abelian group and let α be a continuous endomorphism of X. Let ξ_1 and ξ_2 be independent random variables with values in the group X. If the conditional distribution of the linear form $L_2 = \xi_1 + \alpha\xi_2$ given $L_1 = \xi_1 + \xi_2$ is symmetric, then the linear forms $M_1 = (I+\alpha)\xi_1 + 2\alpha\xi_2$ and $M_2 = 2\xi_1 + (I+\alpha)\xi_2$ are independent.*

REMARK 9.8. Corollary 9.7 implies the following analogue of Heyde's theorem for the group $\mathbb{R}^n$. Let α be a topological automorphism of $\mathbb{R}^n$ satisfying the condition $\mathrm{Ker}(I+\alpha) = \{0\}$. Let ξ_1 and ξ_2 be independent random variables with values in the group $\mathbb{R}^n$ and distributions μ_1 and μ_2. If the conditional distribution of the linear form $L_2 = \xi_1 + \alpha\xi_2$ given $L_1 = \xi_1 + \xi_2$ is symmetric, then μ_j are Gaussian distributions. Indeed, by Corollary 9.7, the linear forms $M_1 = (I+\alpha)\xi_1 + 2\alpha\xi_2$ and $M_2 = 2\xi_1 + (I+\alpha)\xi_2$ are independent. Inasmuch as $\mathrm{Ker}(I+\alpha) = \{0\}$, coefficients of the linear forms M_j are topological automorphisms of the group $\mathbb{R}^n$. By the Ghurye–Olkin theorem, μ_j are Gaussian distributions.

THEOREM 9.9. *Let X be a locally compact Abelian group containing no elements of order 2. Let α be a topological automorphism of X. Let ξ_1 and ξ_2 be independent random variables with values in the group X and distributions μ_1 and μ_2 with nonvanishing characteristic functions. The symmetry of the conditional distribution of the linear form $L_2 = \xi_1 + \alpha\xi_2$ given $L_1 = \xi_1 + \xi_2$ implies that $\mu_1, \mu_2 \in \Gamma(X)$ if and only if α satisfies the condition* $\mathrm{Ker}(I + \alpha) = \{0\}$.

PROOF. Necessity. Put $K = \mathrm{Ker}(I + \alpha)$ and assume $K \neq \{0\}$. Let ξ_1 and ξ_2 be independent identically distributed random variables with values in K and distribution μ. Applying Lemma 9.5 to the group K, we get that the conditional distribution of the linear form $L_2 = \xi_1 - \xi_2$ given $L_1 = \xi_1 + \xi_2$ is symmetric. It is obvious that $\alpha x = -x$ for all $x \in K$. Thus, if we consider the random variables ξ_1 and ξ_2 as random variables with values in X, then the conditional distribution of the linear form $L_2 = \xi_1 + \alpha\xi_2$ given $L_1 = \xi_1 + \xi_2$ is symmetric. Taking into account that μ is an arbitrary distribution, the necessity is proved.

Sufficiency. Denote by Y the character group of the group X. By Corollary 9.2, the characteristic functions $\hat{\mu}_j(y)$ satisfy equation (9.3). By Lemma 9.6, this implies that the characteristic functions $\hat{\mu}_j(y)$ satisfy the equation

$$\begin{aligned}(9.17)\quad &\hat{\mu}_1((I+\tilde{\alpha})u+2v)\hat{\mu}_2(2\tilde{\alpha}u+(I+\tilde{\alpha})v)\\ &\qquad= \hat{\mu}_1((I+\tilde{\alpha})u)\hat{\mu}_2(2\tilde{\alpha}u)\hat{\mu}_1(2v)\hat{\mu}_2((I+\tilde{\alpha})v), \quad u, v \in Y.\end{aligned}$$

Set $\nu_j = \mu_j * \bar{\mu}_j$. This implies that $\hat{\nu}_j(y) = |\hat{\mu}_j(y)|^2 > 0$ for all $y \in Y$, $j = 1, 2$. The characteristic functions $\hat{\nu}_j(y)$ also satisfy equation (9.17).

Put $\psi_j(y) = -\ln \hat{\nu}_j(y)$, $j = 1, 2$. It follows from (9.17) that the functions $\psi_j(y)$ satisfy the equation

$$(9.18)\qquad \psi_1((I+\tilde{\alpha})u+2v)+\psi_2(2\tilde{\alpha}u+(I+\tilde{\alpha})v) = P(u)+Q(v), \quad u, v \in Y,$$

where

$$(9.19)\qquad P(y) = \psi_1((I+\tilde{\alpha})y)+\psi_2(2\tilde{\alpha}y), \quad Q(y) = \psi_1(2y)+\psi_2((I+\tilde{\alpha})y).$$

We use the finite difference method to solve equation (9.18). Let h_1 be an arbitrary element of the group Y. Substitute $u+(I+\tilde{\alpha})h_1$ for u and $v-2\tilde{\alpha}h_1$ for v in equation (9.18). Subtracting equation (9.18) from the resulting equation, we get

$$(9.20)\quad \Delta_{(I-\tilde{\alpha})^2h_1}\psi_1((I+\tilde{\alpha})u+2v) = \Delta_{(I+\tilde{\alpha})h_1}P(u)+\Delta_{-2\tilde{\alpha}h_1}Q(v), \quad u, v \in Y.$$

Let h_2 be an arbitrary element of the group Y. Substitute $u+2h_2$ for u and $v-(I+\tilde{\alpha})h_2$ for v in equation (9.20). Subtracting equation (9.20) from the resulting equation, we obtain

$$(9.21)\qquad \Delta_{2h_2}\Delta_{(I+\tilde{\alpha})h_1}P(u)+\Delta_{-(I+\tilde{\alpha})h_2}\Delta_{-2\tilde{\alpha}h_1}Q(v) = 0, \quad u, v \in Y.$$

Let h be an arbitrary element of the group Y. Substitute $u+h$ for u in equation (9.21). Subtracting equation (9.21) from the resulting equation, we get

$$(9.22)\qquad \Delta_h\Delta_{2h_2}\Delta_{(I+\tilde{\alpha})h_1}P(u) = 0, \quad u \in Y.$$

It is worth noting that we got equation (9.22) using neither X contains no elements of order 2 nor α satisfies the condition $\mathrm{Ker}(I+\alpha) = \{0\}$.

Since the group X has no elements of order 2, by Theorem 1.12, the subgroup $Y^{(2)}$ is dense in Y. In view of $\mathrm{Ker}(I+\alpha) = \{0\}$, it follows from statement (b) of Theorem 1.20 that the subgroup $(I+\tilde{\alpha})(Y)$ is dense in Y. Taking into account that h, h_1, h_2 in (9.22) are arbitrary elements of Y, it follows from (9.22) that the function $P(y)$ satisfies the equation

$$\Delta_h^3 P(y) = 0, \quad y, h \in Y. \tag{9.23}$$

In view of $P(-y) = P(y)$ for all $y \in Y$ and $P(0) = 0$, (9.23) implies that

$$P(u+v) + P(u-v) = 2[P(u) + P(v)], \quad u, v \in Y. \tag{9.24}$$

Let μ be the distribution of the random variable $M_1 = (I+\alpha)\xi_1 + 2\alpha\xi_2$. Put $\nu = \mu * \bar{\mu}$. It is obvious that $\mu = (I+\alpha)(\mu_1) * (2\alpha)(\mu_2)$. It follows from (9.19) that the characteristic function $\hat{\nu}(y)$ is of the form

$$\hat{\nu}(y) = \exp\{-P(y)\}, \quad y \in Y.$$

In view of (9.24), we have $\nu \in \Gamma(X)$. Inasmuch as the group X contains no elements of order 2, X has no subgroups topologically isomorphic to the circle group $\mathbb{T}$. By Theorem 3.14, we get that $(I+\alpha)(\mu_1) \in \Gamma(X)$. Since the endomorphism $I+\alpha$ is a continuous monomorphism, we conclude that $\mu_1 \in \Gamma(X)$.

Let k be an arbitrary element of the group Y. Substitute $v+k$ for v in equation (9.21). Subtracting equation (9.21) from the resulting equation, we get

$$\Delta_k \Delta_{-(I+\tilde{\alpha})h_2} \Delta_{-2\tilde{\alpha}h_1} Q(v) = 0, \quad v \in Y. \tag{9.25}$$

Taking into account (9.19) and (9.25), considering the distribution of the random variable $M_2 = 2\xi_1 + (I+\alpha)\xi_2$, and reasoning similarly as above, we obtain that $\mu_2 \in \Gamma(X)$. □

Let X be a locally compact Abelian group containing no elements of order 2. Let α be a topological automorphism of X. Let ξ_1 and ξ_2 be independent random variables with values in the group X with nonvanishing characteristic functions. Consider the linear forms $L_1 = \xi_1 + \xi_2$ and $L_2 = \xi_1 + \alpha\xi_2$. The rest of the section is devoted to group analogues of Heyde's theorem when we do not assume that α satisfies the condition $\operatorname{Ker}(I+\alpha) = \{0\}$.

LEMMA 9.10. *Let X be a locally compact Abelian group with character group Y. Let H be a closed subgroup of the group Y such that the subgroup $H^{(2)}$ is dense in H. Put $K = A(X, H)$. Let α be a topological automorphism of the group X such that $\tilde{\alpha}(H) = H$. Let ξ_1 and ξ_2 be independent random variables with values in X and distributions μ_1 and μ_2 such that*

$$|\hat{\mu}_1(y)| = |\hat{\mu}_2(y)| = 1, \quad y \in H. \tag{9.26}$$

Assume that the conditional distribution of the linear form $L_2 = \xi_1 + \alpha\xi_2$ given $L_1 = \xi_1 + \xi_2$ is symmetric. Then there are some shifts λ_j of the distributions μ_j such that λ_j are supported in K. Moreover, if η_j are independent random variables with values in X and distributions λ_j, then the conditional distribution of the linear form $N_2 = \eta_1 + \alpha\eta_2$ given $N_1 = \eta_1 + \eta_2$ is symmetric.

PROOF. By Corollary 9.2, the characteristic functions $\hat{\mu}_j(y)$ satisfy equation (9.3). In view of statement (v) of Theorem 2.5, it follows from (9.26) that there are elements $p_j \in X$ such that

$$\hat{\mu}_j(y) = (p_j, y), \quad y \in H, \ j = 1, 2. \tag{9.27}$$

Taking into account that $\tilde{\alpha}(H) = H$, we can consider the restriction of equation (9.3) to H and use (9.27). We get

$$(p_1, u+v)(p_2, u+\tilde{\alpha}v) = (p_1, u-v)(p_2, u-\tilde{\alpha}v), \quad u, v \in H. \tag{9.28}$$

In view of $K = A(X, H)$, it follows from (9.28) that

$$2(p_1 + \alpha p_2) \in K. \tag{9.29}$$

By Theorem 1.9, the character group of the factor-group X/K is topologically isomorphic to the annihilator $A(Y, K)$ and by Theorem 1.8, $A(Y, K) = H$. Therefore, by Theorem 1.12, if the subgroup $H^{(2)}$ is dense in H, then the factor-group X/K contains no elements of order 2. Hence the subgroup K has the property: if $2x \in K$, then $x \in K$. For this reason (9.29) implies that $p_1 + \alpha p_2 \in K$. Put $x_1 = -\alpha p_2$, $x_2 = p_2$. Then $x_1 + \alpha x_2 = 0$. This implies that the characteristic functions $f_j(y) = (-x_j, y)$, $j = 1, 2$, satisfy equation (9.3). Hence the characteristic functions of the distributions $\lambda_j = \mu_j * E_{-x_j}$ also satisfy equation (9.3). Let η_j be independent random variables with values in the group X and distributions λ_j. By Corollary 9.2, the conditional distribution of the linear form $N_2 = \eta_1 + \alpha\eta_2$ given $N_1 = \eta_1 + \eta_2$ is symmetric. Taking into account that $x_1 = p_1 - k$, where $k \in K$, $x_2 = p_2$, and $K = A(X, H)$, we find from (9.27) that $\hat{\lambda}_j(y) = 1$ for all $y \in H$, $j = 1, 2$. By Proposition 2.10, this implies that $\sigma(\lambda_j) \subset K$, $j = 1, 2$. □

THEOREM 9.11. *Let X be a finite Abelian group containing no elements of order 2. Let α be an automorphism of X. Set $K = \mathrm{Ker}(I + \alpha)$. Let ξ_1 and ξ_2 be independent random variables with values in the group X and distributions μ_1 and μ_2 with nonvanishing characteristic functions. If the conditional distribution of the linear form $L_2 = \xi_1 + \alpha\xi_2$ given $L_1 = \xi_1 + \xi_2$ is symmetric, then $\mu_j = \omega * E_{x_j}$, where $\omega \in \mathrm{M}^1(K)$, $x_j \in X$, $j = 1, 2$.*

PROOF. Denote by Y the character group of the group X. Since the group Y is isomorphic to the group X, the group Y also contains no elements of order 2.

1. First we show that the proof of the theorem is reduced to the case when $(I - \alpha) \in \mathrm{Aut}(X)$. By Corollary 9.2, the characteristic functions $\hat{\mu}_j(y)$ satisfy equation (9.3). Set $L = \mathrm{Ker}(I - \widetilde{\alpha})$ and assume that $L \neq \{0\}$. It is obvious that $\widetilde{\alpha}(L) = L$. Since $\widetilde{\alpha}y = y$ for all $y \in L$, the restriction of equation (9.3) to the subgroup L takes the form

$$\hat{\mu}_1(u + v)\hat{\mu}_2(u + v) = \hat{\mu}_1(u - v)\hat{\mu}_2(u - v), \quad u, v \in L. \tag{9.30}$$

Substituting $u = v = y$ in (9.30), we obtain $\hat{\mu}_1(2y)\hat{\mu}_2(2y) = 1$ for all $y \in L$. Inasmuch as the subgroup L is finite and contains no elements of order 2, we have $L^{(2)} = L$. Thus, $\hat{\mu}_1(y)\hat{\mu}_2(y) = 1$. Hence $|\hat{\mu}_1(y)| = |\hat{\mu}_2(y)| = 1$ for all $y \in L$. Put $G = A(X, L)$. By Lemma 9.10, we can replace the distributions μ_j by their shifts λ_j in such a way that $\sigma(\lambda_j) \subset G$, and if η_j are independent random variables with values in the group X and distributions λ_j, then the conditional distribution of the linear form $N_2 = \eta_1 + \alpha\eta_2$ given $N_1 = \eta_1 + \eta_2$ is symmetric. It follows from $\widetilde{\alpha}(L) = L$ that $\alpha(G) = G$, i.e., the restriction of α to G is an automorphism of the group G. Since $\sigma(\lambda_j) \subset G$, we can consider η_j as independent random variables with values in the group G. In so doing, the conditional distribution of the linear form $N_2 = \eta_1 + \alpha_G\eta_2$ given $N_1 = \eta_1 + \eta_2$ is symmetric. Note that G is a proper subgroup of the group X, because $L \neq \{0\}$. Obviously, $\mathrm{Ker}(I + \alpha_G) \subset \mathrm{Ker}(I + \alpha)$. Thus, we may prove the theorem supposing that the independent random variables take values not in the group X but in the group G. Furthermore, if $\mathrm{Ker}(I - \widetilde{\alpha}_G) \neq \{0\}$, then we repeat this argument. Since G is a finite group, after a finite number of steps, each time replacing the distributions by their shifts, we obtain a subgroup F of the group X, an automorphism α_F of the group F satisfying the condition

$\mathrm{Ker}(I-\widetilde{\alpha}_F)=\{0\}$, and independent random variables ζ_j with values in the group F such that the conditional distribution of the linear form $M_2=\zeta_1+\alpha_F\zeta_2$ given $M_1=\zeta_1+\zeta_2$ is symmetric. Furthermore, the distributions of the random variables ζ_j are shifts of μ_j. Inasmuch as F is a finite group, $\mathrm{Ker}(I-\widetilde{\alpha}_F)=\{0\}$ implies that $(I-\alpha_F)\in \mathrm{Aut}(F)$. Thus, we can suppose that $(I-\alpha)\in\mathrm{Aut}(X)$. Observe also that in this argument we did not use the fact that the characteristic functions $\hat{\mu}_j(y)$ do not vanish.

2. Put $H=(I+\widetilde{\alpha})(Y)$ and prove that (9.26) is satisfied. Set $\nu_j=\mu_j*\bar{\mu}_j$. Then $\hat{\nu}_j(y)=|\hat{\mu}_j(y)|^2>0$ for all $y\in Y$, $j=1,2$. The characteristic functions $\hat{\nu}_j(y)$ satisfy equation (9.3). Set $f(y)=\hat{\nu}_1(y)$, $g(y)=\hat{\nu}_2(y)$ and rewrite equation (9.3) for the functions $\hat{\nu}_j(y)$ in this notation

$$f(u+v)g(u+\widetilde{\alpha}v)=f(u-v)g(u-\widetilde{\alpha}v),\quad u,v\in Y. \tag{9.31}$$

Substituting $u=v=y$ in (9.31), we obtain

$$f(2y)g((I+\widetilde{\alpha})y)=g((I-\widetilde{\alpha})y),\quad y\in Y. \tag{9.32}$$

Since $(I-\alpha)\in\mathrm{Aut}(X)$, by statement (c) of Theorem 1.20, we have $(I-\widetilde{\alpha})\in\mathrm{Aut}(Y)$ and find from (9.32) that

$$g(y)=f(2(I-\widetilde{\alpha})^{-1}y)g((I+\widetilde{\alpha})(I-\widetilde{\alpha})^{-1}y),\quad y\in Y. \tag{9.33}$$

Inasmuch as $g(y)<1$ for all $y\in Y$, (9.33) implies the inequality

$$g(y)\le f(2(I-\widetilde{\alpha})^{-1}y),\quad y\in Y. \tag{9.34}$$

Substituting $u=\widetilde{\alpha}y$, $v=y$ in (9.31), we obtain

$$f((I+\widetilde{\alpha})y)g(2\widetilde{\alpha}y)=f(-(I-\widetilde{\alpha})y),\quad y\in Y. \tag{9.35}$$

Since $f(y)<1$ for all $y\in Y$, (9.35) implies the inequality

$$f(y)\le g(-2\widetilde{\alpha}(I-\widetilde{\alpha})^{-1}y),\quad y\in Y. \tag{9.36}$$

From (9.34) and (9.36) we find

$$g(y)\le f(2(I-\widetilde{\alpha})^{-1}y)\le g(-4\widetilde{\alpha}(I-\widetilde{\alpha})^{-2}y),\quad y\in Y. \tag{9.37}$$

Set $\kappa=-4\widetilde{\alpha}(I-\widetilde{\alpha})^{-2}$. Since the group Y contains no elements of order 2, multiplication by 2 is an automorphism of the group Y. This implies that $\kappa\in\mathrm{Aut}(Y)$. The group $\mathrm{Aut}(Y)$ is finite because the group Y is finite. Denote by m the order of the element κ of the group $\mathrm{Aut}(Y)$. We find from (9.37) the inequalities

$$g(y)\le g(\kappa y)\le\dots\le g(\kappa^{m-1}y)\le g(\kappa^m y)=g(y),\quad y\in Y.$$

Hence

$$g(y)=g(\kappa y),\quad y\in Y, \tag{9.38}$$

and the equality

$$g(y)=f(2(I-\widetilde{\alpha})^{-1}y),\quad y\in Y, \tag{9.39}$$

follows from (9.37) and (9.38). Taking into account that the characteristic functions $f(y)$ and $g(y)$ do not vanish and the fact that $(I-\widetilde{\alpha})\in\mathrm{Aut}(Y)$, we find from (9.33) and (9.39) that $g(y)=1$ for all $y\in H$.

Prove an analogous statement for the function $f(y)$. It follows from (9.34) and (9.36) that

$$f(y)\le g(-2\widetilde{\alpha}(I-\widetilde{\alpha})^{-1}y)\le f(\kappa y),\quad y\in Y. \tag{9.40}$$

We get from (9.40) the inequalities

$$f(y) \leq f(\kappa y) \leq \cdots \leq f(\kappa^{m-1}y) \leq f(\kappa^m y) = f(y), \quad y \in Y.$$

Hence

$$f(y) = f(\kappa y), \quad y \in Y. \tag{9.41}$$

We obtain from (9.40) and (9.41) that

$$f(y) = g(-2\widetilde{\alpha}(I - \widetilde{\alpha})^{-1}y), \quad y \in Y. \tag{9.42}$$

Since $(I - \widetilde{\alpha}) \in \mathrm{Aut}(Y)$, we find from (9.35) that

$$f(y) = f(-(I + \widetilde{\alpha})(I - \widetilde{\alpha})^{-1}y)g(-2\widetilde{\alpha}(I - \widetilde{\alpha})^{-1}y), \quad y \in Y. \tag{9.43}$$

Taking into account that the characteristic functions $f(y)$ and $g(y)$ do not vanish and the fact that $(I - \widetilde{\alpha}) \in \mathrm{Aut}(Y)$, we find from (9.42) and (9.43) that $f(y) = 1$ for all $y \in H$. Thus, we proved that $f(y) = g(y) = 1$ for all $y \in H$ and this implies that (9.26) is satisfied.

3. In view of $H = (I + \widetilde{\alpha})(Y)$, it follows from Theorem 1.19 and statement (a) of Theorem 1.20 that $K = A(X, H)$. Inasmuch as the subgroup H contains no elements of order 2, we have $H^{(2)} = H$. It is obvious that $\widetilde{\alpha}(H) = H$. By Lemma 9.10, we can replace the distributions μ_j by their shifts λ_j in such a way that the distributions λ_j are supported in the subgroup K, and if η_j are independent random variables with values in the group X and distributions λ_j, then the conditional distribution of the linear form $N_2 = \eta_1 + \alpha\eta_2$ given $N_1 = \eta_1 + \eta_2$ is symmetric. In view of $K = \mathrm{Ker}(I + \alpha)$, the restriction of the automorphism α to the subgroup K coincides with $-I$. Since $\sigma(\lambda_j) \subset K$, we can consider η_j as independent random variables with values in the group K. In so doing, the conditional distribution of the linear form $N_2 = \eta_1 - \eta_2$ given $N_1 = \eta_1 + \eta_2$ is symmetric. Applying Lemma 9.5 to the group K, we get $\lambda_1 = \lambda_2 = \omega$. This implies the statement of the theorem. □

REMARK 9.12. Let X be a finite Abelian group and let α be an automorphism of the group X. Set $K = \mathrm{Ker}(I + \alpha)$. Let ω be a distribution on X supported in the subgroup K. Let x_1 and x_2 be elements of the group X such that $x_1 + \alpha x_2 = 0$. Put $\mu_j = \omega * E_{x_j}$, $j = 1, 2$. Let ξ_j be independent random variables with values in the group X and distributions μ_j. It is easy to see that the characteristic functions $\hat{\mu}_j(y)$ satisfy equation (9.3). By Corollary 9.2, the conditional distribution of the linear form $L_2 = \xi_1 + \alpha\xi_2$ given $L_1 = \xi_1 + \xi_2$ is symmetric. From what has been said it follows that Theorem 9.11 cannot be strengthened by narrowing the class of distributions which are characterized by the symmetry of the conditional distribution of the linear form L_2 given L_1.

LEMMA 9.13. *Let $X = \mathbb{R} \times K$, where K is a locally compact Abelian group. Denote by (t, k), where $t \in \mathbb{R}$, $k \in K$, elements of X and by L the character group of the group K. Assume that $L^{(2)} = L$. Let α be a topological automorphism of the group X of the form $\alpha(t, k) = (at, -k)$, where $a \neq -1$. Let ξ_1 and ξ_2 be independent random variables with values in X and distributions μ_1 and μ_2 with nonvanishing characteristic functions. If the conditional distribution of the linear form $L_2 = \xi_1 + \alpha\xi_2$ given $L_1 = \xi_1 + \xi_2$ is symmetric, then $\mu_j = \gamma_j * \omega$, where $\gamma_j \in \Gamma(\mathbb{R})$ and $\omega \in \mathrm{M}^1(K)$.*

PROOF. Denote by Y the character group of the group X. The group Y is topologically isomorphic to the group $\mathbb{R} \times L$. Denote by $y = (s, l)$, where $s \in \mathbb{R}$,

$l \in L$, elements of the group Y. By Corollary 9.2, the characteristic functions $\hat{\mu}_j(s,l)$ satisfy equation (9.3) which takes the form

$$(9.44)\quad \hat{\mu}_1(s_1+s_2,l_1+l_2)\hat{\mu}_2(s_1+as_2,l_1-l_2) = \hat{\mu}_1(s_1-s_2,l_1-l_2)\hat{\mu}_2(s_1-as_2,l_1+l_2), \quad s_j\in\mathbb{R},\ l_j\in L.$$

Substitute $l_1=l_2=0$ in (9.44). In view of Corollary 9.2 and Heyde's theorem, the obtained equation implies that

$$\hat{\mu}_j(s,0)=\exp\{-\sigma_j s^2+i\beta_j s\}, \quad s\in\mathbb{R},$$

where $\sigma_j\geq 0$, $\beta_j\in\mathbb{R}$, $j=1,2$. Moreover,

$$(9.45)\qquad \beta_1+a\beta_2=0$$

and

$$(9.46)\qquad \sigma_1+a\sigma_2=0.$$

In view of (9.45), we can replace the distributions μ_j by their shifts $\mu_j * E_{-\beta_j}$ and suppose, without loss of generality, that $\beta_1=\beta_2=0$. It follows from (9.46) that either $\sigma_1=\sigma_2=0$ or $\sigma_1>0$ and $\sigma_2>0$.

If $\sigma_1=\sigma_2=0$, then $\hat{\mu}_1(s,0)=\hat{\mu}_2(s,0)=1$ for all $s\in\mathbb{R}$. By Proposition 2.10, it follows from this that $\sigma(\mu_j)\subset A(X,\mathbb{R})=K$, $j=1,2$. Hence we may consider ξ_j as independent random variables with values in the group K. If so, the conditional distribution of the linear form $L_2=\xi_1-\xi_2$ given $L_1=\xi_1+\xi_2$ is symmetric. Since $L^{(2)}=L$, by Theorem 1.12, the group K contains no elements of order 2. The statement of the lemma follows in this case from Corollary 9.2 and Lemma 9.5, applied to the group K.

Thus, we assume

$$(9.47)\qquad \hat{\mu}_j(s,0)=\exp\{-\sigma_j s^2\}, \quad s\in\mathbb{R},$$

where $\sigma_j>0$, $j=1,2$. It follows from (9.46) that $a<0$. In view of (9.47), by Proposition 2.13, for each fixed $l\in L$ the functions $\hat{\mu}_j(s,l)$ can be extended to the complex plane $\mathbb{C}$ as entire functions in s. It is easy to see that equation (9.44) holds for all $s_j\in\mathbb{C}$, $l_j\in L$. It also follows from (9.44) that the functions $\hat{\mu}_j(s,l)$ do not vanish for all $s\in\mathbb{C}$, $l\in L$. Indeed, consider equation (9.44), assuming that $s_j\in\mathbb{C}$, $l_j\in L$ and substitute $l_1=l_2=l$ in it. In view of (9.47) and taking into account that $L^{(2)}=L$, rewrite the obtained equation in the form

$$(9.48)\quad \hat{\mu}_1(s_1+s_2,l)\exp\{-\sigma_2(s_1+as_2)^2\} = \exp\{-\sigma_1(s_1-s_2)^2\}\hat{\mu}_2(s_1-as_2,l), \quad s_j\in\mathbb{C},\ l\in L.$$

Suppose that $\hat{\mu}_1(s_0,l_0)=0$ for some $s_0\in\mathbb{C}$, $l_0\in L$. Taking into account that $a\neq -1$, substitute $s_1=s_0a(a+1)^{-1}$, $s_2=s_0(a+1)^{-1}$, $l=l_0$ in (9.48). Inasmuch as $\hat{\mu}_2(0,l_0)\neq 0$, the right-hand side of the obtained equality is nonzero, whereas the left-hand side is equal to zero. From the obtained contradiction it follows that the function $\hat{\mu}_1(s,l)$ does not vanish for all $s\in\mathbb{C}$, $l\in L$. Similarly, if $\hat{\mu}_2(s_0,l_0)=0$ for some $s_0\in\mathbb{C}$, $l_0\in L$, then substituting $s_1=s_0(a+1)^{-1}$, $s_2=-s_0(a+1)^{-1}$, $l=l_0$ in (9.48) and taking into account that $\hat{\mu}_1(0,l_0)\neq 0$, we obtain the contradiction. Thus, the function $\hat{\mu}_2(s,l)$ also does not vanish for all $s\in\mathbb{C}$, $l\in L$. It follows from Proposition 2.13 and (9.47) that the functions $\hat{\mu}_j(s,l)$ for each fixed $l\in L$ are entire functions in s of order at most 2. By the Hadamard theorem on the representation of an entire function of finite order, taking into account that the functions $\hat\mu_j(s,l)$ do not vanish, we obtain the representations

$$\hat\mu_j(s,l) = \exp\{a_{lj}s^2 + b_{lj}s + c_{lj}\}, \quad s \in \mathbb{R},\ l \in L,\ j = 1,2, \tag{9.49}$$

where a_{lj}, b_{lj}, c_{lj} are some complex constants. Actually, the representation (9.49) is valid when $s \in \mathbb{C}$, but we need it only when $s \in \mathbb{R}$. Set

$$\psi_j(s,l) = a_{lj}s^2 + b_{lj}s + c_{lj}, \quad s \in \mathbb{R},\ j = 1,2,$$

and substitute (9.49) into (9.44). It follows from the obtained equation that the functions $\psi_j(s,l)$ satisfy the equation

$$\begin{aligned}\psi_1(s_1+s_2, l_1+l_2) + \psi_2(s_1+as_2, l_1-l_2) \\ = \psi_1(s_1-s_2, l_1-l_2) + \psi_2(s_1-as_2, l_1+l_2) + 2\pi i n(l_1,l_2), \quad s_j \in \mathbb{R},\ l_j \in L,\end{aligned} \tag{9.50}$$

where $n(l_1,l_2)$ is an integer. Substituting $l_1 = l_2 = l$ in (9.50), we find

$$\begin{aligned}a_{2l1}(s_1+s_2)^2 + b_{2l1}(s_1+s_2) + c_{2l1} + a_{02}(s_1+as_2)^2 + b_{02}(s_1+as_2) \\ + c_{02} = a_{01}(s_1-s_2)^2 + b_{01}(s_1-s_2) + c_{01} + a_{2l2}(s_1-as_2)^2 \\ + b_{2l2}(s_1-as_2) + c_{2l2} + 2\pi i n(l,l), \quad s_j \in \mathbb{R},\ l \in L.\end{aligned} \tag{9.51}$$

Equating the coefficients of s_1^2 and s_2^2 on each side of (9.51), we obtain

$$a_{2l1} + a_{02} = a_{01} + a_{2l2}, \quad a_{2l1} + a^2 a_{02} = a_{01} + a^2 a_{2l2}. \tag{9.52}$$

It follows from (9.47) that

$$a_{01} = -\sigma_1, \quad a_{02} = -\sigma_2. \tag{9.53}$$

Taking into account that $L^{(2)} = L$ and $a < 0$, $a \neq -1$, it follows from (9.52) and (9.53) that

$$a_{l1} = -\sigma_1, \quad a_{l2} = -\sigma_2, \quad l \in L. \tag{9.54}$$

Equating the coefficients of s_1 and s_2 on each side of (9.51), we find

$$b_{2l1} + b_{02} = b_{01} + b_{2l2}, \quad b_{2l1} + ab_{02} = -b_{01} - ab_{2l2}. \tag{9.55}$$

It follows from (9.47) that

$$b_{01} = b_{02} = 0. \tag{9.56}$$

Taking into account that $L^{(2)} = L$ and $a < 0$, $a \neq -1$, it follows from (9.55) and (9.56) that

$$b_{l1} = b_{l2} = 0, \quad l \in L. \tag{9.57}$$

Equating the constant terms on each side of (9.51), we obtain

$$c_{2l1} + c_{02} = c_{01} + c_{2l2} + 2\pi i n(l,l), \quad l \in L. \tag{9.58}$$

In view of (9.47), we can assume that $c_{01} = c_{02} = 0$. Taking this into account and the fact that $L^{(2)} = L$, the equality $\exp\{c_{l1}\} = \exp\{c_{l2}\}$ for all $l \in L$ follows from (9.58). We note that (9.49) implies that $\exp\{c_{lj}\} = \hat\mu_j(0,l)$, $l \in L$, $j = 1,2$. Denote by ω the distribution on the group K with the characteristic function

$$\hat\omega(l) = \exp\{c_{l1}\} = \exp\{c_{l2}\}, \quad l \in L. \tag{9.59}$$

Taking into account (9.54), (9.57) and (9.59), the representation

$$\hat\mu_j(s,l) = \exp\{-\sigma_j s^2\}\hat\omega(l), \quad s \in \mathbb{R},\ l \in L,\ j = 1,2, \tag{9.60}$$

follows from (9.49). Denote by γ_j the Gaussian distributions on $\mathbb{R}$ with the characteristic functions

$$\hat{\gamma}_j(s) = \exp\{-\sigma_j s^2\}, \quad j = 1, 2.$$

It follows from (9.60) that $\mu_j = \gamma_j * \omega$, $j = 1, 2$. □

Let us now formulate the following corollary of Lemma 9.13. We keep the notation used in the proof of Lemma 9.13.

COROLLARY 9.14. *Assume that the conditions of Lemma* 9.13 *are fulfilled. Then all solutions of equation* (9.44) *in the class of characteristic functions are of the form* (9.60).

Let $X = \mathbb{R} \times G$, where G is a finite Abelian group. Denote by $x = (t, g)$, where $t \in \mathbb{R}$, $g \in G$, elements of the group X. Let H be the character group of the group G. The group Y is topologically isomorphic to the group $\mathbb{R} \times H$. Denote by $y = (s, h)$, where $s \in \mathbb{R}$, $h \in H$, elements of the group Y. Let α be a topological automorphism of the group X. It is obvious that $\alpha(\mathbb{R}) = \mathbb{R}$ and $\alpha(G) = G$. Therefore, α acts on elements of X as follows: $\alpha(t, g) = (\alpha_{\mathbb{R}} t, \alpha_G g)$, where $t \in \mathbb{R}$, $g \in G$. In so doing, $\alpha_{\mathbb{R}}$ is a multiplication by a nonzero real number a. We will identify $\alpha_{\mathbb{R}}$ with the number a. The adjoint automorphism $\widetilde{\alpha}_{\mathbb{R}}$ is also identified with a. Thus, $\alpha(t, g) = (at, \alpha_G g)$, and we write α and $\widetilde{\alpha}$ in the form $\alpha = (a, \alpha_G)$, $\widetilde{\alpha} = (a, \widetilde{\alpha}_G)$.

THEOREM 9.15. *Let $X = \mathbb{R} \times G$, where G is a finite Abelian group containing no elements of order* 2. *Let $\alpha = (a, \alpha_G)$ be a topological automorphism of X. Set $K = \mathrm{Ker}(I + \alpha_G)$. Let ξ_1 and ξ_2 be independent random variables with values in the group X and distributions μ_1 and μ_2 with nonvanishing characteristic functions. Assume that the conditional distribution of the linear form $L_2 = \xi_1 + \alpha\xi_2$ given $L_1 = \xi_1 + \xi_2$ is symmetric. If $a \neq -1$, then $\mu_j = \gamma_j * \omega * E_{x_j}$, where $\gamma_j \in \Gamma(\mathbb{R})$, $\omega \in \mathrm{M}^1(K)$, $x_j \in X$. If $a = -1$, then $\mu_j = \omega * E_{x_j}$, where $\omega \in \mathrm{M}^1(\mathbb{R} \times K)$, $x_j \in X$.*

PROOF. By Corollary 9.2, the characteristic functions $\hat{\mu}_j(s, h)$ satisfy equation (9.3) which takes the form

$$\begin{aligned}(9.61)\quad & \hat{\mu}_1(s_1 + s_2, h_1 + h_2)\hat{\mu}_2(s_1 + as_2, h_1 + \widetilde{\alpha}_G h_2) \\ & \qquad = \hat{\mu}_1(s_1 - s_2, h_1 - h_2)\hat{\mu}_2(s_1 - as_2, h_1 - \widetilde{\alpha}_G h_2), \quad s_j \in \mathbb{R},\ h_j \in H.\end{aligned}$$

Denote by ω_j distributions on the group G with the characteristic functions $\hat{\omega}_j(h) = \hat{\mu}_j(0, h)$, $j = 1, 2$. Substituting $s_1 = s_2 = 0$ in (9.61), we obtain

$$(9.62)\qquad \hat{\omega}_1(h_1 + h_2)\hat{\omega}_2(h_1 + \widetilde{\alpha}_G h_2) = \hat{\omega}_1(h_1 - h_2)\hat{\omega}_2(h_1 - \widetilde{\alpha}_G h_2), \quad h_j \in H.$$

Taking into account Corollary 9.2 and equation (9.62), it follows from Theorem 9.11 applying to the group G and the automorphism α_G, that there are elements $g_j \in G$ such that the distribution $\omega = \omega_1 * E_{-g_1} = \omega_2 * E_{-g_2}$ is supported in the subgroup K. Moreover, if ζ_j are independent identically distributed random variables with values in the group G and distribution ω, then the conditional distribution of the linear form $M_2 = \zeta_1 + \alpha_G\zeta_2$ given $M_1 = \zeta_1 + \zeta_2$ is symmetric. Consider the distributions $\lambda_j = \mu_j * E_{-g_j}$. Let η_j be independent random variables with values in the group X and distributions λ_j. Then the conditional distribution of the linear form $N_2 = \eta_1 + \alpha\eta_2$ given $N_1 = \eta_1 + \eta_2$ is also symmetric. It is obvious that $\sigma(\lambda_j) \subset \mathbb{R} \times K$. Since $K = \mathrm{Ker}(I + \alpha_G)$, the restriction of the automorphism α_G to the subgroup K coincides with $-I$. We can consider η_j as independent random variables with values in the subgroup $\mathbb{R}\times K$. From what has been said it follows that proving the theorem we can assume that $X = \mathbb{R} \times K$, where K is a finite Abelian group containing no elements of order 2, and the topological automorphism α is of the form $\alpha(t, k) = (at, -k)$, where $t \in \mathbb{R}$, $k \in K$.

Let $a \neq -1$. Denote by L the character group of the group K. Since the group K contains no elements of order 2 and L is isomorphic to K, the group L also contains no elements of order 2. This implies that $L^{(2)} = L$. Then the statement of the theorem follows from Lemma 9.13.

Let $a = -1$. Then $\alpha = -I$ and by Lemma 9.5, we obtain that $\mu_1 = \mu_2 = \omega$. □

REMARK 9.16. Let $X = \mathbb{R} \times G$, where G is a finite Abelian group. Let $\alpha = (a, \alpha_G)$ be a topological automorphism of the group X. Put $K = \mathrm{Ker}(I + \alpha_G)$.

Assume that $a \neq -1$. Let γ_j be Gaussian distributions on $\mathbb{R}$ with the characteristic functions $\hat\gamma_j(s) = \exp\{-\sigma_j s^2\}$, $s \in \mathbb{R}$, where $\sigma_1 + a\sigma_2 = 0$. Let ω be a distribution on X supported in the subgroup K and let x_j be elements of the group X such that $x_1 + \alpha x_2 = 0$. Put $\mu_j = \gamma_j * \omega * E_{x_j}$, $j = 1, 2$. Let ξ_j be independent random variables with values in the group X and distributions μ_j. It is easy to see that the characteristic functions $\hat\mu_j(y)$ satisfy equation (9.3). By Corollary 9.2, the conditional distribution of the linear form $L_2 = \xi_1 + \alpha\xi_2$ given $L_1 = \xi_1 + \xi_2$ is symmetric.

Assume that $a = -1$. Then we argue in the same way as in Remark 9.12 and construct independent random variables ξ_j with values in the group X and distributions $\mu_j = \omega * E_{x_j}$, where ω is a distribution supported in $\mathbb{R} \times K$, $x_j \in X$, $j = 1, 2$, such that the conditional distribution of the linear form $L_2 = \xi_1 + \alpha\xi_2$ given $L_1 = \xi_1 + \xi_2$ is symmetric.

From what has been said it follows that Theorem 9.15 cannot be strengthened by narrowing the class of distributions which are characterized by the symmetry of the conditional distribution of the linear form L_2 given L_1.

Let $\Sigma_{\boldsymbol{a}}$ be the $\boldsymbol{a}$-adic solenoid with character group $H_{\boldsymbol{a}}$. We consider $H_{\boldsymbol{a}}$ as a subset of $\mathbb{R}$. Let $\alpha \in \mathrm{Aut}(\Sigma_{\boldsymbol{a}})$. Then $\alpha = f_p f_q^{-1}$ for some mutually prime p and q, where $f_p, f_q \in \mathrm{Aut}(\Sigma_{\boldsymbol{a}})$. We identify $\alpha = f_p f_q^{-1}$ with the real number $\frac{p}{q}$. If $\alpha = f_p f_q^{-1}$, then $\widetilde\alpha = f_p f_q^{-1}$ and we also identify $\widetilde\alpha$ with the real number $\frac{p}{q}$. We note that if $\alpha \neq -I$, then either $\mathrm{Ker}(I + \alpha) = \{0\}$ or $\mathrm{Ker}(I + \alpha) \cong \mathbb{Z}(m)$ for some $m = 2, 3, 4, \ldots$. Put $G = (\Sigma_{\boldsymbol{a}})_{(2)}$. It is easy to verify that there are only two possibilities for G: either $G = \{0\}$ or $G \cong \mathbb{Z}(2)$. It is obvious that $G = \{0\}$ if and only if $f_2 \in \mathrm{Aut}(\Sigma_{\boldsymbol{a}})$.

For $\boldsymbol{a}$-adic solenoids containing no elements of order 2 we shall describe distributions which are characterised by the symmetry of the conditional distribution of one linear form of two independent random variables given another with no restrictions on a topological automorphism α. Recall, as noted in Example (3.9), the characteristic function of a Gaussian distribution γ on an $\boldsymbol{a}$-adic solenoid $\Sigma_{\boldsymbol{a}}$ is of the form (3.14).

We need the following lemma. Its proof is actually contained in the proof of Theorem 9.3. Nevertheless, we present it here.

LEMMA 9.17. *Let Y be an Abelian group and let β be an endomorphism of Y. Assume that the functions $\varphi_j(y)$ satisfy the equation*

$$\varphi_1(u + v) + \varphi_2(u + \beta v) - \varphi_1(u - v) - \varphi_2(u - \beta v) = 0, \quad u, v \in Y. \tag{9.63}$$

Then the function $\varphi_1(y)$ satisfies the equation

$$\Delta_{(I-\beta)k_3}\Delta_{2k_2}\Delta_{(I+\beta)k_1}\varphi_1(y) = 0, \quad y \in Y, \tag{9.64}$$

and the function $\varphi_2(y)$ satisfies the equation

$$\Delta_{-(I-\beta)k_3}\Delta_{(I+\beta)k_2}\Delta_{2\beta k_1}\varphi_2(y) = 0, \quad y \in Y, \tag{9.65}$$

where k_j are arbitrary elements of the group Y.

PROOF. We use the finite difference method to solve equation (9.63). Let k_1 be an arbitrary element of Y. Put $h_1 = \beta k_1$. Hence $h_1 - \beta k_1 = 0$. Substitute $u + h_1$ for u and $v + k_1$ for v in equation (9.63). Subtracting equation (9.63) from the resulting equation, we obtain

$$\Delta_{(I+\beta)k_1}\varphi_1(u+v) + \Delta_{2\beta k_1}\varphi_2(u+\beta v) - \Delta_{(\beta-I)k_1}\varphi_1(u-v) = 0, \quad u, v \in Y. \tag{9.66}$$

Let k_2 be an arbitrary element of the group Y. Put $h_2 = k_2$. Hence $h_2 - k_2 = 0$. Substitute $u + h_2$ for u and $v + k_2$ for v in equation (9.66). Subtracting equation (9.66) from the resulting equation, we find

$$\Delta_{2k_2}\Delta_{(I+\beta)k_1}\varphi_1(u+v) + \Delta_{(I+\beta)k_2}\Delta_{2\beta k_1}\varphi_2(u+\beta v) = 0, \quad u, v \in Y. \tag{9.67}$$

Let k_3 be an arbitrary element of Y. Put $h_3 = -\beta k_3$. Hence $h_3 + \beta k_3 = 0$. Substitute $u + h_3$ for u and $v + k_3$ for v in equation (9.67). Subtracting equation (9.67) from the resulting equation, we get

$$\Delta_{(I-\beta)k_3}\Delta_{2k_2}\Delta_{(I+\beta)k_1}\varphi_1(u+v) = 0, \quad u, v \in Y.$$

Substituting $v = 0$ in this equation, we obtain (9.64).

Reasoning similarly we receive from (9.67) that the function $\varphi_2(y)$ satisfies equation (9.65). □

THEOREM 9.18. *Let $\Sigma_{\boldsymbol{a}}$ be an $\boldsymbol{a}$-adic solenoid containing no elements of order 2. Let α be a topological automorphism of $\Sigma_{\boldsymbol{a}}$. Put $K = \mathrm{Ker}(I + \alpha)$. Let ξ_1 and ξ_2 be independent random variables with values in the group $\Sigma_{\boldsymbol{a}}$ and distributions μ_1 and μ_2 with nonvanishing characteristic functions. Assume that the conditional distribution of the linear form $L_2 = \xi_1 + \alpha\xi_2$ given $L_1 = \xi_1 + \xi_2$ is symmetric. If $\alpha \neq -I$, then μ_j are represented in the form $\mu_j = \gamma_j * \omega$, where $\gamma_j \in \Gamma(\Sigma_{\boldsymbol{a}})$ and $\omega \in \mathrm{M}^1(K)$. Moreover, if $\alpha > 0$, then $\mu_j = E_{x_j} * \omega$, where $x_j \in \Sigma_{\boldsymbol{a}}$, $j = 1, 2$. If $\alpha = -I$, then $\mu_1 = \mu_2$.*

PROOF. Let $H_{\boldsymbol{a}}$ be the character group of the group $\Sigma_{\boldsymbol{a}}$. Let $\alpha = f_p f_q^{-1}$ for some mutually prime p and q, where $f_p, f_q \in \mathrm{Aut}(\Sigma_{\boldsymbol{a}})$. By Corollary 9.2, the characteristic functions $\hat{\mu}_j(y)$ satisfy equation (9.3). Put $\nu_j = \mu_j * \bar{\mu}_j$. This implies that $\hat{\nu}_j(y) = |\hat{\mu}_j(y)|^2 > 0$ for all $y \in H_{\boldsymbol{a}}$. The characteristic functions $\hat{\nu}_j(y)$ also satisfy equation (9.3) which takes the form

$$\hat{\nu}_1(u+v)\hat{\nu}_2(u+\widetilde{\alpha}v) = \hat{\nu}_1(u-v)\hat{\nu}_2(u-\widetilde{\alpha}v), \quad u, v \in H_{\boldsymbol{a}}. \tag{9.68}$$

Describe the scheme of the proof of the theorem. First we get some representation for the characteristic functions $\hat{\nu}_j(y)$. By the topological automorphism α we find some natural numbers m and n and consider the group $\mathbb{R} \times \mathbb{Z}(mn)$. Next we construct distributions $M_j \in \mathrm{M}^1(\mathbb{R} \times \mathbb{Z}(mn))$ and a continuous monomorphism $\pi : \mathbb{R} \times \mathbb{Z}(mn) \to \Sigma_{\boldsymbol{a}}$ such that $\nu_j = \pi(M_j)$. In so doing, the characteristic functions of the distributions M_j satisfy some Heyde's functional equation. Next we solve the obtained Heyde's functional equation and obtain the representation for the characteristic functions of the distributions M_j. Finally, we find the desired representation for the distributions μ_j.

First prove the theorem assuming that $\alpha \neq \pm I$. Put $\varphi_j(y) = \ln \hat{\nu}_j(y)$, $j = 1, 2$. It follows from (9.68) that the functions $\varphi_j(y)$ satisfy the equation

$$\varphi_1(u+v) + \varphi_2(u+\widetilde{\alpha}v) - \varphi_1(u-v) - \varphi_2(u-\widetilde{\alpha}v) = 0, \quad u, v \in H_{\boldsymbol{a}}. \tag{9.69}$$

Applying Lemma 9.17, we get from (9.69) that the function $\varphi_1(y)$ satisfies the equation

$$\Delta_{(I-\widetilde{\alpha})k_3}\Delta_{2k_2}\Delta_{(I+\widetilde{\alpha})k_1}\varphi_1(u) = 0, \quad u \in H_{\boldsymbol{a}}, \tag{9.70}$$

and the function $\varphi_2(y)$ satisfies the equation

$$\Delta_{-(I-\widetilde{\alpha})k_3}\Delta_{(I+\widetilde{\alpha})k_2}\Delta_{2\widetilde{\alpha}k_1}\varphi_2(u) = 0, \quad u \in H_{\boldsymbol{a}}, \tag{9.71}$$

where k_j are arbitrary elements of $H_{\boldsymbol{a}}$.

Put

$$H = (I+\widetilde{\alpha})(H_{\boldsymbol{a}}) \cap H_{\boldsymbol{a}}^{(2)} \cap (I-\widetilde{\alpha})(H_{\boldsymbol{a}}). \tag{9.72}$$

It follows from (9.70) and (9.71) that each of the functions $\varphi_j(y)$ satisfies the equation

$$\Delta_h^3\varphi_j(y) = 0, \quad h \in H, \ y \in H_{\boldsymbol{a}}, \ j = 1, 2. \tag{9.73}$$

It is useful to remark that we obtained equation (9.73) without using the fact that the $\boldsymbol{a}$-adic solenoid $\Sigma_{\boldsymbol{a}}$ contains no elements of order 2.

Since the group $\Sigma_{\boldsymbol{a}}$ contains no elements of order 2, by Theorem 1.12, we have $H_{\boldsymbol{a}}^{(2)} = H_{\boldsymbol{a}}$. It is obvious that $(I+\widetilde{\alpha})(H_{\boldsymbol{a}}) = H_{\boldsymbol{a}}^{(m)}$ for some natural m. We can assume that m is the minimum possible. Similarly, $(I-\widetilde{\alpha})(H_{\boldsymbol{a}}) = H_{\boldsymbol{a}}^{(n)}$ for some natural n. We also assume that n is the minimum possible. It follows from $H_{\boldsymbol{a}}^{(2)} = H_{\boldsymbol{a}}$ that m and n are odd. Since m and n are divisors of $p+q$ and $p-q$ respectively and p and q are mutually prime, m and n are also mutually prime. This implies that $H = H_{\boldsymbol{a}}^{(m)} \cap H_{\boldsymbol{a}}^{(n)} = H_{\boldsymbol{a}}^{(mn)}$. Consider the factor-group $H_{\boldsymbol{a}}/H$. It is obvious that $H_{\boldsymbol{a}}/H \cong \mathbb{Z}(mn)$. Take an element $y_0 \in H_{\boldsymbol{a}}$ in such a way that the coset $y_0 + H$ is a generator of the factor-group $H_{\boldsymbol{a}}/H$. Then

$$H_{\boldsymbol{a}} = H \cup (y_0 + H) \cup (2y_0 + H) \cup \cdots \cup ((mn-1)y_0 + H) \tag{9.74}$$

is a H-coset decomposition of the group $H_{\boldsymbol{a}}$. Put

$$\psi_{lj}(y) = \varphi_j(ly_0 + y), \quad y \in H, \ l = 0, 1, \dots, mn-1, \ j = 1, 2.$$

It follows from (9.73) that each of the functions $\psi_{lj}(y)$ satisfies the equation

$$\Delta_h^3\psi_{lj}(y) = 0, \quad h, y \in H, \ l = 0, 1, \dots, mn-1, \ j = 1, 2.$$

This implies that for any coset $ly_0 + H$ there exists the polynomial

$$P_{lj}(s) = a_{lj}s^2 + b_{lj}s + c_{lj}, \quad j = 1, 2, \tag{9.75}$$

on the real line with real coefficients such that

$$P_{lj}(y) = \varphi_j(y), \quad y \in ly_0 + H, \ l = 0, 1, \dots, mn-1, \ j = 1, 2.$$

It follows from this that

$$\hat{\nu}_j(y) = \exp\{a_{lj}y^2 + b_{lj}y + c_{lj}\}, \ y \in ly_0 + H, \ l = 0, 1, \dots, mn-1, \ j = 1, 2. \tag{9.76}$$

Consider the group $\mathbb{R} \times \mathbb{Z}(mn)$. Observe that the character group of the group $\mathbb{R} \times \mathbb{Z}(mn)$ is topologically isomorphic to $\mathbb{R} \times \mathbb{Z}(mn)$. In order not to complicate the notation, suppose that the character group of the group $\mathbb{R} \times \mathbb{Z}(mn)$ coincides with $\mathbb{R} \times \mathbb{Z}(mn)$. Denote by (s, l), where $s \in \mathbb{R}$, $l \in \mathbb{Z}(mn)$, its elements. Define the mapping

$$\tau : H_{\boldsymbol{a}} \to \mathbb{R} \times \mathbb{Z}(mn)$$

as follows:

$$\tau y = (y, l) \text{ if } y \in l y_0 + H, \quad l = 0, 1, \ldots, mn - 1.$$

Then τ is a homomorphism and the subgroup $\tau(H_{\boldsymbol{a}})$ is dense in $\mathbb{R} \times \mathbb{Z}(mn)$. Since m and n are divisors of $p+q$ and $p-q$ respectively and p and q are mutually prime, p and q are mutually prime with m and with n. Hence $f_p, f_q \in \mathrm{Aut}(\mathbb{Z}(mn))$.

Let $y \in l y_0 + H$. If we consider l as an element of the group $\mathbb{Z}(mn)$ and f_p and f_q as automorphisms of the group $\mathbb{Z}(mn)$, then it is easy to see that $\widetilde{\alpha} y \in (f_p f_q^{-1} l) y_0 + H$. Therefore, on the one hand,

$$\tau \widetilde{\alpha} y = (\widetilde{\alpha} y, f_p f_q^{-1} l) = (f_p f_q^{-1} y, f_p f_q^{-1} l), \quad y \in H_{\boldsymbol{a}}.$$

On the other hand, since $\tau y = (y, l)$ and $f_p f_q^{-1}(y, l) = (f_p f_q^{-1} y, f_p f_q^{-1} l)$, we have

$$\tau \widetilde{\alpha} y = f_p f_q^{-1} \tau y, \quad y \in H_{\boldsymbol{a}}. \tag{9.77}$$

Define on the set $\tau(H_{\boldsymbol{a}})$ the functions $g_j(s, l)$ in the following way. Let $(s, l) = \tau y \in \tau(H_{\boldsymbol{a}})$. Put $g_j(s, l) = \hat{\nu}_j(y)$, $j = 1, 2$. Inasmuch as τ is a monomorphism, the functions $g_j(s, l)$ are well defined. It follows from (9.76) that the functions $g_j(s, l)$ are continuous on the subgroup $\tau(H_{\boldsymbol{a}})$ in the topology induced on $\tau(H_{\boldsymbol{a}})$ by the topology of the group $\mathbb{R} \times \mathbb{Z}(mn)$. Taking into account that the subgroup $\tau(H_{\boldsymbol{a}})$ is dense in $\mathbb{R} \times \mathbb{Z}(mn)$, the functions $g_j(s, l)$ can be extended by continuity to some continuous positive definite functions $\bar{g}_j(s, l)$ on the group $\mathbb{R} \times \mathbb{Z}(mn)$. By Bochner's theorem, there exist the distributions $M_j \in \mathrm{M}^1(\mathbb{R} \times \mathbb{Z}(mn))$ such that $\widehat{M}_j(s, l) = \bar{g}_j(s, l)$, $s \in \mathbb{R}$, $l \in \mathbb{Z}(mn)$. It is obvious that

$$\widehat{M}_j(\tau y) = \hat{\nu}(y), \quad y \in H_{\boldsymbol{a}}, \ j = 1, 2. \tag{9.78}$$

It follows from (9.76) and (9.78) that the characteristic functions $\widehat{M}_j(s, l)$ do not vanish and hence $\widehat{M}_j(s, l) > 0$, $j = 1, 2$.

Put

$$\pi = \widetilde{\tau}, \quad \pi : \mathbb{R} \times \mathbb{Z}(mn) \to \Sigma_{\boldsymbol{a}}.$$

Since the subgroup $\tau(H_{\boldsymbol{a}})$ is dense in $\mathbb{R} \times \mathbb{Z}(mn)$, by statement (b) of Theorem 1.20, π is a continuous monomorphism. By Corollary 2.4, π induces an isomorphism of the semigroups $\mathrm{M}^1(\mathbb{R} \times \mathbb{Z}(mn))$ and $\mathrm{M}^1(\pi(\mathbb{R} \times \mathbb{Z}(mn)))$. We also denote this isomorphism by π. It follows from (9.78) that

$$\nu_j = \pi(M_j), \quad j = 1, 2. \tag{9.79}$$

Let us find representations for the characteristic functions $\widehat{M}_j(s, l)$. It follows from (9.68) and (9.78) that

$$\widehat{M}_1(\tau u + \tau v) \widehat{M}_2(\tau u + \tau \widetilde{\alpha} v) = \widehat{M}_1(\tau u - \tau v) \widehat{M}_2(\tau u - \tau \widetilde{\alpha} v), \quad u, v \in H_{\boldsymbol{a}}. \tag{9.80}$$

Taking into account (9.77), we find from (9.80) that

$$\begin{aligned} \widehat{M}_1(\tau u + \tau v) \widehat{M}_2(\tau u + f_p f_q^{-1} \tau v) \\ = \widehat{M}_1(\tau u - \tau v) \widehat{M}_2(\tau u - f_p f_q^{-1} \tau v), \quad u, v \in H_{\boldsymbol{a}}. \end{aligned} \tag{9.81}$$

Since the subgroup $\tau(H_{\boldsymbol{a}})$ is dense in $\mathbb{R}\times\mathbb{Z}(mn)$, (9.81) implies that the characteristic functions $\widehat{M}_j(s,l)$ on the group $\mathbb{R}\times\mathbb{Z}(mn)$ satisfy the following Heyde's functional equation

$$
\begin{aligned}
(9.82)\quad & \widehat{M}_1(s_1+s_2,l_1+l_2)\widehat{M}_2(s_1+pq^{-1}s_2,l_1+f_pf_q^{-1}l_2)\\
&= \widehat{M}_1(s_1-s_2,l_1-l_2)\widehat{M}_2(s_1-pq^{-1}s_2,l_1-f_pf_q^{-1}l_2),\quad s_j\in\mathbb{R},\ l_j\in\mathbb{Z}(mn).
\end{aligned}
$$

Note that $\mathbb{Z}(mn)=\mathbb{Z}(m)\times\mathbb{Z}(n)$, and we represent an element $l\in\mathbb{Z}(mn)$ in the form $l=(a,b)$, $a\in\mathbb{Z}(m)$, $b\in\mathbb{Z}(n)$. Since m is a divisor of $p+q$ and n is a divisor of $p-q$, it is easy to see that the automorphism $f_pf_q^{-1}$ acts on the group $\mathbb{Z}(m)\times\mathbb{Z}(n)$ in the following way: $f_pf_q^{-1}(a,b)=(-a,b)$. Hence we can write equation (9.82) in the form

$$
\begin{aligned}
(9.83)\quad & \widehat{M}_1(s_1+s_2,a_1+a_2,b_1+b_2)\widehat{M}_2(s_1+pq^{-1}s_2,a_1-a_2,b_1+b_2)\\
&= \widehat{M}_1(s_1-s_2,a_1-a_2,b_1-b_2)\widehat{M}_2(s_1-pq^{-1}s_2,a_1+a_2,b_1-b_2),\\
&\qquad s_j\in\mathbb{R},\ a_j\in\mathbb{Z}(m),\ b_j\in\mathbb{Z}(n).
\end{aligned}
$$

Putting $s_1=s_2=0$, $a_1=a_2=0$ in (9.83), we get

$$
\begin{aligned}
(9.84)\quad & \widehat{M}_1(0,0,b_1+b_2)\widehat{M}_2(0,0,b_1+b_2)\\
&= \widehat{M}_1(0,0,b_1-b_2)\widehat{M}_2(0,0,b_1-b_2),\quad b_j\in\mathbb{Z}(n).
\end{aligned}
$$

Putting $b_1=b_2=b$ in (9.84), we obtain $\widehat{M}_j(0,0,2b)=1$ for all $b\in\mathbb{Z}(n)$. Taking into account that n is odd, we conclude that $\widehat{M}_j(0,0,b)=1$, $j=1,2$, for all $b\in\mathbb{Z}(n)$. By Proposition 2.10, this implies that $\widehat{M}_j(s,a,b)=\widehat{M}_j(s,a,0)$ for all $(s,a,b)\in\mathbb{R}\times\mathbb{Z}(m)\times\mathbb{Z}(n)$ and $\sigma(M_j)\subset A(\mathbb{R}\times\mathbb{Z}(m)\times\mathbb{Z}(n),\mathbb{Z}(n))=\mathbb{R}\times\mathbb{Z}(m)$, $j=1,2$. Therefore, we can write the characteristic functions $\widehat{M}_j(s,a,b)$ in the form $\widehat{M}_j(s,a)$. Then equation (9.83) takes the form

$$
\begin{aligned}
(9.85)\quad & \widehat{M}_1(s_1+s_2,a_1+a_2)\widehat{M}_2(s_1+pq^{-1}s_2,a_1-a_2)\\
&= \widehat{M}_1(s_1-s_2,a_1-a_2)\widehat{M}_2(s_1-pq^{-1}s_2,a_1+a_2),\quad s_j\in\mathbb{R},\ a_j\in\mathbb{Z}(m).
\end{aligned}
$$

Inasmuch as m is odd, we can apply Corollary 9.14 to the group $\mathbb{R}\times\mathbb{Z}(m)$. We get from (9.85) that the functions $\widehat{M}_j(s,a)$ are represented in the form

$$
(9.86)\qquad \widehat{M}_j(s,a)=\exp\{-\sigma_js^2\}\widehat{\Omega}(a),\quad s\in\mathbb{R},\ a\in\mathbb{Z}(m),\ j=1,2,
$$

where Ω is a distribution on the group $\mathbb{Z}(m)$.

It follows from (9.79) that the distributions ν_j are concentrated on the subgroup $\pi(\mathbb{R}\times\mathbb{Z}(mn))$. By Proposition 2.1, we can substitute the distributions μ_j by their shifts and suppose, without loss of generality, that μ_j are also concentrated on the subgroup $\pi(\mathbb{R}\times\mathbb{Z}(mn))$. Since π is an isomorphism of the semigroups $\mathrm{M}^1(\mathbb{R}\times\mathbb{Z}(mn))$ and $\mathrm{M}^1(\pi(\mathbb{R}\times\mathbb{Z}(mn)))$, we have

$$
(9.87)\qquad M_j=\pi^{-1}(\nu_j)=\pi^{-1}(\mu_j)*\pi^{-1}(\bar{\mu}_j)=\pi^{-1}(\mu_j)*\overline{\pi^{-1}(\mu_j)},\quad j=1,2.
$$

Put $N_j=\pi^{-1}(\mu_j)$, $j=1,2$. We find from (9.87) that

$$
(9.88)\qquad \widehat{M}_j(s,a)=\widehat{N}_j(s,a)\overline{\widehat{N}_j(s,a)},\quad s\in\mathbb{R},\ a\in\mathbb{Z}(m),\ j=1,2.
$$

Substituting $a = 0$ in equation (9.88) and taking into account (9.86), by Cramér's decomposition theorem for the Gaussian distribution on the real line, we get

$$(9.89)\qquad \widehat{N}_j(s,0) = \exp\left\{-\frac{\sigma_j}{2}s^2 + i\beta_{0j}s\right\}, \quad s \in \mathbb{R},\ j = 1,2,$$

where $\beta_{0j} \in \mathbb{R}$. We can assume, without loss of generality, that $\beta_{0j} = 0$. In view of (9.89), it follows from Proposition 2.13 that $\widehat{N}_j(s,a)$ are entire functions for any $a \in \mathbb{Z}(m)$. Since (9.88) holds for all $s \in \mathbb{C}$, $a \in \mathbb{Z}(m)$, it follows from (9.86) that they do not vanish for all $s \in \mathbb{C}$, $a \in \mathbb{Z}(m)$. By Proposition 2.13, (9.89) implies that each of the entire functions $\widehat{N}_j(s,a)$ is of at most order 2 and type $\frac{\sigma_j}{2}$. Applying the Hadamard theorem on the representation of an entire function of finite order, (9.86) and (9.88), we obtain that

$$(9.90)\qquad \widehat{N}_j(s,a) = \exp\left\{-\frac{\sigma_j}{2}s^2 + i\beta_{aj}s + d_{aj}\right\}, \quad s \in \mathbb{R},\ a \in \mathbb{Z}(m),\ j = 1,2,$$

where $\beta_{aj} \in \mathbb{R}$. It follows from (9.90) that

$$\hat{N}_j(-iy+x,a)/\hat{N}_j(-iy,0) = \exp\left\{-\frac{\sigma_j}{2}x^2 + \beta_{aj}y + \sigma_j ixy + \beta_{aj}ix + d_{aj}\right\}.$$

By Proposition 2.13, the function $\widehat{N}_j(-iy+x,a)/\widehat{N}_j(-iy,0)$ for each fixed y is a characteristic function in a variable $(x,a) \in \mathbb{R} \times \mathbb{Z}(m)$. Thus, in particular, for each y the inequality $|\widehat{N}_j(-iy+x,a)/\widehat{N}_j(-iy,0)| \le 1$ holds true. It follows from this that $\beta_{aj} = 0$, $j = 1,2$, for all $a \in \mathbb{Z}(m)$ and (9.90) implies that

$$(9.91)\qquad \widehat{N}_j(s,a) = \exp\left\{-\frac{\sigma_j}{2}s^2 + d_{aj}\right\}, \quad (s,a) \in \mathbb{R} \times \mathbb{Z}(m), \quad j = 1,2.$$

Let P_j be the Gaussian distributions on $\mathbb{R}$ with the characteristic functions

$$\widehat{P}_j(s) = \exp\{-\frac{\sigma_j}{2}s^2\}, \quad s \in \mathbb{R}.$$

Denote by Q_j the distributions on the group $\mathbb{Z}(m)$ with the characteristic functions

$$\widehat{Q}_j(a) = \exp\{d_{aj}\}, \quad a \in \mathbb{Z}(m).$$

It follows from (9.91) that $N_j = P_j * Q_j$, $j = 1,2$, and hence

$$\mu_j = \pi(N_j) = \pi(P_j) * \pi(Q_j).$$

It is obvious that $\pi(P_j) \in \Gamma(\Sigma_{\boldsymbol{a}})$. Let us verify that $\pi(\mathbb{Z}(m)) = K$.

We have

$$(9.92)\qquad A(H_{\boldsymbol{a}}, K) = (I + \widetilde{\alpha})(H_{\boldsymbol{a}}) = H_{\boldsymbol{a}}^{(m)}.$$

Observe that (9.74) implies the equality

$$H_{\boldsymbol{a}}^{(m)} = H \cup (my_0 + H) \cup (2my_0 + H) \cup \cdots \cup ((n-1)my_0 + H),$$

and hence the equivalence

$$(9.93)\qquad y \in H_{\boldsymbol{a}}^{(m)} \iff \tau y \in \mathbb{R} \times \{0\} \times \mathbb{Z}(n).$$

We have

$$y \in A(H_{\boldsymbol{a}}, \pi(\mathbb{Z}(m)) \iff (g, \tau y) = 1 \text{ for all } g \in \mathbb{Z}(m) \iff \tau y \in \mathbb{R} \times \{0\} \times \mathbb{Z}(n).$$

Taking into account (9.92) and (9.93), this implies the equality $A(H_{\boldsymbol{a}}, \pi(\mathbb{Z}(m)) = A(H_{\boldsymbol{a}}, K)$ and hence $\pi(\mathbb{Z}(m)) = K$. Thus, $\pi(Q_j) \in \mathrm{M}^1(K)$. Put $\lambda_j = \pi(P_j)$, $\rho_j = \pi(Q_j)$. So, we proved that $\mu_j = \lambda_j * \rho_j$, where $\lambda_j \in \Gamma(\Sigma_{\boldsymbol{a}})$, $\rho_j \in \mathrm{M}^1(K)$, $j = 1,2$.

Let the characteristic functions $\hat{\lambda}_j(y)$ be of the form

$$\hat{\lambda}_j(y) = (x_j, y)\exp\{-\sigma_j y^2\}, \quad y \in H_{\boldsymbol{a}},\ j = 1, 2. \tag{9.94}$$

Put $L = A(H_{\boldsymbol{a}}, K)$ and note that $\widetilde{\alpha}(L) = L$. Since $\rho_j \in \mathrm{M}^1(K)$, we have $\hat{\rho}_j(y) = 1$ for all $y \in L$, $j = 1, 2$. Consider the restriction of equation (9.3) for the characteristic functions $\hat{\mu}_j(y)$ to L. Taking into account (9.94), we find from the obtained equation that

$$\sigma_1 + pq^{-1}\sigma_2 = 0 \tag{9.95}$$

and hence

$$(x_1, u + v)(x_2, u + \widetilde{\alpha}v) = (x_1, u - v)(x_2, u - \widetilde{\alpha}v), \quad u, v \in L. \tag{9.96}$$

It follows from (9.96) that $2(x_1 + \alpha x_2) \in K$. Inasmuch as $f_2 \in \mathrm{Aut}(\Sigma_{\boldsymbol{a}})$, we have $x_1 + \alpha x_2 \in K$. Put $x_0 = x_1 + \alpha x_2$, $\gamma_1 = \lambda_1 * E_{-x_0}$, $\omega_1 = \rho_1 * E_{x_0}$, $\gamma_2 = \lambda_2$, $\omega_2 = \rho_2$. It is obvious that $\mu_j = \gamma_j * \omega_j$ and $\omega_j \in \mathrm{M}^1(K)$, $j = 1, 2$. It is easy to see that the characteristic functions $\hat{\gamma}_j(y)$ satisfy equation (9.3). Hence the characteristic functions $\hat{\omega}_j(y)$ also satisfy equation (9.3). We have

$$\hat{\omega}_1(u + v)\hat{\omega}_2(u + \widetilde{\alpha}v) = \hat{\omega}_1(u - v)\hat{\omega}_2(u - \widetilde{\alpha}v), \quad u, v \in H_{\boldsymbol{a}}. \tag{9.97}$$

We consider here ω_j as distributions on the $\boldsymbol{a}$-adic solenoid $\Sigma_{\boldsymbol{a}}$. Denote by M the character group of the group K. Since $\omega_j \in \mathrm{M}^1(K)$, $j = 1, 2$, and $\alpha x = -x$ for all $x \in K$, we can consider ω_j as distributions on the group K. Then equation (9.97) takes the form

$$\hat{\omega}_1(u + v)\hat{\omega}_2(u - v) = \hat{\omega}_1(u - v)\hat{\omega}_2(u + v), \quad u, v \in M.$$

Substituting here $u = v = y$, we get $\hat{\omega}_1(2y) = \hat{\omega}_2(2y)$ for all $y \in M$. Since $K \cong \mathbb{Z}(m)$ and m is odd, this implies that $\hat{\omega}_1(y) = \hat{\omega}_2(y)$ for all $y \in M$. Hence $\omega_1 = \omega_2 = \omega$. Thus, $\mu_j = \gamma_j * \omega$, where $\gamma_j \in \Gamma(\Sigma_{\boldsymbol{a}})$, $\omega \in \mathrm{M}^1(K)$, $j = 1, 2$.

Assume now that $\alpha > 0$. Then (9.95) implies that $\sigma_1 = \sigma_2 = 0$, i.e., γ_j are degenerate distributions. Hence the distributions μ_j are of the required form. Thus, we proved the theorem assuming that $\alpha \neq \pm I$.

It still remains the case when $\alpha = \pm I$. Since the group $\Sigma_{\boldsymbol{a}}$ contains no elements of order 2, by Theorem 1.12, we have $H_{\boldsymbol{a}}^{(2)} = H_{\boldsymbol{a}}$. Let $\alpha = I$. Inasmuch as $\Sigma_{\boldsymbol{a}}$ has no elements of order 2, we have $K = \{0\}$. Substituting $\alpha = I$, $u = v = y$ in equation (9.68), we arrive at $\hat{\nu}_1(2y) = \hat{\nu}_2(2y) = 1$ for all $y \in H_{\boldsymbol{a}}$. It follows from this that $\hat{\nu}_1(y) = \hat{\nu}_2(y) = 1$ for all $y \in H_{\boldsymbol{a}}$. Hence $\nu_1 = \nu_2 = E_0$. Therefore, μ_j are degenerate distributions. In this case the theorem is also proved. Let $\alpha = -I$. Since $\Sigma_{\boldsymbol{a}}$ contains no elements of order 2, the statement of the theorem in this case follows directly from Lemma 9.5. □

REMARK 9.19. Assume that an $\boldsymbol{a}$-adic solenoid $\Sigma_{\boldsymbol{a}}$ contains no elements of order 2. Let $\alpha = f_p f_q^{-1}$ be a topological automorphism of the group $\Sigma_{\boldsymbol{a}}$. Put $K = \mathrm{Ker}(I + \alpha)$. Theorem 9.18 cannot be strengthened by narrowing the class of distributions which are characterized by the symmetry of the conditional distribution of the linear form L_2 given L_1. Indeed, let γ_j be the Gaussian distributions on the group $\Sigma_{\boldsymbol{a}}$ with the characteristic functions of the form

$$\hat{\gamma}_j(y) = (x_j, y)\exp\{-\sigma_j y^2\}, \quad y \in H_{\boldsymbol{a}},\ j = 1, 2,$$

where $x_j \in \Sigma_{\boldsymbol{a}}$ such that $x_1 + \alpha x_2 = 0$. Assume also that $\sigma_1 + pq^{-1}\sigma_2 = 0$. Let $\omega \in \mathrm{M}^1(K)$. Put $\mu_j = \gamma_j * \omega$, $j = 1, 2$. It easily follows from Corollary 9.2

that if ξ_1 and ξ_2 are independent random variables with values in the group $\Sigma_{\boldsymbol{a}}$ and distributions μ_1 and μ_2, then the conditional distribution of the linear form $L_2 = \xi_1 + \alpha\xi_2$ given $L_1 = \xi_1 + \xi_2$ is symmetric.

In Section 11 in Proposition 11.19 we shall prove that generally speaking, Theorem 9.18 is false if an $\boldsymbol{a}$-adic solenoid $\Sigma_{\boldsymbol{a}}$ contains an element of order 2.

LEMMA 9.20. *Let $X = \Sigma_{\boldsymbol{a}} \times K$, where $\boldsymbol{a} = (2, 3, 4, \dots)$ and K is a locally compact Abelian group. Denote by (z, k), where $z \in \Sigma_{\boldsymbol{a}}$, $k \in K$, elements of X. Denote by L the character group of the group K and suppose $L^{(2)} = L$. Let α be a topological automorphism of the group X of the form $\alpha(z, k) = (az, -k)$, where $a \neq -1$. Let ξ_1 and ξ_2 be independent random variables with values in X and distributions μ_1 and μ_2 with nonvanishing characteristic functions. If the conditional distribution of the linear form $L_2 = \xi_1 + \alpha\xi_2$ given $L_1 = \xi_1 + \xi_2$ is symmetric, then $\mu_j = \gamma_j * \omega$, where $\gamma_j \in \Gamma(\Sigma_{\boldsymbol{a}})$ and $\omega \in \mathrm{M}^1(K)$.*

PROOF. Denote by Y the character group of the group X. The group Y is topologically isomorphic to the group $\mathbb{Q} \times L$. Denote by $y = (r, l)$, where $r \in \mathbb{Q}$, $l \in L$, elements of the group Y. By Corollary 9.2, the characteristic functions $\hat{\mu}_j(r, l)$ satisfy equation (9.3) which takes the form

$$\text{(9.98)} \quad \hat{\mu}_1(r_1 + r_2, l_1 + l_2)\hat{\mu}_2(r_1 + ar_2, l_1 - l_2) \\ = \hat{\mu}_1(r_1 - r_2, l_1 - l_2)\hat{\mu}_2(r_1 - ar_2, l_1 + l_2), \quad r_j \in \mathbb{Q},\ l_j \in L.$$

Substitute $l_1 = l_2 = 0$ in (9.98). Taking into account that multiplication by a nonzero rational number is a topological automorphism of the group $\Sigma_{\boldsymbol{a}}$ and Corollary 9.2, it follows from the obtained equation and Theorem 9.9, applied to the group $\Sigma_{\boldsymbol{a}}$ and the topological automorphism a, that

$$\hat{\mu}_j(r, 0) = (z_j, r)\exp\{-\sigma_j r^2\}, \quad r \in \mathbb{Q},$$

where $\sigma_j \geq 0$, $z_j \in \Sigma_{\boldsymbol{a}}$, $j = 1, 2$. Since $z_1 + az_2 = 0$, we can replace the distributions μ_j by their shifts $\mu_j * E_{-z_j}$ and suppose, without loss of generality, that $z_1 = z_2 = 0$. We also note that $\sigma_1 + a\sigma_2 = 0$. This implies that either $\sigma_1 = \sigma_2 = 0$ or $\sigma_1 > 0$ and $\sigma_2 > 0$. The case when $\sigma_1 = \sigma_2 = 0$, is considered in the same way as in the proof of Lemma 9.13. Thus, we will assume that

$$\text{(9.99)} \qquad \hat{\mu}_j(r, 0) = \exp\{-\sigma_j r^2\}, \quad r \in \mathbb{Q},$$

where $\sigma_j > 0$, $j = 1, 2$. Embed in the natural way $(r, l) \to (r, l)$ the group $\mathbb{Q} \times L$ into the group $\mathbb{R} \times L$. It follows from (9.99) and statement (vii) of Theorem 2.5 that the characteristic functions $\hat{\mu}_j(r, l)$, $j = 1, 2$, are uniformly continuous on the subgroup $\mathbb{Q} \times L$ in the topology induced on $\mathbb{Q} \times L$ by the topology of the group $\mathbb{R} \times L$. Hence the characteristic functions $\hat{\mu}_j(r, l)$ can be extended by continuity from the subgroup $\mathbb{Q} \times L$ to the group $\mathbb{R} \times L$. We retain the notation $\hat{\mu}_j(s, l)$, where $s \in \mathbb{R}$, $l \in L$ for the extended functions. It follows from (9.98) that the characteristic functions $\hat{\mu}_j(s, l)$ satisfy equation (9.44). By Corollary 9.14, the representation (9.60) holds. Hence

$$\hat{\mu}_j(r, l) = \exp\{-\sigma_j r^2\}\hat{\omega}(l), \quad r \in \mathbb{Q},\ l \in L,\ j = 1, 2.$$

The statement of the lemma follows from this. □

Consider the group $X = \Sigma_{\boldsymbol{a}} \times G$, where $\boldsymbol{a} = (2, 3, 4, \dots)$ and G is a finite Abelian group. Denote by $x = (z, g)$, where $z \in \Sigma_{\boldsymbol{a}}$, $g \in G$, elements of the group X. Let Y be the character group of the group X, and let H be the character group of the group G. The group Y is topologically isomorphic to the group $\mathbb{Q}\times H$. Denote by $y=(r,h)$, where $r\in\mathbb{Q}$, $h\in H$, elements of the group Y. Let α be a topological automorphism of the group X. Since $\Sigma_{\boldsymbol{a}}$ is the connected component of the zero of X and G is the torsion subgroup of X, we have $\alpha(\Sigma_{\boldsymbol{a}})=\Sigma_{\boldsymbol{a}}$ and $\alpha(G)=G$. Therefore, α acts on elements of X as follows: $\alpha(z,g)=(\alpha_{\Sigma_{\boldsymbol{a}}}z,\alpha_G g)$, where $z\in\Sigma_{\boldsymbol{a}}$, $g\in G$. In so doing, $\alpha_{\Sigma_{\boldsymbol{a}}}$ is a multiplication by a nonzero rational number a. We will identify $\alpha_{\Sigma_{\boldsymbol{a}}}$ with the rational number a. Thus, $\alpha(z,g)=(az,\alpha_G g)$ and we will write α in the form $\alpha=(a,\alpha_G)$. We shall also identify the adjoint automorphism $\widetilde{\alpha}_{\Sigma_{\boldsymbol{a}}}$ of the group $\mathbb{Q}$ with a. Thus, $\widetilde{\alpha}=(a,\widetilde{\alpha}_G)$.

THEOREM 9.21. *Let $X=\Sigma_{\boldsymbol{a}}\times G$, where $\boldsymbol{a}=(2,3,4,\dots)$ and G is a finite Abelian group containing no elements of order 2. Let $\alpha=(a,\alpha_G)$ be a topological automorphism of X. Set $K=\operatorname{Ker}(I+\alpha_G)$. Let ξ_1 and ξ_2 be independent random variables with values in the group X and distributions μ_1 and μ_2 with nonvanishing characteristic functions. Assume that the conditional distribution of the linear form $L_2=\xi_1+\alpha\xi_2$ given $L_1=\xi_1+\xi_2$ is symmetric. If $a\neq-1$, then $\mu_j=\gamma_j*\omega*E_{x_j}$, where $\gamma_j\in\Gamma(\Sigma_{\boldsymbol{a}})$, $\omega\in\mathrm{M}^1(K)$, $x_j\in X$. If $a=-1$, then $\mu_j=\omega*E_{x_j}$, where $\omega\in\mathrm{M}^1(\Sigma_{\boldsymbol{a}}\times K)$, $x_j\in X$.*

PROOF. Denote by Y the character group of the group X and by H the character group of the group G. Then the group Y is topologically isomorphic to the group $\mathbb{Q}\times H$. Denote by $y=(r,h)$, where $r\in\mathbb{Q}$, $h\in H$, elements of the group Y. By Corollary 9.2, the characteristic functions $\hat\mu_j(r,h)$ satisfy equation (9.3) which takes the form

$$\begin{aligned}&\hat\mu_1(r_1+r_2,h_1+h_2)\hat\mu_2(r_1+ar_2,h_1+\widetilde{\alpha}_G h_2)\\&\qquad=\hat\mu_1(r_1-r_2,h_1-h_2)\hat\mu_2(r_1-ar_2,h_1-\widetilde{\alpha}_G h_2),\quad r_j\in\mathbb{Q},\ h_j\in H.\end{aligned}$$

The proof of the theorem boils down to solving this equation and repeats the proof of Theorem 9.15. We only change $\mathbb{R}$ for $\Sigma_{\boldsymbol{a}}$ and use Lemma 9.20 instead of Lemma 9.13.

Note that arguing in the same way as in Remark 9.16, it is easy to verify that Theorem 9.21 cannot be strengthened by narrowing the class of distributions which are characterized by the symmetry of the conditional distribution of the linear form L_2 given L_1. □

10. Discrete and compact totally disconnected Abelian groups

Let X be a second countable locally compact Abelian group. In this section we continue to study Heyde's theorem for two independent random variables. We assume X contains no elements of order 2 and consider independent random variables ξ_j with values in the group X and distributions μ_j. Without loss of generality, we can suppose that $L_1=\xi_1+\xi_2$ and $L_2=\xi_1+\alpha\xi_2$, where $\alpha\in\operatorname{Aut}(X)$. We prove that if X is a countable discrete Abelian group, then the symmetry of the conditional distribution of the linear form L_2 given L_1 implies that μ_j are shifts of idempotent distributions. We describe compact totally disconnected Abelian groups X for which the symmetry of the conditional distribution of the linear form L_2 given L_1 implies that μ_j are shifts of idempotent distributions. This class of groups turns out to be narrower than the class of compact totally disconnected Abelian groups for which the Skitovich–Darmois theorem is true. Unlike in Section 9, the characteristic functions $\hat\mu_j(y)$ can vanish, but we suppose that α satisfies the condition

$\mathrm{Ker}(I+\alpha)=\{0\}$. This condition is necessary to characterize shifts of idempotent distributions. We describe a wide class of locally compact Abelian groups where an analogue of Heyde's theorem holds. We also prove that in the same manner as in the case of the Skitovich–Darmois theorem, Heyde's theorem is not valid on compact connected Abelian groups.

We start by studying the group analogue of Heyde's theorem on discrete Abelian groups.

LEMMA 10.1. *Let X be a locally compact Abelian group containing no elements of order 2. Let α be a topological automorphism of X. Let ξ_1 and ξ_2 be independent random variables with values in the group X and distributions $\mu_1=m_{K_1}$ and $\mu_2=m_{K_2}$, where K_1 and K_2 are finite subgroups of X. If the conditional distribution of the linear form $L_2=\xi_1+\alpha\xi_2$ given $L_1=\xi_1+\xi_2$ is symmetric, then $K_1=K_2=K$ and $\alpha(K)=K$.*

PROOF. Denote by Y the character group of the group X. Put $f(y)=\widehat{m}_{K_1}(y)$, $g(y)=\widehat{m}_{K_2}(y)$, $E_j=A(Y,K_j)$, $j=1,2$. It follows from (2.3) that

$$f(y)=\begin{cases}1, & \text{if } y\in E_1,\\ 0, & \text{if } y\notin E_1,\end{cases}\qquad g(y)=\begin{cases}1, & \text{if } y\in E_2,\\ 0, & \text{if } y\notin E_2.\end{cases} \tag{10.1}$$

By Corollary 9.2, the characteristic functions of the random variables ξ_1 and ξ_2 satisfy equation (9.3) which takes the form (9.31). Substituting $u=v=y$ in equation (9.31), we get (9.32). Assume that

$$(I-\widetilde{\alpha})y\in E_2. \tag{10.2}$$

Then, it follows from (10.1) and (9.32) that

$$2y\in E_1 \tag{10.3}$$

and

$$(I+\widetilde{\alpha})y\in E_2. \tag{10.4}$$

Since group X has no elements of order 2 and K_2 is a finite group, we have

$$(K_2)^{(2)}=K_2. \tag{10.5}$$

It follows from (10.2) and (10.4) that

$$2y\in E_2. \tag{10.6}$$

Taking into account (10.5) and applying Lemma 7.10 to $\alpha=f_2$ and $K=K_2$, we get from (10.6) that $y\in E_2$. So, we proved that (10.2) implies that $y\in E_2$. By Lemma 7.10, applying to the compact subgroup K_2 and the continuous endomorphism $I-\alpha$, we get that $(I-\alpha)(K_2)\supset K_2$. Since K_2 is a finite group, it follows from this that

$$(I-\alpha)(K_2)=K_2 \tag{10.7}$$

and hence

$$\alpha(K_2)=K_2. \tag{10.8}$$

Assume that $y\in E_2$. It is obvious that (10.7) implies the inclusion $(I-\widetilde{\alpha})(E_2)\subset E_2$. Hence (10.2) holds and then (10.3) follows from (10.1) and (9.32). Since group X contains no elements of order 2 and K_1 is a finite group, we have $(K_1)^{(2)}=K_1$.

Taking into account (10.3) and applying Lemma 7.10 to $\alpha = f_2$ and $K = K_1$, it follows from $(K_1)^{(2)} = K_1$ that $y \in E_1$. So, we proved that

$$E_2 \subset E_1. \tag{10.9}$$

Substituting $u = \widetilde{\alpha}y$, $v = y$ in equation (9.31), we get (9.35). Arguing similarly as above, we obtain that (9.35) implies the inclusion

$$(I - \widetilde{\alpha})(E_1) \subset E_1. \tag{10.10}$$

Let $y \in E_1$. It follows from (10.10) that $(I - \widetilde{\alpha})y \in E_1$. In view of (10.1), we get from (9.35) that

$$2\widetilde{\alpha}y \in E_2. \tag{10.11}$$

It is easy to see that (10.5) and (10.8) imply that $2\alpha(K_2) = K_2$. Taking into account (10.11) and applying Lemma 7.10 to the compact subgroup K_2 and the continuous endomorphism 2α, we get that $y \in E_2$.

So, we proved that $E_1 \subset E_2$. In view of (10.9), this implies that $E_1 = E_2$. By Theorem 1.8, $K_1 = K_2 = K$. The equality $\alpha(K) = K$ follows from (10.8). □

THEOREM 10.2. *Let X be a finite Abelian group containing no elements of order 2. Let α be an automorphism of X satisfying the condition*

$$\mathrm{Ker}(I + \alpha) = \{0\}. \tag{i}$$

Let ξ_1 and ξ_2 be independent random variables with values in the group X and distributions μ_1 and μ_2. If the conditional distribution of the linear form $L_2 = \xi_1 + \alpha\xi_2$ given $L_1 = \xi_1 + \xi_2$ is symmetric, then $\mu_j = m_K * E_{x_j}$, where K is a subgroup of X and $x_j \in X$, $j = 1, 2$. Moreover, $\alpha(K) = K$.

PROOF. Since X is a finite group, it follows from condition (i) that $(I+\alpha)$ is an automorphism of X. By Corollary 9.7, the linear forms $M_1 = (I+\alpha)\xi_1 + 2\alpha\xi_2$ and $M_2 = 2\xi_1 + (I+\alpha)\xi_2$ are independent. Since the group X has no elements of order 2, multiplication by 2 is an automorphism of X. Hence all coefficients of the linear forms M_j are automorphisms of the group X. By Theorem 7.1, $\mu_1, \mu_2 \in I(X)$. This means that $\mu_j = m_{K_j} * E_{x_j}$, where K_j are subgroups of X, $x_j \in X$, $j = 1, 2$.

By Corollary 9.2, the characteristic functions $\hat{\mu}_j(y)$ satisfy equation (9.3). This implies that the characteristic functions $\widehat{m}_{K_j}(y)$ also satisfy equation (9.3). Let ζ_j be independent random variables with values in the group X and distributions m_{K_j}, $j = 1, 2$. By Corollary 9.2, the conditional distribution of the linear form $N_2 = \zeta_1 + \alpha\zeta_2$ given $N_1 = \zeta_1 + \zeta_2$ is symmetric. By Lemma 10.1, $K_1 = K_2 = K$ and $\alpha(K) = K$. □

We will prove that Theorem 10.2 holds true for discrete Abelian groups. We need some lemmas.

LEMMA 10.3. *Let X be a discrete torsion Abelian group with character group Y. Assume that X contains no elements of order 2. Let α be an automorphism of the group X satisfying condition* (i) *of Theorem* 10.2. *Let ξ_1 and ξ_2 be independent random variables with values in X and distributions μ_1 and μ_2 such that $\hat{\mu}_j(y) \geq 0$ for all $y \in Y$, $j = 1, 2$. If the conditional distribution of the linear form $L_2 = \xi_1 + \alpha\xi_2$ given $L_1 = \xi_1 + \xi_2$ is symmetric, then $\hat{\mu}_1(y) = \hat{\mu}_2(y) = 1$ for all $y \in B$, where B is an open subgroup of the group Y.*

PROOF. By Corollary 9.2, the characteristic functions $\hat{\mu}_j(y)$ satisfy equation (9.3). By Lemma 9.6, this implies that the characteristic functions $\hat{\mu}_j(y)$ satisfy equation (9.17). By Theorems 1.2 and 1.5, Y is a compact totally disconnected Abelian group. By Theorem 1.17, any neighborhood of the zero in the group Y contains an open subgroup. Hence we can choose an open subgroup U of the group Y such that $\hat{\mu}_j(y) > 0$ for all $y \in U$, $j = 1, 2$. Put $\psi_j(y) = -\ln \hat{\mu}_j(y)$, $y \in U$, $j = 1, 2$. Since $\hat{\mu}_j(y) \leq 1$, it follows from this that

$$\psi_j(y) \geq 0, \quad y \in U, \; j = 1, 2. \tag{10.12}$$

Put $V = U \cap \widetilde{\alpha}^{-1}(U) \cap \widetilde{\alpha}^{-2}(U)$. Then V is an open subgroup of Y. This implies that V is a closed subgroup and hence compact. It follows from (9.17) that the functions $\psi_j(y)$ satisfy equation (9.18) on the group V and (9.19) holds. Arguing as in the proof of Theorem 9.9 and retaining the same notation, we obtain that the function $P(y)$ satisfies equation (9.22) on the group V. Put $W_P = V \cap V^{(2)} \cap (I + \widetilde{\alpha})(V)$. Since h, h_1 and h_2 in equation (9.22) are arbitrary elements of the group V, it follows from (9.22) that the function $P(y)$ satisfies the equation

$$\Delta_h^3 P(y) = 0, \quad y, h \in W_P. \tag{10.13}$$

Put $W_Q = V \cap (I + \widetilde{\alpha})(V) \cap (\widetilde{\alpha}(V))^{(2)}$. Reasoning similarly, we obtain that the function $Q(y)$ satisfies the equation

$$\Delta_h^3 Q(y) = 0, \quad y, h \in W_Q. \tag{10.14}$$

Set

$$W = W_P \cap W_Q = V^{(2)} \cap (I + \widetilde{\alpha})(V) \cap (\widetilde{\alpha}(V))^{(2)}.$$

Since V is compact, W is a compact subgroup of the group Y. It follows from (10.13) and (10.14) that $P(y)$ and $Q(y)$ are continuous polynomials on the group W. In view of $P(0) = Q(0) = 0$, by Proposition 1.30, we have

$$P(y) = Q(y) = 0, \quad y \in W. \tag{10.15}$$

Put $B = (I + \widetilde{\alpha})(W)$. It follows from (9.19) and (10.15) that $\psi_1(y) = \psi_2(y) = 0$ for all $y \in B$. Hence

$$\hat{\mu}_1(y) = \hat{\mu}_1(y) = 1, \quad y \in B.$$

In view of condition (i) of Theorem 10.2, it follows from statement (b) of Theorem 1.20 that the subgroup $(I + \widetilde{\alpha})(Y)$ is dense in Y. Taking into account that Y is a compact group, we have

$$(I + \widetilde{\alpha})(Y) = Y.$$

By Theorem 1.23, it follows from this that the continuous endomorphism $I + \widetilde{\alpha}$ is open. Hence $(I + \widetilde{\alpha})(V)$ is an open subgroup of Y. Since X is a torsion group containing no elements of order 2, we have $f_2 \in \mathrm{Aut}(X)$. By statement (c) of Theorem 1.20, $f_2 \in \mathrm{Aut}(Y)$. Hence $V^{(2)}$ and $(\widetilde{\alpha}(V))^{(2)}$ are open subgroups of Y. This implies that W is an open subgroup of Y. Hence B is also open subgroup of Y. □

LEMMA 10.4. *Let Y be a compact connected Abelian group and let W be a neighborhood of the zero in Y. Let $P(y)$ be a continuous function satisfying the equation*

$$\Delta_h^k P(y) = 0, \quad y, h \in W, \tag{10.16}$$

for some natural k and the condition $P(0)=0$. Let $\hat{\mu}(y)$ be a characteristic function on the group Y such that the representation

$$\hat{\mu}(y)=e^{P(y)}, \quad y\in W,$$

holds. Then $\mu=E_0$.

PROOF. Since Y is a compact Abelian group, by Theorem 1.18, every neighborhood of the zero W contains a compact subgroup H of Y such that the factor-group Y/H is topologically isomorphic to a group of the form $\mathbb{T}^m\times F$, where $m\ge 0$ and F is a finite Abelian group. Since Y is a connected group, $F=\{0\}$, i.e., $Y/H\cong\mathbb{T}^m$. Let $p_1: Y\to Y/H$ be the natural homomorphism and let $p_2: Y/H\to\mathbb{T}^m$ be the above-mentioned isomorphism. Let $p=p_2p_1$ be the continuous epimorphism $p: Y\to\mathbb{T}^m$. By Theorem 1.23, p is an open epimorphism. Hence $p(W)$ is a neighborhood of the zero in the group $\mathbb{T}^m$. Denote by $t=(t_1,\dots,t_m)$, where $-\pi\le t_j<\pi$, elements of the group $\mathbb{T}^m$. The operation in $\mathbb{T}^m$ is the coordinate-wise addition modulo 2π. Consider the restriction of equation (10.16) to the subgroup H. The function $P(y)$ is a continuous polynomial on the group H. Since H is a compact group, by Proposition 1.30, $P(y)=\text{const}$ for all $y\in H$. Hence $P(y)=0$ for all $y\in H$ and therefore $\hat{\mu}(y)=1$ for all $y\in H$. By Proposition 2.10, it follows from this that $\hat{\mu}(y+h)=\hat{\mu}(y)$ for all $y\in Y$, $h\in H$. Thus, the function $\hat{\mu}(y)$ induces a positive definite function $f(t)$ on the group $\mathbb{T}^m$ by the formula $f(t)=\hat{\mu}(y)$, where $t=py$. In view of $\mathbb{T}^m\cong(\mathbb{Z}^m)^*$, by Bochner's theorem, there exists a distribution $\lambda\in \mathrm{M}^1(\mathbb{Z}^m)$ such that $\hat{\lambda}(t)=f(t)$ for all $t\in\mathbb{T}^m$. Moreover, in the neighborhood of the zero $p(W)$ in the group $\mathbb{T}^m$ we have the representation

$$f(t)=\exp\{\widetilde{P}(t)\}, \quad t\in p(W), \tag{10.17}$$

where $\widetilde{P}(t)=P(y)$, $t=py$. It follows from (10.16) that $\widetilde{P}(t)$ is an ordinary polynomial of m variables $t_1,\dots,t_m$. Inasmuch as $\mathbb{Z}^m\subset\mathbb{R}^m$, we can consider λ as a distribution on $\mathbb{R}^m$ supported in $\mathbb{Z}^m$, i.e., we assume that the function $f(t)$ is defined on $\mathbb{R}^m$ and $f(t)$ is a 2π-periodic function in each variable. It follows from (10.17) that the left-hand side of (10.17) is an entire function, and representation (10.17) is valid for all $t\in\mathbb{R}^m$ [**73**, Chapter VI, §1]. Since $\widetilde{P}(t)$ is a continuous polynomial and $\widetilde{P}(t)$ is a 2π-periodic function in each variable, we obtain that $\widetilde{P}(t)=\text{const}$ for all $t\in\mathbb{R}^m$. Hence $\widetilde{P}(t)=0$ for all $t\in\mathbb{R}^m$. This implies that $f(t)=1$ for all $t\in\mathbb{R}^m$ and therefore, $\hat{\mu}(y)=1$ for all $y\in Y$. Thus $\mu=E_0$. □

LEMMA 10.5. *Let X be a discrete torsion-free Abelian group. Let α be an automorphism of X satisfying condition* (i) *of Theorem* 10.2. *Let ξ_1 and ξ_2 be independent random variables with values in the group X and distributions μ_1 and μ_2. If the conditional distribution of the linear form $L_2=\xi_1+\alpha\xi_2$ given $L_1=\xi_1+\xi_2$ is symmetric, then μ_1 and μ_2 are degenerate distributions.*

PROOF. Denote by Y the character group of the group X. By Corollary 9.2, the characteristic functions $\hat{\mu}_j(y)$ satisfy equation (9.3). Put $\nu_j=\mu_j*\bar{\mu}_j$. This implies that $\hat{\nu}_j(y)=|\hat{\mu}_j(y)|^2\ge 0$ for all $y\in Y$, $j=1,2$. The characteristic functions $\hat{\nu}_j(y)$ also satisfy equation (9.3). Thus in the course of the proof of the lemma, we may suppose without loss of generality that $\hat{\mu}_j(y)\ge 0$ for all $y\in Y$, $j=1,2$. We will prove that in this case $\mu_1=\mu_2=E_0$.

Let U be a neighborhood of the zero in the group Y such that $\hat{\mu}_j(y)>0$ for $y\in U$, $j=1,2$. Put $\varphi_j(y)=-\ln\hat{\mu}_j(y)$, $y\in U$. Let V be a symmetric neighborhood of the zero in Y such that for any automorphism $\lambda_j \in \{I, \widetilde{\alpha}\}$ the inclusion

$$\sum_{j=1}^{8} \lambda_j(V) \subset U \tag{10.18}$$

holds. It follows from equation (9.3) that the functions $\varphi_j(y)$ satisfy equation (9.63), where $\beta = \widetilde{\alpha}$, when $u, v \in V$. By Lemma 9.17, the functions $\varphi_1(y)$ and $\varphi_2(y)$ satisfy the equations

$$\Delta_{(I-\widetilde{\alpha})k_3}\Delta_{2k_2}\Delta_{(I+\widetilde{\alpha})k_1}\varphi_1(u) = 0, \quad u \in V, \tag{10.19}$$

and

$$\Delta_{-(I-\widetilde{\alpha})k_3}\Delta_{(I+\widetilde{\alpha})k_2}\Delta_{2\widetilde{\alpha}k_1}\varphi_2(u) = 0, \quad u \in V. \tag{10.20}$$

Inasmuch as (10.18) holds, arguments of the functions $\varphi_j(y)$ are in U.

By Theorems 1.2 and 1.3, Y is a compact connected Abelian group. In view of condition (i) of Theorem 10.2, it follows from statement (b) of Theorem 1.20 that the subgroup $(I + \widetilde{\alpha})(Y)$ is dense in Y. Taking into account that Y is a compact group, we have $(I + \widetilde{\alpha})(Y) = Y$. By Theorem 1.23, it follows from this that the continuous epimorphism $I + \widetilde{\alpha}$ is open. Hence $(I + \widetilde{\alpha})(V)$ is a neighborhood of the zero in Y. Since Y is a connected group, by Theorem 1.13, $Y^{(n)} = Y$ for all natural n. In particular, $Y^{(2)} = Y$. By Theorem 1.23, this implies that $f_2 : Y \to Y$ is an open epimorphism. Hence $V^{(2)}$ is also a neighborhood of the zero in the group Y.

Put $L = (I - \widetilde{\alpha})(Y)$. Then L a compact connected subgroup of the group Y. By Theorem 1.23, the continuous homomorphism $I - \widetilde{\alpha}$, as an epimorphism from Y to L, is open. It follows from this that $(I - \widetilde{\alpha})(V)$ is a neighborhood of the zero in the group L. Taking into account (10.19), it follows from what has been said that there exists a neighborhood of the zero W in the group L such that the function $\varphi_1(y)$ satisfies the equation

$$\Delta_h^3\varphi_1(y) = 0, \quad y, h \in W. \tag{10.21}$$

Considering the group L instead of Y and applying Lemma 10.4 to the group L, we obtain from (10.21) that $\hat{\mu}_1(y) = 1$ for all $y \in L$. Hence by Proposition 2.10, $\sigma(\mu_1) \subset A(X, L)$. Reasoning similarly, we get from (10.20) that $\sigma(\mu_2) \subset A(X, L)$.

Put $G = A(X, L)$. Since $L = (I - \widetilde{\alpha})(Y)$, it follows from Theorem 1.19 and statement (a) of Theorem 1.20 that $G = \mathrm{Ker}(I - \alpha)$. It is obvious that $\alpha(G) = G$. Inasmuch as $\sigma(\mu_j) \subset G$, $j = 1, 2$, the random variables ξ_1 and ξ_2 take values in G. Denote by N the character group of the group G. The characteristic functions of the distributions μ_j, considered as distributions on G, satisfy equation (9.3) on N. Moreover, since the restriction of α to G is the identity automorphism, equation (9.3) on the group N takes the form

$$\hat{\mu}_1(u + v)\hat{\mu}_2(u + v) = \hat{\mu}_1(u - v)\hat{\mu}_2(u - v), \quad u, v \in N. \tag{10.22}$$

Substituting $u = v = y$ in equation (10.22) and taking into account that $\hat{\mu}_j(y) \geq 0$, we obtain

$$\hat{\mu}_1(2y) = \hat{\mu}_2(2y) = 1, \quad y \in N. \tag{10.23}$$

Since G is a discrete torsion-free Abelian group, by Theorem 1.3, N is a compact connected Abelian group. Hence by Theorem 1.13, $N^{(n)} = N$ for all natural n. In particular, $N^{(2)} = N$. It follows from (10.23) that then $\hat{\mu}_1(y) = \hat{\mu}_2(y) = 1$ for all $y \in N$. This implies that $\mu_1 = \mu_2 = E_0$. □

Taking into account that for a discrete Abelian group X the conditions $\mathrm{Ker}(I+\alpha)=\{0\}$ and $(I+\widetilde{\alpha})(Y)=Y$ are equivalent, the following statement follows from Corollary 9.2 and Lemma 10.5.

COROLLARY 10.6. *Let Y be a compact connected Abelian group. Let $\widetilde{\alpha}$ be a topological automorphism of Y satisfying the condition*

$$\text{(i)} \qquad (I+\widetilde{\alpha})(Y)=Y.$$

If $\hat{\mu}_j(y)$ are characteristic functions on the group Y satisfying equation (9.3), then $\hat{\mu}_j(y)=(x_j,y)$, where $x_j\in X$, $j=1,2$.

LEMMA 10.7. *Let X be a discrete Abelian group. Let α be an automorphism of X satisfying the condition*

$$\text{(i)} \qquad \mathrm{Ker}(I+\alpha)\subset b_X.$$

Let ξ_1 and ξ_2 be independent random variables with values in the group X and distributions μ_1 and μ_2. Assume that the conditional distribution of the linear form $L_2=\xi_1+\alpha\xi_2$ given $L_1=\xi_1+\xi_2$ is symmetric. Then we can replace the distributions μ_j by their shifts λ_j such that the distributions λ_j are supported in the subgroup b_X. Moreover, if η_j are independent random variables with values in the group b_X and distributions λ_j, then the conditional distribution of the linear form $M_2=\eta_1+\alpha\eta_2$ given $M_1=\eta_1+\eta_2$ is symmetric.

PROOF. Denote by Y the character group of the group X. By Theorem 1.2, Y is a compact Abelian group. Inasmuch as X is a discrete group, b_X is the subgroup consisting of all elements of finite order of the group X. Consider the factor-group X/b_X and denote by $[x]$ its elements. By Theorem 1.9, the character group of the group X/b_X is topologically isomorphic to the annihilator $A(Y,b_X)$. By Theorem 1.10, $A(Y,b_X)=c_Y$. It follows from $\alpha(b_X)=b_X$ that α induces an automorphism $\hat{\alpha}$ of the factor-group X/b_X by the formula $\hat{\alpha}[x]=[\alpha x]$. We note that $\widetilde{\alpha}(c_Y)=c_Y$, and the topological automorphism of the group c_Y which is adjoint to $\hat{\alpha}$ is the restriction of $\widetilde{\alpha}$ to c_Y. Verify that

$$(10.24) \qquad (I+\widetilde{\alpha}_{c_Y})(c_Y)=c_Y.$$

Taking into account that for a discrete Abelian group X condition (i) of Theorem 10.2 and condition (i) of Corollary 10.6 are equivalent, it suffices to verify that

$$(10.25) \qquad \mathrm{Ker}(I+\hat{\alpha})=\{0\}.$$

Let $x_0\in X$ and $[x_0]\in\mathrm{Ker}(I+\hat{\alpha})$. Then $[(I+\alpha)x_0]=0$ and this implies that $(I+\alpha)x_0\in b_X$, i.e., $n(I+\alpha)x_0=0$ for some nonzero integer n. Hence $(I+\alpha)nx_0=0$. Taking into account (i), this implies that $nx_0\in b_X$. Thus, $x_0\in b_X$. So, (10.25) holds and hence (10.24) is proved.

By Corollary 9.2, the characteristic functions $\hat{\mu}_j(y)$ satisfy equation (9.3). Consider the restriction of equation (9.3) to c_Y. Applying Corollary 10.6 to the group c_Y, we get that the restrictions of the characteristic functions $\hat{\mu}_j(y)$ to the subgroup c_Y are characters of the subgroup c_Y. By Theorem 1.9, there exist elements $x_j\in X$, $j=1,2$, such that

$$(10.26) \qquad \hat{\mu}_j(y)=(x_j,y), \quad y\in c_Y,\ j=1,2,$$

hold. By Theorem 1.10, $A(X, c_Y) = b_X$. Substituting (10.26) into equation (9.3), we get

$$2(x_1 + \alpha x_2) \in b_X. \tag{10.27}$$

Since b_X consists of all elements of finite order of the group X, it follows from (10.27) that

$$x_1 + \alpha x_2 \in b_X. \tag{10.28}$$

Put $\tilde{x}_1 = -\alpha x_2$, $\tilde{x}_2 = x_2$. It is easy to see that (10.26) and (10.28) imply that

$$\hat{\mu}_j(y) = (\tilde{x}_j, y), \quad y \in c_Y, \ j = 1, 2. \tag{10.29}$$

Inasmuch as $\tilde{x}_1 + \alpha\tilde{x}_2 = 0$, the characteristic functions $f_j(y) = (-\tilde{x}_j, y)$, $j = 1, 2$, on the group Y satisfy equation (9.3).

Put

$$\lambda_j = \mu_j * E_{-\tilde{x}_j}. \quad j = 1, 2. \tag{10.30}$$

Then the characteristic functions $\hat{\lambda}_j(y)$ satisfy equation (9.3). Let η_j be independent random variables with values in the group X and distributions λ_j. Then by Corollary 9.2, the conditional distribution of the linear form $M_2 = \eta_1 + \alpha\eta_2$ given $M_1 = \eta_1 + \eta_2$ is symmetric. Moreover, (10.29) and (10.30) imply that $\hat{\lambda}_j(y) = 1$ for all $y \in c_Y$. It follows from Proposition 2.10 that $\sigma(\lambda_j) \subset A(X, c_Y)$. By Theorem 1.10, $A(X, c_Y) = b_X$.

It is easy to see that the lemma is false if condition (i) is omitted. □

THEOREM 10.8. *Let X be a discrete Abelian group containing no elements of order 2. Let α be an automorphism of X satisfying condition* (i) *of Theorem* 10.2. *Let ξ_1 and ξ_2 be independent random variables with values in the group X and distributions μ_1 and μ_2. If the conditional distribution of the linear form $L_2 = \xi_1 + \alpha\xi_2$ given $L_1 = \xi_1 + \xi_2$ is symmetric, then $\mu_j = m_K * E_{x_j}$, where K is a finite subgroup of X and $x_j \in X$, $j = 1, 2$. Moreover, $\alpha(K) = K$.*

PROOF. Denote by Y the character group of the group X. Taking into account that $\alpha(b_X) = b_X$ and condition (i) of Theorem 10.2 is satisfied, we can apply Lemma 10.7 and assume X is a torsion group. By Corollary 9.2, the characteristic functions $\hat{\mu}_j(y)$ satisfy equation (9.3). Put $\nu_j = \mu_j * \bar{\mu}_j$. This implies that $\hat{\nu}_j(y) = |\hat{\mu}_j(y)|^2 \geq 0$ for all $y \in Y$. The characteristic functions $\hat{\nu}_j(y)$ also satisfy equation (9.3). If we prove that $\nu_1 = \nu_2 = m_K$, where K is a finite subgroup of the group X, this easily implies that $\mu_j = m_K * E_{x_j}$, where $x_j \in X$, $j = 1, 2$. Therefore, we can solve equation (9.3) supposing that $\hat{\mu}_j(y) \geq 0$. So, we will assume that.

Put $f(y) = \hat{\mu}_1(y)$, $g(y) = \hat{\mu}_2(y)$. Then equation (9.3) takes the form (9.31). We will prove that in this case $f(y) = g(y) = \hat{m}_K(y)$, where K is a finite subgroup of the group X.

Set $E_f = \{y \in Y : f(y) = 1\}$, $E_g = \{y \in Y : g(y) = 1\}$, $F = A(X, E_f)$, $G = A(X, E_g)$. It follows from Proposition 2.10 that $\sigma(\mu_1) \subset F$ and $\sigma(\mu_2) \subset G$. Since condition (i) of Theorem 10.2 is satisfied, it follows from Lemma 10.3 that there exists an open subgroup B of the group Y such that $B \subset E_f \cap E_g$. Put $S = A(X, B)$. Then F and G are subgroups of S. By Theorem 1.11, since B is an open subgroup, S is a compact subgroup. Taking into account that X is a discrete group, this implies that S is a finite subgroup. Hence F and G are also finite subgroups.

It follows from (9.31) that the equation

$$f^n(u+v)g^n(u+\widetilde{\alpha}v) = f^n(u-v)g^n(u-\widetilde{\alpha}v), \quad u,v \in Y \tag{10.31}$$

is true for all natural n. It is obvious that there exist limits

$$\bar{f}(y) = \lim_{n\to\infty} f^n(y) = \begin{cases} 1, & \text{if } y \in E_f, \\ 0, & \text{if } y \notin E_f, \end{cases} \tag{10.32}$$

$$\bar{g}(y) = \lim_{n\to\infty} g^n(y) = \begin{cases} 1, & \text{if } y \in E_g, \\ 0, & \text{if } y \notin E_g. \end{cases}$$

By Theorem 1.8, $E_f = A(Y,F)$ and $E_g = A(Y,G)$. In view of (2.3), we have

$$\widehat{m}_F(y) = \begin{cases} 1, & \text{if } y \in E_f, \\ 0, & \text{if } y \notin E_f, \end{cases} \qquad \widehat{m}_G(y) = \begin{cases} 1, & \text{if } y \in E_g, \\ 0, & \text{if } y \notin E_g. \end{cases}$$

Hence

$$\widehat{m}_F(y) = \bar{f}(y), \quad \widehat{m}_G(y) = \bar{g}(y).$$

Let ζ_1 and ζ_2 be independent random variables with values in the group X and distributions m_F and m_G. It follows from (10.31) and (10.32) that the characteristic functions $\bar{f}(y)$ and $\bar{g}(y)$ also satisfy equation (9.31). By Corollary 9.2, this implies that the conditional distribution of the linear form $L_2 = \zeta_1+\alpha\zeta_2$ given $L_1 = \zeta_1+\zeta_2$ is symmetric. Applying Lemma 10.1, we obtain that $F = G$ and $\alpha(F) = F$.

We now return to the random variables ξ_1 and ξ_2 and the linear forms $L_1 = \xi_1 + \xi_2$ and $L_2 = \xi_1 + \alpha\xi_2$. Inasmuch as $\sigma(\mu_j) \subset F$, the random variables ξ_j take values in the finite group F. Taking into account that $\alpha(F) = F$ and that α satisfies condition (i) of Theorem 10.2, we can apply Theorem 10.2 to the group F. Since $\hat{\mu}_j(y) \geq 0$, by Theorem 10.2, $\mu_1 = \mu_2 = m_K$, where K is a finite subgroup of the group X, and $\alpha(K) = K$. □

We complement Theorem 10.8 with the following statement.

Proposition 10.9. *Let X be a locally compact Abelian group containing no elements of order 2, let K be a finite subgroup of X, and let α be a topological automorphism of X. Let ξ_1 and ξ_2 be independent identically distributed random variables with values in the group X and distribution m_K. Then the following statements are equivalent:*

(i) *the conditional distribution of the linear form $L_2 = \xi_1 + \alpha\xi_2$ given $L_1 = \xi_1 + \xi_2$ is symmetric;*
(ii) *$(I-\alpha)(K) = K$.*

Proof. Denote by Y the character group of the group X. Assume that (i) is true. The characteristic function $\hat{\mu}(y)$ is of the form (2.3). By Corollary 9.2, the characteristic function $\hat{\mu}(y)$ satisfies the equation

$$\hat{\mu}(u+v)\hat{\mu}(u+\widetilde{\alpha}v) = \hat{\mu}(u-v)\hat{\mu}(u-\widetilde{\alpha}v), \quad u,v \in Y. \tag{10.33}$$

Put $u = v = y$ in (10.33). We get

$$\hat{\mu}(2y)\hat{\mu}((I+\widetilde{\alpha})y) = \hat{\mu}((I-\widetilde{\alpha})y), \quad y \in Y. \tag{10.34}$$

Assume that $(I-\widetilde{\alpha})y \in A(Y,K)$. Then it follows from (2.3) and (10.34) that $2y \in A(Y,K)$. Since the group X contains no elements of order 2 and K is a finite subgroup, K is a Corwin group. This easily implies that $y \in A(Y,K)$. By Lemma

7.10 applying to $I-\alpha$, we obtain $(I-\alpha)(K)\supset K$. Inasmuch in K is a finite group, we receive (ii).

Assume that (ii) is true. We shall verify that $\hat\mu(y)$ satisfies equation (10.33). Assume that for some $u, v\in Y$ the left-hand side of equation (10.33) is equal to 1. Then $u+v, u+\widetilde{\alpha}v\in A(Y,K)$. This implies that $(I-\widetilde{\alpha})v\in A(Y,K)$. Taking into account (ii), by Lemma 7.10, applying to $I-\alpha$, we get $v\in A(Y,K)$. Hence, $u\in A(Y,K)$ and then $\widetilde{\alpha}v\in A(Y,K)$. It follows from this that $u-v, u-\widetilde{\alpha}v\in A(Y,K)$. We get that the right-hand side of equation (10.33) is also equal to 1. Reasoning similarly we check that if the right-hand side of equation (10.33) is equal to 1, then the left-hand side of equation (10.33) is equal to 1. Thus the characteristic function $\hat\mu(y)$ satisfies equation (10.33). By Corollary 9.2, (i) holds.

Note that the proof of the statement (ii)$\Rightarrow$(i) is based on Lemma 7.10 and Corollary 9.2 only. Therefore, the statement (ii)$\Rightarrow$ (i) also holds in the case when K is a compact subgroup of X. □

Now we describe compact totally disconnected Abelian groups X for which for two independent random variables, Heyde's theorem holds. We need some lemmas.

LEMMA 10.10. *Let p, where $p>2$, be a prime number. Consider the group*

$$X=\mathop{\mathbf{P}}_{n=1}^{\infty}\mathbb{Z}(p^{k_n}),\quad k_n\le k_{n+1},\quad n=1,2,\dots \tag{10.35}$$

Then there exist a topological automorphism α of the group X satisfying condition (i) of Theorem 10.2 and independent identically distributed random variables ξ_1 and ξ_2 with values in X and distribution μ such that the conditional distribution of the linear form $L_2=\xi_1+\alpha\xi_2$ given $L_1=\xi_1+\xi_2$ is symmetric, while $\mu\notin I(X)$.

PROOF. Denote by Y the character group of the group X. By Theorem 1.7, the group Y is topologically isomorphic to a weak direct product of the groups $\mathbb{Z}(p^{k_n})$. In order not to complicate the notation suppose

$$Y=\mathop{\mathbf{P}^*}_{n=1}^{\infty}\mathbb{Z}(p^{k_n}). \tag{10.36}$$

We denote by $t=\{t_n\}_{n=1}^\infty$, where $t_n\in\mathbb{Z}(p^{k_n})$, elements of the group X, and by $s=\{s_n\}_{n=1}^\infty$, where $s_n\in\mathbb{Z}(p^{k_n})$, elements of the group Y. Let $s=\{s_n\}_{n=1}^\infty\in Y$. We say that $s\in\mathbb{Z}(p^{k_1})$ if $s_2=s_3=\dots=0$.

Let $i\le j$. Denote by $\alpha_{i,j}$ the homomorphism $\alpha_{i,j}:\mathbb{Z}(p^{k_i})\to\mathbb{Z}(p^{k_j})$ of the form

$$\alpha_{i,j}t_i=p^{k_j-k_i}t_i,\quad t_i\in\mathbb{Z}(p^{k_i}).$$

It is easy to verify that the adjoint homomorphism $\widetilde{\alpha}_{i,j}:\mathbb{Z}(p^{k_j})\to\mathbb{Z}(p^{k_i})$ is of the form

$$\widetilde{\alpha}_{i,j}s_j=s_j(\text{mod }p^{k_i}),\quad s_j\in\mathbb{Z}(p^{k_j}).$$

Define a continuous homomorphism $\alpha: X\to X$ by the formula $\alpha\{t_n\}_{n=1}^\infty=\{h_n\}_{n=1}^\infty$, where

$$h_n=\begin{cases}(p^{k_1}-t_1)(\text{mod }p^{k_1}), & \text{if } n=1,\\ (\alpha_{n-1,n}t_{n-1}+p^{k_n}-t_n)(\text{mod }p^{k_n}), & \text{if } n\ge 2.\end{cases} \tag{10.37}$$

It is easy to see that α is a topological automorphism of the group X satisfying condition (i) of Theorem 10.2. We find from (10.37) that the adjoint automorphism

$\widetilde{\alpha}: Y \to Y$ is of the form $\widetilde{\alpha}\{s_n\}_{n=1}^{\infty} = \{u_n\}_{n=1}^{\infty}$, where

(10.38) $$u_n = (\widetilde{\alpha}_{n,n+1}s_{n+1} + p^{k_n} - s_n)(\text{mod } p^{k_n}), \quad n \geq 1.$$

It follows from (10.35) and (10.36) that $X = \mathbb{Z}(p^{k_1}) \times G$, where $G = \mathop{\mathbf{P}}_{n=2}^{\infty} \mathbb{Z}(p^{k_n})$ and $Y = \mathbb{Z}(p^{k_1}) \times H$, where $H = \mathop{\mathbf{P}^*}_{n=2}^{\infty} \mathbb{Z}(p^{k_n})$. Obviously, $A(Y, G) = \mathbb{Z}(p^{k_1})$. Take $0 < a < 1$. Consider the distribution $\mu = am_X + (1-a)m_G$ on the group X. It follows from (2.3) that the characteristic function $\hat{\mu}(y)$ is of the form

(10.39) $$\hat{\mu}(y) = \begin{cases} 1, & \text{if } \ y = 0, \\ 1 - a, & \text{if } \ y \in \mathbb{Z}(p^{k_1})\backslash\{0\}, \\ 0, & \text{if } \ y \notin \mathbb{Z}(p^{k_1}). \end{cases}$$

Let ξ_1 and ξ_2 be independent identically distributed random variables with values in the group X and distribution μ. Let us check that the conditional distribution of the linear form $L_2 = \xi_1 + \alpha\xi_2$ given $L_1 = \xi_1 + \xi_2$ is symmetric. By Corollary 9.2, it suffices to verify that the characteristic functions of the random variables ξ_j satisfy equation (9.3) which takes the form

(10.40) $$\hat{\mu}(u + v)\hat{\mu}(u + \widetilde{\alpha}v) = \hat{\mu}(u - v)\hat{\mu}(u - \widetilde{\alpha}v), \quad u, v \in Y.$$

Let $u, v \in Y$. Consider three cases:

1. $u, v \in \mathbb{Z}(p^{k_1})$. It follows from (10.38) that $\widetilde{\alpha}y = -y$ when $y \in \mathbb{Z}(p^{k_1})$. Hence the restriction of equation (10.40) to the subgroup $\mathbb{Z}(p^{k_1})$ takes the form

$$\hat{\mu}(u + v)\hat{\mu}(u - v) = \hat{\mu}(u - v)\hat{\mu}(u + v), \quad u, v \in \mathbb{Z}(p^{k_1}).$$

It is obvious that in this case (10.40) is fulfilled.

2. Either $u \in \mathbb{Z}(p^{k_1})$, $v \notin \mathbb{Z}(p^{k_1})$ or $u \notin \mathbb{Z}(p^{k_1})$, $v \in \mathbb{Z}(p^{k_1})$. Then $u \pm v \notin \mathbb{Z}(p^{k_1})$. It follows from (10.39) that $\hat{\mu}(u \pm v) = 0$. Then both sides of equation (10.40) are equal to zero.

3. $u, v \notin \mathbb{Z}(p^{k_1})$. Assume that the left-hand side in equation (10.40) is not equal to zero. Then (10.39) implies that $u + v, u + \widetilde{\alpha}v \in \mathbb{Z}(p^{k_1})$. It follows from this that

(10.41) $$(I - \widetilde{\alpha})v \in \mathbb{Z}(p^{k_1}).$$

Let $v = \{v_n\}_{n=1}^{\infty}$, $v_n \in \mathbb{Z}(p^{k_n})$. We find from (10.38) and (10.41) that

(10.42) $$(2v_n + p^{k_n} - \widetilde{\alpha}_{n,n+1}v_{n+1})(\text{mod } p^{k_n}) = 0, \ n \geq 2.$$

It follows from $p > 2$ that if $v_n \neq 0$, then $2v_n(\text{mod } p^{k_n}) \neq 0$, $n \geq 1$. Assume that $v_2 \neq 0$. Then we find from (10.42) that $v_n \neq 0$ for each $n \geq 2$. But this contradicts the fact that $v_n = 0$ for all but a finite set of indices n. Hence $v_2 = 0$. Reasoning by induction we prove that $v_3 = 0$, then $v_4 = 0$, and so on. Thus, we get that $v = \{v_n\}_{n=1}^{\infty} \in \mathbb{Z}(p^{k_1})$, contrary to the assumption. The obtained contradiction shows that the left-hand side of equation (10.40) is equal to zero. Similarly, we prove that when $u, v \notin \mathbb{Z}(p^{k_1})$ the right-hand side of equation (10.40)) is also equal to zero. So, both sides of equation (10.40) are equal to zero.

Thus, we proved that equation (10.40) becomes an equality for all $u, v \in Y$. □

LEMMA 10.11. *There exist a topological automorphism α of the group of p-adic integers Δ_p satisfying condition* (i) *of Theorem* 10.2 *and independent random variables ξ_1 and ξ_2 with values in Δ_p and distributions μ_1 and μ_2 such that the* *conditional distribution of the linear form* $L_2 = \xi_1 + \alpha\xi_2$ *given* $L_1 = \xi_1 + \xi_2$ *is symmetric, while* $\mu_j \notin I(\Delta_p)$, $j = 1, 2$.

The assertion of the lemma follows from Theorem 13.3. The proof of Theorem 13.3 does not depend on Lemma 10.11.

THEOREM 10.12. *Let* X *be a compact totally disconnected Abelian group containing no elements of order* 2. *Let* α *be a topological automorphism of* X *satisfying condition* (i) *of Theorem* 10.2. *Let* ξ_1 *and* ξ_2 *be independent random variables with values in the group* X *and distributions* μ_1 *and* μ_2. *The symmetry of the conditional distribution of the linear form* $L_2 = \xi_1 + \alpha\xi_2$ *given* $L_1 = \xi_1 + \xi_2$ *implies that* $\mu_j \in I(X)$, $j = 1, 2$, *if and only if the group* X *is topologically isomorphic to a group of the form*

$$\mathop{\mathbf{P}}_{p\in\mathcal{P},\ p>2} X_p, \tag{10.43}$$

where X_p *is a finite* p*-group. Moreover, if* X *is of the form* (10.43), *then* $\mu_j = m_K * E_{x_j}$, *where* K *is a compact subgroup of* X, $x_j \in X$, $j = 1, 2$ *and* $\alpha(K) = K$.

PROOF. Necessity. Assume that a compact totally disconnected Abelian group X contains no elements of order 2, and X is not topologically isomorphic to a group of the form (10.43). It follows from Lemma 7.21 that in this case for some prime number p, $p > 2$, there exists a compact subgroup K of the group X such that K is a topological direct factor of X, and K is topologically isomorphic to either the group of p-adic integers Δ_p or a group of the form (10.35). We have $X = K \times G$. Denote by $x = (k, g)$, where $k \in K$, $g \in G$, elements of the group X. Applying either Lemma 10.10 or Lemma 10.11, we find a topological automorphism α_K of the group K satisfying condition (i) of Theorem 10.2 and independent random variables ξ_1 and ξ_2 with values in K and distributions μ_1 and μ_2 such that the conditional distribution of the linear form $M_2 = \xi_1 + \alpha_K\xi_2$ given $M_1 = \xi_1 + \xi_2$ is symmetric, whereas $\mu_j \notin I(K)$, $j = 1, 2$. Put $\alpha(k, g) = (\alpha_K k, g)$. Since the group X contains no elements of order 2, α is a topological automorphism of X satisfying condition (i) of Theorem 10.2. If we consider ξ_j as independent random variables with values in the group X, then the conditional distribution of the linear form $L_2 = \xi_1 + \alpha\xi_2$ given $L_1 = \xi_1 + \xi_2$ is symmetric, whereas $\mu_j \notin I(X)$, $j = 1, 2$. The necessity is proved.

Sufficiency. Assume that a compact totally disconnected Abelian group X is topologically isomorphic to a group of the form (10.43). Let α be a topological automorphism of the group X satisfying condition (i) of Theorem 10.2. Let ξ_1 and ξ_2 be independent random variables with values in X and distributions μ_1 and μ_2. Denote by Y the character group of the group X. By Corollary 9.2, the symmetry of the conditional distribution of the linear form L_2 given L_1 implies that the characteristic functions $\hat\mu_j(y)$ satisfy equation (9.3). Put $\nu_j = \mu_j * \bar\mu_j$. This implies that $\hat\nu_j(y) = |\hat\mu_j(y)|^2 \geq 0$ for all $y \in Y$. The characteristic functions $\hat\nu_j(y)$ also satisfy equation (9.3).

By Theorem 1.7, it follows from (10.43) that

$$Y = \mathop{\mathbf{P}^*}_{p\in\mathcal{P},\ p>2} Y_p, \tag{10.44}$$

where the group Y_p is isomorphic to the character group of the group X_p. Put

$$A_m = \mathop{\mathbf{P}}_{p\in\mathcal{P},\ 2<p\leq m} X_p, \quad B_m = \mathop{\mathbf{P}}_{p\in\mathcal{P},\ 2<p\leq m} Y_p, \quad m = 3, 4, \ldots.$$

Then A_m and B_m are finite Abelian groups, B_m is isomorphic to the character group of the group A_m, $B_m \subset B_{m+1}$ and $Y = \bigcup_{m=3}^{\infty} B_m$. It is obvious that A_m and B_m are characteristic subgroups. Denote by α_m the restriction of α to the subgroup A_m. Then $\alpha_m \in \mathrm{Aut}(A_m)$ and α_m satisfies condition (i) of Theorem 10.2. Note that the restriction of $\widetilde{\alpha}$ to subgroup B_m coincides with $\widetilde{\alpha}_m$. Consider the restriction of equation (9.3) for the characteristic functions $\hat{\nu}_j(y)$ to the subgroup B_m. We have

$$\hat{\nu}_1(u+v)\hat{\nu}_2(u+\widetilde{\alpha}_m v) = \hat{\nu}_1(u-v)\hat{\nu}_2(u-\widetilde{\alpha}_m v), \quad u, v \in B_m. \tag{10.45}$$

Taking into account Corollary 9.2 and applying Theorem 10.2 to the finite group A_m, we obtain from (10.45) that $\hat{\nu}_1(y) = \hat{\nu}_2(y)$ for all $y \in B_m$, and the characteristic functions $\hat{\nu}_j(y)$ take only values 0 and 1 on B_m. This implies that $\hat{\nu}_1(y) = \hat{\nu}_2(y)$ for all $y \in Y$, and the characteristic functions $\hat{\nu}_j(y)$ take only values 0 and 1 on Y. Put $E = \{y \in Y : \hat{\nu}_j(y) = 1\}$. Then E is a subgroup of Y. Put $K = A(X, E)$. By Theorem 1.8, $E = A(Y, K)$. It follows from (2.3) and statement (ii) of Theorem 1.8 that $\nu_1 = \nu_2 = m_K$. This implies that $\mu_j = m_K * E_{x_j}$, where $x_j \in X$, $j = 1, 2$. Furthermore, we can assert that $\alpha(K) = K$. Indeed, put $E_m = E \cap B_m$. It follows from Theorem 10.2 that $\widetilde{\alpha}_m(E_m) = E_m$. Hence $\widetilde{\alpha}(E) = E$, so that $\alpha(K) = K$. The sufficiency is proved. □

Now we shall prove a theorem which generalizes both Theorems 10.8 and 10.12 and describe a wide class of locally compact Abelian groups where for two independent random variables an analogue of Heyde's theorem holds. We need the following lemma.

LEMMA 10.13. *Let $X = \mathbb{R}^n \times G$, where $n \geq 0$ and G is a compact totally disconnected Abelian group of the form* (10.43). *Let α be a topological automorphism of the group X satisfying condition* (i) *of Theorem* 10.2. *Let ξ_1 and ξ_2 be independent random variables with values in X and distributions μ_1 and μ_2. If the conditional distribution of the linear form $L_2 = \xi_1 + \alpha\xi_2$ given $L_1 = \xi_1 + \xi_2$ is symmetric, then $\mu_j = \gamma_j * m_K * E_{x_j}$, where $\gamma_j \in \Gamma(\mathbb{R}^n)$, K is a compact subgroup of G, $x_j \in G$, $j = 1, 2$. Moreover, $\alpha(K) = K$.*

PROOF. Denote by $x = (t, g)$, where $t \in \mathbb{R}^n$, $g \in G$, elements of the group X. Let Y be the character group of the group X. We can suppose that $Y = \mathbb{R}^n \times H$, where H is a group topologically isomorphic to the character group of the group G. Denote by $y = (s, h)$, where $s \in \mathbb{R}^n$, $h \in H$, elements of the group Y. Since $\mathbb{R}^n$ is the connected component of the zero of the group X, we have $\alpha(\mathbb{R}^n) = \mathbb{R}^n$. Inasmuch as G is the subgroup of X consisting of all compact elements of the group X, we have $\alpha(G) = G$. This implies that the restriction of α to each of the subgroups $\mathbb{R}^n$ and G is a topological automorphism of the corresponding subgroup. This means that α can be written in the form $\alpha(t, g) = (\alpha_{\mathbb{R}^n} t, \alpha_G g)$, $(t, g) \in X$. Observe that the restrictions of $\widetilde{\alpha}$ to the subgroups $\mathbb{R}^n$ and H coincide with $\widetilde{\alpha}_{\mathbb{R}^n}$ and $\widetilde{\alpha}_G$ respectively. Hence $\widetilde{\alpha}$ can be written in the form $\widetilde{\alpha}(s, h) = (\widetilde{\alpha}_{\mathbb{R}^n} s, \widetilde{\alpha}_G h)$, $(s, h) \in Y$.

By Corollary 9.2, the characteristic functions $\hat{\mu}_j(s, h)$ satisfy equation (9.3) which takes the form

$$\begin{aligned} &\hat{\mu}_1(s_1 + s_2, h_1 + h_2)\hat{\mu}_2(s_1 + \widetilde{\alpha}_{\mathbb{R}^n} s_2, h_1 + \widetilde{\alpha}_G h_2) \\ &\quad = \hat{\mu}_1(s_1 - s_2, h_1 - h_2)\hat{\mu}_2(s_1 - \widetilde{\alpha}_{\mathbb{R}^n} s_2, h_1 - \widetilde{\alpha}_G h_2), \quad s_j \in \mathbb{R}^n, \ h_j \in H. \end{aligned} \tag{10.46}$$

It is obvious that the topological automorphism $\alpha_{\mathbb{R}^n}$ satisfies condition (i) of Theorem 10.2.

Substituting $h_1 = h_2 = 0$ in equation (10.46) and taking into account Corollary 9.2 and Remark 9.8, we obtain from the resulting equation

$$\hat{\mu}_j(s,0) = \exp\{i\langle t_j, s\rangle - \langle A_j s, s\rangle\}, \quad s \in \mathbb{R}^n,\ j = 1,2, \tag{10.47}$$

where $t_j \in \mathbb{R}^n$ and A_j are symmetric positive semi-definite $n\times n$ matrices. Similarly, substituting $s_1 = s_2 = 0$ in equation (10.46), taking into account Corollary 9.2, and applying Theorem 10.12 to the group G, we obtain from the resulting equation that

$$\hat{\mu}_j(0,h) = \widehat{m}_K(h)(g_j, h), \quad h \in H,\ j = 1,2, \tag{10.48}$$

where K is a compact subgroup in G and $g_j \in G$. Moreover, $\alpha_G(K) = K$.

Substituting (10.48) into equation (10.46) and taking into account (2.3), we find that $2(g_1 + \alpha_G g_2) \in K$. Since G is a group of the form (10.43) and K is a compact subgroup of G, multiplication by 2 is a topological automorphism of K. Hence

$$g_1 + \alpha_G g_2 \in K. \tag{10.49}$$

Put $x_1 = -\alpha_G g_2$, $x_2 = g_2$. Then $x_1 + \alpha_G x_2 = 0$. It follows from (2.3), (10.48), and (10.49) that

$$\hat{\mu}_j(0,h) = \widehat{m}_K(h)(x_j, h), \quad h \in H,\ j = 1,2. \tag{10.50}$$

Consider the new independent random variables $\eta_j = \xi_j - x_j$. Denote by λ_j the distribution of the random variable η_j, $j = 1,2$. Since

$$\lambda_j = \mu_j * E_{-x_j}, \quad j = 1,2 \tag{10.51}$$

and $x_1 + \alpha x_2 = 0$, the characteristic functions $\hat{\lambda}_j(y)$ also satisfy equation (9.3). By Corollary 9.2, the conditional distribution of the linear form $N_2 = \eta_1 + \alpha\eta_2$ given $N_1 = \eta_1 + \eta_2$ is symmetric. It follows from (2.3), (10.50) and (10.51) that $\hat{\lambda}_j(0,h) = 1$ for all $h \in A(H,K)$. By Proposition 2.10, this implies that $\sigma(\lambda_j) \subset A(X, A(H,K)) = \mathbb{R}^n \times K$. Inasmuch as $\alpha(\mathbb{R}^n \times K) = \mathbb{R}^n \times K$, we can suppose that the random variables η_j take values in the group $X = \mathbb{R}^n \times K$. We also suppose $Y = \mathbb{R}^n \times L$, where L is a group topologically isomorphic to the character group of the group K. Denote by $y = (s,l)$, where $s \in \mathbb{R}^n$, $l \in L$, elements of the group Y. In so doing,

$$\hat{\lambda}_j(0,l) = \begin{cases} 1, & \text{if } l = 0,\\ 0, & \text{if } l \neq 0, \quad j = 1,2. \end{cases} \tag{10.52}$$

Consider equation (9.3) for the characteristic functions $\hat{\lambda}_j(s,l)$ on the group $Y = \mathbb{R}^n \times L$. We have

$$\begin{aligned} &\hat{\lambda}_1(s_1+s_2, l_1+l_2)\hat{\lambda}_2(s_1+\widetilde{\alpha}_{\mathbb{R}^n}s_2, l_1+\widetilde{\alpha}_K l_2)\\ &\quad = \hat{\lambda}_1(s_1-s_2, l_1-l_2)\hat{\lambda}_2((s_1-\widetilde{\alpha}_{\mathbb{R}^n}s_2, l_1-\widetilde{\alpha}_K l_2), \quad s_j \in \mathbb{R}^n,\ l_j \in L. \end{aligned} \tag{10.53}$$

Substitute $s_1 = s_2 = s$, $l_1 = l$, and $l_2 = -l$ in equation (10.53). We obtain

$$\begin{aligned} &\hat{\lambda}_1(2s,0)\hat{\lambda}_2((I+\widetilde{\alpha}_{\mathbb{R}^n})s, (I-\widetilde{\alpha}_K)l)\\ &\quad = \hat{\lambda}_1(0,2l)\hat{\lambda}_2((I-\widetilde{\alpha}_{\mathbb{R}^n})s, (I+\widetilde{\alpha}_K)l), \quad s \in \mathbb{R}^n,\ l \in L. \end{aligned} \tag{10.54}$$

Assume that $l \neq 0$. Then $2l \neq 0$, and (10.52) implies that $\hat{\lambda}_1(0, 2l) = 0$. Therefore, it follows from (10.54) that $\hat{\lambda}_1(2s, 0)\hat{\lambda}_2((I + \widetilde{\alpha}_{\mathbb{R}^n})s, (I - \widetilde{\alpha}_K)l) = 0$. Taking into account (10.47), this implies that

$$\hat{\lambda}_2((I + \widetilde{\alpha}_{\mathbb{R}^n})s, (I - \widetilde{\alpha}_K)l) = 0.$$

Since $\alpha(\mathbb{R}^n) = \mathbb{R}^n$ and the topological automorphism $\alpha_{\mathbb{R}^n}$ satisfies condition (i) of Theorem 10.2, we have $I + \alpha_{\mathbb{R}^n} \in \mathrm{Aut}(\mathbb{R}^n)$. Hence $I + \widetilde{\alpha}_{\mathbb{R}^n} \in \mathrm{Aut}(\mathbb{R}^n)$ and

$$\hat{\lambda}_2(s, (I - \widetilde{\alpha}_K)l) = 0, \quad s \in \mathbb{R}^n,\ l \in L,\ l \neq 0. \tag{10.55}$$

In particular,

$$\hat{\lambda}_2(0, (I - \widetilde{\alpha}_K)l) = 0, \quad l \in L,\ l \neq 0. \tag{10.56}$$

It follows from (10.52) and (10.56) that

$$\mathrm{Ker}(I - \widetilde{\alpha}_K) = \{0\}. \tag{10.57}$$

Since K is a compact subgroup of the group G, the group K is topologically isomorphic to a compact totally disconnected Abelian group of the form (10.43). By Theorem 1.7, this implies that the group L is topologically isomorphic to a group of the form (10.44). Then, it follows from (10.57) that $I - \widetilde{\alpha}_K \in \mathrm{Aut}(L)$. Hence we find from (10.55) that

$$\hat{\lambda}_2(s, l) = 0, \quad s \in \mathbb{R}^n,\ l \in L,\ l \neq 0. \tag{10.58}$$

Arguing similarly, we obtain

$$\hat{\lambda}_1(s, l) = 0, \quad s \in \mathbb{R}^n,\ l \in L,\ l \neq 0. \tag{10.59}$$

It follows from (10.47), (10.58), and (10.59) that

$$\hat{\lambda}_j(s, l) = \begin{cases} \exp\{i\langle t_j, s\rangle - \langle A_j s, s\rangle\}, & \text{if } l = 0, \\ 0, & \text{if } l \neq 0, \quad j = 1, 2. \end{cases} \tag{10.60}$$

Taking into account (10.60) and returning from the distributions λ_j to the distributions μ_j, we prove the lemma. □

THEOREM 10.14. *Let*

$$X = \mathbb{R}^n \times D \times G, \tag{10.61}$$

where $n \geq 0$, D is a discrete Abelian group containing no elements of order 2, and G is a compact totally disconnected Abelian group of the form (10.43). Let α be a topological automorphism of the group X satisfying condition (i) of Theorem 10.2. Let ξ_1 and ξ_2 be independent random variables with values in X and distributions μ_1 and μ_2. If the conditional distribution of the linear form $L_2 = \xi_1 + \alpha\xi_2$ given $L_1 = \xi_1 + \xi_2$ is symmetric, then $\mu_j = \gamma_j * m_K * E_{x_j}$, where $\gamma_j \in \Gamma(\mathbb{R}^n)$, K is a compact subgroup of X, $x_j \in X$, $j = 1, 2$. Moreover, $\alpha(K) = K$.

PROOF. Denote by Y the character group of the group X. We can suppose that $Y = \mathbb{R}^n \times L \times H$, where L and H are groups topologically isomorphic to the character groups of the groups D and G respectively. Let

$$G = \mathop{\mathbf{P}}_{p \in \mathcal{P},\ p > 2} G_p,$$

where G_p is a finite p-group. Taking into account Theorem 1.7, suppose that

$$H = \mathop{\mathbf{P}^*}_{p \in \mathcal{P},\ p > 2} H_p,$$

where H_p is a group topologically isomorphic to the character group of the group G_p. Put

$$A_m = \mathop{\mathbf{P}}_{p\in\mathcal{P},\ 2<p\le m} G_p, \quad B_m = \mathop{\mathbf{P}}_{p\in\mathcal{P},\ 2<p\le m} H_p,$$
$$C_m = \mathop{\mathbf{P}}_{p\in\mathcal{P},\ p>m} G_p, \quad m = 3, 4, \dots.$$

Inasmuch as $\mathbb{R}^n$ is the connected component of the zero of the group X, we have $\alpha(\mathbb{R}^n) = \mathbb{R}^n$. It follows from the definition of the product topology that $\alpha(C_m) \subset G$ for some m. Taking into account that C_m is the only subgroup of G which is topologically isomorphic to C_m, we have $\alpha(C_m) = C_m$. Hence $\alpha(\mathbb{R}^n \times C_m) = \mathbb{R}^n \times C_m$.

Represent the group X in the form $X = \mathbb{R}^n \times D \times A_m \times C_m$ and consider the factor-group $X/(\mathbb{R}^n \times C_m)$. Denote by $[x]$ its elements. Note that the factor-group $X/(\mathbb{R}^n \times C_m)$ is topologically isomorphic to the group $D \times A_m$. Since $\alpha(\mathbb{R}^n \times C_m) = \mathbb{R}^n \times C_m$, the topological automorphism α induces a topological automorphism $[\alpha]$ on the factor-group $X/(\mathbb{R}^n \times C_m)$ by the formula $[\alpha][x] = [\alpha x]$. It is easy to see that any continuous monomorphism of the group $\mathbb{R}^n \times C_m$ to itself is a topological automorphism. In view of condition (i) of Theorem 10.2, this implies that

$$\mathrm{Ker}(I + [\alpha]) = \{0\}. \tag{10.62}$$

By Theorem 1.9, the character group of the factor-group $X/(\mathbb{R}^n \times C_m)$ is topologically isomorphic to the annihilator $A(Y, \mathbb{R}^n \times C_m) = L \times B_m$. Since $\alpha(\mathbb{R}^n \times C_m) = \mathbb{R}^n \times C_m$, we have $\widetilde{\alpha}(L \times B_m) = (L \times B_m)$. Observe also that the restriction of $\widetilde{\alpha}$ to the subgroup $L \times B_m$ is the adjoint automorphism to $[\alpha]$.

By Corollary 9.2, the characteristic functions $\hat\mu_j(y)$ satisfy equation (9.3). Let η_1 and η_2 be independent random variables with values in the factor-group $X/(\mathbb{R}^n \times C_m)$ such that the characteristic functions of their distributions coincide with the restrictions to the subgroup $L \times B_m$ of the characteristic functions $\hat\mu_j(y)$. By Corollary 9.2, the conditional distribution of the linear form $M_2 = \eta_1 + [\alpha]\eta_2$ given $M_1 = \eta_1 + \eta_2$ is symmetric. Obviously, the factor-group $X/(\mathbb{R}^n \times C_m)$ is discrete and contains no elements of order 2. Therefore, taking into account (10.62) and applying Theorem 10.8 to the discrete Abelian group $X/(\mathbb{R}^n \times C_m)$, we obtain that the restrictions of the characteristic functions $\hat\mu_j(y)$ to the subgroup $L \times B_m$ are of the form

$$\hat\mu_j(y) = \begin{cases} (g_j, y), & \text{if } y \in E, \\ 0, & \text{if } y \notin E, \quad j = 1, 2, \end{cases}$$

where $g_j \in D \times A_m$, E is an open subgroup in $L \times B_m$, which is the annihilator of a finite subgroup of the group $X/(\mathbb{R}^n \times C_m)$. Moreover,

$$\widetilde{\alpha}(E) = E. \tag{10.63}$$

Using the same reasoning as in the proof of Lemma 10.13, we can replace the independent random variables ξ_j by the independent random variables $\zeta_j = \xi_j - x_j$ with distributions λ_j in such a way that the restrictions of the characteristic functions $\hat\lambda_j(y)$ to the subgroup $L \times B_m$ are of the form

$$\hat\lambda_j(y) = \begin{cases} 1, & \text{if } y \in E, \\ 0, & \text{if } y \notin E, \quad j = 1, 2. \end{cases} \tag{10.64}$$

Moreover, the conditional distribution of the linear form $N_2 = \zeta_1 + \alpha\zeta_2$ given $N_1 = \zeta_1 + \zeta_2$ is symmetric.

By Proposition 2.10, (10.64) implies that $\sigma(\lambda_j) \subset A(X, E)$. It is easy to see that $A(X, E) = \mathbb{R}^n \times F$, where F is a compact totally disconnected Abelian group of the form (10.43). It follows from (10.63) that $\alpha(\mathbb{R}^n \times F) = \mathbb{R}^n \times F$. Applying Lemma 10.13 to the group $\mathbb{R}^n \times F$, we get the required representation for the distributions λ_j. Hence we obtain the required representation for the distributions μ_j too. □

An analogue of Heyde's theorem for two independent random variables is valid not only for locally compact Abelian groups of the form (10.61). The following statement holds true.

PROPOSITION 10.15. *There exists a locally compact Abelian group X containing no elements of order 2 such that X is not topologically isomorphic to any group of the form* (10.61) *and the following statement holds: Let α be a topological automorphism of the group X satisfying condition* (i) *of Theorem* 10.2. *Let ξ_1 and ξ_2 be independent random variables with values in X and distributions μ_1 and μ_2. If the conditional distribution of the linear form $L_2 = \xi_1 + \alpha\xi_2$ given $L_1 = \xi_1 + \xi_2$ is symmetric, then $\mu_j = m_K * E_{x_j}$, where K is a compact subgroup of X, $x_j \in X$, $j = 1, 2$. Moreover, $\alpha(K) = K$.*

PROOF. Denote by p_n the nth prime number. Put

$$S = \mathop{\mathbf{P}}_{n=2}^{\infty} \mathbb{Z}(p_n^2)$$

and consider S only as an "algebraic" group. We do not consider S as a topological group equipped with the product topology.

Consider a subgroup

$$K_n = \{0, p_n, 2p_n, \dots, (p_n - 1)p_n\}$$

of the group $\mathbb{Z}(p_n^2)$. It is obvious that K_n is isomorphic to $\mathbb{Z}(p_n)$. Denote by Y a subgroup of S consisting of all elements $y = (y_2, \dots, y_n, \dots)$, $y_n \in \mathbb{Z}(p_n^2)$, such that $y_n \notin K_n$ only for a finite number of indexes n. Put

$$F = \mathop{\mathbf{P}}_{n=2}^{\infty} K_n.$$

Then F is a subgroup of Y. Consider the group F in the product topology. Obviously, F is a second countable compact Abelian group. Define a topology on Y as follows. A subset U in Y is called open if for each $y \in U$ there is a neighborhood of the zero V_y in F such that $y + V_y \subset U$. Then Y is a second countable locally compact Abelian group. In so doing F is an open subgroup of Y. It is obvious that Y is a nondiscrete noncompact totally disconnected Abelian group. Denote by X the character group of the group Y.

Since multiplication by 2 is a topological automorphism of the group Y, the group X has no elements of order 2. To see that the group X is not topologically isomorphic to a group of the form (10.61), suppose the contrary. It is obvious that then in (10.61) $n = 0$, and then the group Y can be represented in the form $Y = L \times H$, where the group L is compact and the group H is discrete. It is easy to see that there exists m such that

$$\mathop{\mathbf{P}}_{n=m}^{\infty} K_n \subset L.$$

This implies that the group H can contain only a finite number of distinct subgroups K_n. Hence H is a finite group. It follows from this that Y is a compact group. The obtained contradiction shows that the group X is not topologically isomorphic to any group of the form (10.61).

We shall verify that

$$(I+\widetilde{\alpha})(F)=F. \tag{10.65}$$

By statement (b) of Theorem 1.20, α satisfies condition (i) of Theorem 10.2 if and only if

$$\overline{(I+\widetilde{\alpha})(Y)}=Y. \tag{10.66}$$

Put

$$B=\mathop{\mathbf{P}^*}_{n=2}^{\infty}\mathbb{Z}(p_n^2),$$

and consider B as a subgroup of the group Y in the topology induced on B by the topology of Y. Denote by $\widetilde{\alpha}_B$ the restriction of $\widetilde{\alpha}$ to the subgroup B. Suppose $\mathrm{Ker}(I+\widetilde{\alpha}_B)\neq\{0\}$. This implies that $(I+\widetilde{\alpha}_B)(B)$ is a proper subgroup of B. It follows from this that $(I+\widetilde{\alpha})(B)$ is contained in a proper closed subgroup of the group Y. Inasmuch as B is a dense subgroup of Y, this implies that $(I+\widetilde{\alpha})(Y)$ is also contained in a proper closed subgroup of the group Y. This contradicts (10.66). Hence $\mathrm{Ker}(I+\widetilde{\alpha}_B)=\{0\}$. Then $I+\widetilde{\alpha}_B$ is an automorphism of the group B. Put

$$C=\mathop{\mathbf{P}^*}_{n=2}^{\infty}K_n,$$

and consider C as a subgroup of the group F in the topology induced on C by the topology of Y. Then C is a dense subgroup of F. Since $I+\widetilde{\alpha}_B$ is an automorphism of the group B, we have

$$(I+\widetilde{\alpha})(C)=C. \tag{10.67}$$

Take $y_0\in F$. The subgroup C is dense in F. This implies that there exists a sequence $y_j\in C$ such that $y_j\to y_0$. It follows from (10.67) that $y_j=(I+\widetilde{\alpha})z_j$, where $z_j\in C$. Since F is a compact group, there exist a sequence $\{j_k\}$ and an element $z\in F$ such that $z_{j_k}\to z$. Hence $y_0=(I+\widetilde{\alpha})z$. Thus, (10.65) is proved.

By Corollary 9.2, the characteristic functions $\hat{\mu}_j(y)$ satisfy equation (9.3). Put $\nu_j=\mu_j*\bar{\mu}_j$. This implies that $\hat{\nu}_j(y)=|\hat{\mu}_j(y)|^2\geq 0$ for all $y\in Y$. The characteristic functions $\hat{\nu}_j(y)$ also satisfy equation (9.3). It is easy to see that the proposition will be proved if we prove that $\nu_1=\nu_2=m_K$, where K is a compact subgroup of the group X and $\alpha(K)=K$. Therefore, we can solve equation (9.3) assuming that $\hat{\mu}_j(y)\geq 0$ for all $y\in Y$. It is obvious that $\widetilde{\alpha}_F\in\mathrm{Aut}(F)$. Consider the restriction of equation (9.3) to the subgroup F. We have

$$\hat{\mu}_1(u+v)\hat{\mu}_2(u+\widetilde{\alpha}_F v)=\hat{\mu}_1(u-v)\hat{\mu}_2(u-\widetilde{\alpha}_F v),\quad u,v\in F. \tag{10.68}$$

The group F is topologically isomorphic to the character group of the discrete group $\mathop{\mathbf{P}^*}_{n=2}^{\infty}K_n$, and $\widetilde{\alpha}_F$ is the adjoint automorphism to an automorphism $\bar{\alpha}$ of the group $\mathop{\mathbf{P}^*}_{n=2}^{\infty}K_n$. Note that (10.65) implies that $\bar{\alpha}$ satisfies condition (i) of Theorem 10.2. Taking into account Corollary 9.2 and applying Theorem 10.8 to the discrete group $\mathop{\mathbf{P}^*}_{n=2}^{\infty}K_n$, we conclude from (10.68) that the restrictions of the characteristic functions $\hat\mu_j(y)$ to the subgroup F are of the form

$$\hat\mu_j(y) = \begin{cases} 1, & \text{if } y \in E, \\ 0, & \text{if } y \notin E, \quad j = 1, 2, \end{cases}$$

where E is an open subgroup of F. In so doing

$$\widetilde\alpha_F(E) = E. \tag{10.69}$$

Put $M = A(X, E)$. It follows from (10.69) that $\alpha(M) = M$. By Proposition 2.10, $\sigma(\mu_j) \subset M$, $j = 1, 2$. Thus, we reduced the proof of the proposition to the case when the random variables ξ_j take values in the group M. The character group of the group M is topologically isomorphic to the factor-group Y/E. Since E is an open subgroup of F, there is some natural number m such that the group Y/E is topologically isomorphic to a group of the form

$$\mathop{\mathbf{P}}_{n=2}^{m} \mathbb{Z}(p_n^{l_n}) \times \mathop{\mathbf{P}^*}_{n=m+1}^{\infty} \mathbb{Z}(p_n),$$

where $l_n \in \{0, 1, 2\}$. Hence the group M is topologically isomorphic to a group of the form (10.61). The proposition follows now from Theorem 10.12. □

Consider now the case when X is a compact connected Abelian group. To conclude this section, we prove that on these groups there is not even a weak analogue of Heyde's theorem. The following theorem is valid.

THEOREM 10.16. *Let X be a compact connected Abelian group containing no elements of order* 2. *Then there exist a topological automorphism α of the group X satisfying condition* (i) *of Theorem* 10.2, *and independent random variables ξ_1 and ξ_2 with values in X and distributions μ_1 and μ_2 such that the conditional distribution of the linear form $L_2 = \xi_1 + \alpha\xi_2$ given $L_1 = \xi_1 + \xi_2$ is symmetric, whereas $\mu_j \notin \Gamma(X) * I(X)$, $j = 1, 2$.*

PROOF. Denote by Y the character group of the group X. By Theorems 1.2 and 1.3, Y is a discrete torsion-free Abelian group. Two cases are possible: the group X contains a nonzero element of finite order or X is a torsion-free group.

1. Assume that the group X contains a nonzero element of finite order. Denote by p the minimum of orders of nonzero elements of X. Then p is a prime and $p \geq 3$, because X contains no elements of order 2. Since X is a connected Abelian group, by Theorem 1.13, $X^{(n)} = X$ for all natural n. This implies that if $\operatorname{Ker} f_n = \{0\}$, then $f_n \in \operatorname{Aut}(X)$. Set $\alpha = -f_{p-1}$. Then $\alpha \in \operatorname{Aut}(X)$. Inasmuch as $I + \alpha = -f_{p-2}$, we have $I + \alpha \in \operatorname{Aut}(X)$. Thus, α satisfies condition (i) of Theorem 10.2. By Theorem 1.12, $Y^{(p)} = A(Y, \operatorname{Ker} f_p)$. Hence $Y^{(p)}$ is a proper subgroup in the group Y. Take $y_0 \in Y$ such that $y_0 \notin Y^{(p)}$. Since $p \geq 3$, the numbers 2 and p are mutually prime. Therefore, there exist some integers m and n such that $2m + pn = 1$. It follows from this that $y_0 = 2my_0 + pny_0$. This implies that $2y_0 \notin Y^{(p)}$.

Consider on the group X the function

$$r(x) = 1 + \operatorname{Re}(x, y_0), \quad x \in X.$$

Then $r(x) \geq 0$ for all $x \in X$, and

$$\int\limits_X r(x)dm_X(x) = 1.$$

Denote by μ the distribution on X with the density $r(x)$ with respect to m_X. It is easy to see that the characteristic function $\hat{\mu}(y)$ is of the form

$$\hat{\mu}(y) = \begin{cases} 1, & \text{if } \ y = 0, \\ \frac{1}{2}, & \text{if } \ y = \pm y_0, \\ 0, & \text{if } \ y \notin \{0, \pm y_0\}. \end{cases}$$

It is obvious that $\mu \notin \Gamma(X) * I(X)$.

Let ξ_1 and ξ_2 be independent identically distributed random variables with values in the group X and distribution μ. Let us verify that the conditional distribution of the linear form $L_2 = \xi_1 + \alpha\xi_2$ given $L_1 = \xi_1 + \xi_2$ is symmetric. By Corollary 9.2, it suffices to verify that the characteristic functions of the random variables ξ_j satisfy equation (9.3) which takes the form

$$\hat{\mu}(u+v)\hat{\mu}(u+(1-p)v) = \hat{\mu}(u-v)\hat{\mu}(u-(1-p)v), \quad u, v \in Y. \tag{10.70}$$

Equation (10.70) holds true if $v = 0$. So, assume that $v \neq 0$. Then the left-hand side of equation (10.70) vanishes. Otherwise, taking into account (10.70), we get $u+v \in \{0, \pm y_0\}$ and $u+(1-p)v \in \{0, \pm y_0\}$. This implies that $pv \in \{0, \pm y_0, \pm 2y_0\}$, contradicting the fact that $y_0, 2y_0 \notin Y^{(p)}$. Reasoning similarly we obtain that if $v \neq 0$, then the right-hand side of equation (10.70) vanishes too. Thus, in the case when the group X contains a nonzero element of finite order, the theorem is proved.

2. Assume that X is a torsion-free group. Taking into account Theorem 1.15 and the fact that X is a connected group, we get that the group X is topologically isomorphic to a group of the form $(\Sigma_{\boldsymbol{a}})^{\mathfrak{n}}$, where $\boldsymbol{a} = (2, 3, 4, \dots)$, $\mathfrak{n} \leq \aleph_0$. Arguing as in the proof of the necessity in Theorem 10.12 it suffices to prove the theorem for the group $\Sigma_{\boldsymbol{a}}$. Since $(\Sigma_{\boldsymbol{a}})^* \cong \mathbb{Q}$, in order not to complicate the notation, suppose $(\Sigma_{\boldsymbol{a}})^* = \mathbb{Q}$.

Let μ be the distribution on the group $\Sigma_{\boldsymbol{a}}$ such that $\mu \notin \Gamma(\Sigma_{\boldsymbol{a}}) * I(\Sigma_{\boldsymbol{a}})$, constructed in the proof of case 2 in Theorem 7.24. Put $\alpha = -f_2^{-2}$. Then $\alpha \in \mathrm{Aut}(\Sigma_{\boldsymbol{a}})$ and we have $I+\alpha = f_3 f_2^{-2}$. Obviously, $I+\alpha \in \mathrm{Aut}(\Sigma_{\boldsymbol{a}})$. Thus, α satisfies condition (i) of Theorem 10.2.

Let ξ_1 and ξ_2 be independent identically distributed random variables with values in the group $\Sigma_{\boldsymbol{a}}$ and distribution μ. Verify that the conditional distribution of the linear form $L_2 = \xi_1 + \alpha\xi_2$ given $L_1 = \xi_1 + \xi_2$ is symmetric. This is equivalent to the fact that the conditional distribution of the linear form $N_2 = 4\xi_1 - \xi_2$ given $N_1 = \xi_1 + \xi_2$ is symmetric. By Corollary 9.2, it suffices to verify that the characteristic functions of the random variables ξ_j satisfy equation (9.3) which takes the form

$$f(u+4v)f(u-v) = f(u-4v)f(u+v), \quad u, v \in \mathbb{Q}. \tag{10.71}$$

It follows from (7.34) that for $u, v \in H$ equation (10.71) becomes an equality. Observe that if either $u \in H$, $v \notin H$ or $v \in H$, $u \notin H$, then $u \pm v \notin H$. It follows from (7.35) that in this case both sides of equation (10.71) are equal to zero. Suppose $u, v \notin H$. Then the left-hand side in (10.71) is equal to zero. In the opposite case (7.35) implies that $u + 4v, u - v \in H$. It follows from this that $5v \in H$. Hence $v \in H$. The obtained contradiction shows that the left-hand side in (10.71) is equal to zero. Arguing similarly, we verify that if $u, v \notin H$, then the right-hand side in (10.71) is also equal to zero. Thus, for all $u, v \in \mathbb{Q}$ equation (10.71) becomes an equality. □

We recall that if $\mu \in \Gamma(X) * I(X)$, then μ is invariant with respect to a compact subgroup K of the group X. Moreover, μ induces a Gaussian distribution on the factor-group X/K under the natural homomorphism $X \to X/K$. Thus, Theorem 10.16 shows that even a weak analogue Heyde's theorem is not valid for compact connected Abelian groups.

We will complement Theorem 10.16 with two propositions. The first proposition shows that, generally speaking, we cannot replace in Theorem 10.16 the statement:

(A) *Then there exist a topological automorphism α of the group X satisfying condition* (i) *of Theorem* 10.2 *and independent random variables ξ_1 and ξ_2 with values in X and distributions μ_1 and μ_2 such that the conditional distribution of the linear form $L_2 = \xi_1 + \alpha\xi_2$ given $L_1 = \xi_1 + \xi_2$ is symmetric, whereas $\mu_j \notin \Gamma(X) * I(X)$, $j = 1, 2$;*

by the statement:

(B) *Then for each topological automorphism α of the group X, $\alpha \neq I$, satisfying condition* (i) *of Theorem* 10.2 *there exist independent random variables ξ_1 and ξ_2 with values in X and distributions μ_1 and μ_2 such that the conditional distribution of the linear form $L_2 = \xi_1 + \alpha\xi_2$ given $L_1 = \xi_1 + \xi_2$ is symmetric, whereas $\mu_j \notin \Gamma(X) * I(X)$, $j = 1, 2$.*

The second proposition shows that there is a compact connected Abelian group X containing no elements of order 2 when we can replace in Theorem 10.16 statement (A) by statement (B). We should exclude the case when $\alpha = I$, because if a compact connected Abelian group X contains no elements of order 2, then by Theorems 1.12 and 1.13, we have $Y^{(2)} = Y$. Then by Corollary 9.2, the symmetry of the conditional distribution of the linear form $L_2 = \xi_1 + \xi_2$ given $L_1 = \xi_1 + \xi_2$ implies that ξ_1 and ξ_2 have degenerate distributions.

PROPOSITION 10.17. *There exist a compact connected Abelian group X containing no elements of order* 2 *and a topological automorphism α of the group X, $\alpha \neq I$, satisfying condition* (i) *of Theorem* 10.2 *such that the following statement holds: if ξ_1 and ξ_2 are independent random variables with values in X and distributions μ_1 and μ_2 such that the conditional distribution of the linear form $L_2 = \xi_1 + \alpha\xi_2$ given $L_1 = \xi_1 + \xi_2$ is symmetric, then $\mu_j \in \Gamma(X) * I(X)$, $j = 1, 2$.*

PROOF. Let $\boldsymbol{a} = (2, 3, 4, \dots)$ and let $\Sigma_{\boldsymbol{a}}$ be the $\boldsymbol{a}$-adic solenoid with character group $\mathbb{Q}$. Consider the group $\Sigma_{\boldsymbol{a}}^2$. The character group of $\Sigma_{\boldsymbol{a}}^2$ is topologically isomorphic to the group $\mathbb{Q}^2$. It is easy to see that $f_n \in \mathrm{Aut}(\Sigma_{\boldsymbol{a}}^2)$ for all natural n. This implies in particular, that the group $\Sigma_{\boldsymbol{a}}^2$ has no elements of order 2. Denote by $x = (a, b)$, where $a, b \in \Sigma_{\boldsymbol{a}}$, elements of the group $\Sigma_{\boldsymbol{a}}^2$. Define the mapping $\alpha : \Sigma_{\boldsymbol{a}}^2 \to \Sigma_{\boldsymbol{a}}^2$ by the formula

$$\alpha(a, b) = (-3a + 4b, 2a - 3b), \quad (a, b) \in \Sigma_{\boldsymbol{a}}^2.$$

Then α is a topological automorphism of the group $\Sigma_{\boldsymbol{a}}^2$. Since $I + \alpha \in \mathrm{Aut}(\Sigma_{\boldsymbol{a}}^2)$, condition (i) of Theorem 10.2 is fulfilled.

By Corollary 9.7, the linear forms $M_1 = (I + \alpha)\xi_1 + 2\alpha\xi_2$ and $M_2 = 2\xi_1 + (I + \alpha)\xi_2$ are independent. Note that coefficients of the linear forms M_1 and M_2 are topological automorphisms of the group $\Sigma_{\boldsymbol{a}}^2$. Consider the new independent random variables $\eta_1 = (I + \alpha)\xi_1$ and $\eta_2 = 2\alpha\xi_2$. The independence of the linear forms M_1 and M_2 implies that the linear forms $N_1 = \eta_1 + \eta_2$ and $N_2 = 2(I + \alpha)^{-1}\eta_1 + f_2^{-1}\alpha^{-1}(I+\alpha)\eta_2$ are independent. Hence the linear forms $P_1 = \eta_1 + \eta_2$ and $P_2 = \eta_1 + f_2^{-2}\alpha^{-1}(I+\alpha)^2\eta_2$ are also independent. It is easy to verify that $f_2^{-2}\alpha^{-1}(I+\alpha)^2 = -I$. Hence $P_2 = \eta_1 - \eta_2$. Denote by λ_j the distribution of the random variable η_j. Since X is a connected Abelian group containing no elements of order 2, by Corollary 5.3, $\lambda_j \in \Gamma(\Sigma_{\boldsymbol{a}}^2) * I(\Sigma_{\boldsymbol{a}}^2)$, $j = 1, 2$. In view of $I + \alpha, 2\alpha \in \mathrm{Aut}(\Sigma_{\boldsymbol{a}}^2)$, it follows from $\lambda_1 = (I+\alpha)\mu_1$ and $\lambda_2 = 2\alpha\mu_2$ that $\mu_j \in \Gamma(\Sigma_{\boldsymbol{a}}^2) * I(\Sigma_{\boldsymbol{a}}^2)$, $j = 1, 2$. ☐

PROPOSITION 10.18. *There exists a compact connected Abelian group X containing no elements of order* 2 *such that for each topological automorphism α of the group X, $\alpha \neq I$, satisfying condition* (i) *of Theorem* 10.2 *there exist independent random variables ξ_1 and ξ_2 with values in X and distributions μ_1 and μ_2 such that the conditional distribution of the linear form $L_2 = \xi_1 + \alpha\xi_2$ given $L_1 = \xi_1 + \xi_2$ is symmetric, while $\mu_j \notin \Gamma(X) * I(X)$, $j = 1, 2$.*

PROOF. Let $\boldsymbol{a} = (2, 2, 2, \dots)$. Consider the $\boldsymbol{a}$-adic solenoid $\Sigma_{\boldsymbol{a}}$ with character group

$$H_{\boldsymbol{a}} = \left\{ \frac{m}{2^n} : n = 0, 1, \dots;\ m \in \mathbb{Z} \right\}.$$

Since $H_{\boldsymbol{a}}^{(2)} = H_{\boldsymbol{a}}$, by Theorem 1.13, the group $\Sigma_{\boldsymbol{a}}$ contains no elements of order 2. It is easy to see that $\mathrm{Aut}(\Sigma_{\boldsymbol{a}}) = \{\pm f_{2^k}, \pm f_{2^k}^{-1} : k = 0, 1, 2, \dots\}$. In so doing, there exist only two topological automorphisms α, $\alpha \neq I$, of the group $\Sigma_{\boldsymbol{a}}$ satisfying condition (i) of Theorem 10.2. Namely, $\alpha = -f_2$ and $\alpha = -f_2^{-1}$. Let $\alpha = -f_2$. It is easy to see that $f_3 \notin \mathrm{Aut}(\Sigma_{\boldsymbol{a}})$. Then, as has been proved in case 1 of Theorem 10.16, there exist independent random variables ξ_1 and ξ_2 with values in $\Sigma_{\boldsymbol{a}}$ and distributions μ_1 and μ_2 such that the conditional distribution of the linear form $L_2 = \xi_1 - 2\xi_2$ given $L_1 = \xi_1 + \xi_2$ is symmetric, while $\mu_j \notin \Gamma(\Sigma_{\boldsymbol{a}}) * I(\Sigma_{\boldsymbol{a}})$, $j = 1, 2$.

The proof of the existence of independent random variables ξ_1 and ξ_2 with values in the group $\Sigma_{\boldsymbol{a}}$ and distributions μ_1 and μ_2 such that the conditional distribution of the linear form $L_2 = \xi_1 - f_2^{-1}\xi_2$ given $L_1 = \xi_1 + \xi_2$ is symmetric, while $\mu_j \notin \Gamma(\Sigma_{\boldsymbol{a}}) * I(\Sigma_{\boldsymbol{a}})$, $j = 1, 2$, is reduced to the previous case. ☐

11. Locally compact Abelian groups containing an element of order 2

Let X be a second countable locally compact Abelian group. In this section we study Heyde's theorem in the case when the group X contains an element of order 2, i.e., $X_{(2)} \neq \{0\}$. Obviously, $X_{(2)}$ is a characteristic subgroup of the group X. We pay attention to the following. Let ξ_j, $j = 1, 2, \dots, n$, $n \geq 2$, be independent random variables with values in the subgroup $X_{(2)}$ and distributions μ_j. Consider the linear forms $L_1 = \alpha_1\xi_1 + \cdots + \alpha_n\xi_n$ and $L_2 = \beta_1\xi_1 + \cdots + \beta_n\xi_n$, where $\alpha_j, \beta_j \in \mathrm{Aut}(X)$. It is easily seen that for any μ_j the conditional distribution of the linear form L_2 given L_1 is symmetric. Thus, the presence of elements of order 2 in the group X complicates the description of distributions that are characterized by the symmetry of the conditional distribution of one linear form given another.

The simplest locally compact Abelian group on which nondegenerate Gaussian distributions exist and which contains an element of order 2 is the group $\mathbb{R} \times \mathbb{Z}(2)$. Although we are not using this, we note that the group $\mathbb{R} \times \mathbb{Z}(2)$ is topologically isomorphic to the multiplicative group of nonzero real numbers. We start by studying Heyde's theorem for this group. We prove that the class of distributions on the group $\mathbb{R} \times \mathbb{Z}(2)$ which is characterized by the symmetry of the conditional distribution of one linear form of independent random variables given another is much wider than the class $\Gamma(\mathbb{R}) * \mathrm{M}^1(\mathbb{Z}(2))$. Then, we study Heyde's theorem for two independent random variables with values in the groups $\mathbb{R} \times \mathbb{T}$ and $\Sigma_{\boldsymbol{a}} \times \mathbb{T}$, where the $\boldsymbol{a}$-adic solenoid $\Sigma_{\boldsymbol{a}}$ contains no elements of order 2. Next, we study Heyde's theorem for $\boldsymbol{a}$-adic solenoids $\Sigma_{\boldsymbol{a}}$ containing an element of order 2. Finally, we study Heyde's theorem for discrete Abelian groups containing an element of order 2.

Consider the group $\mathbb{R}\times\mathbb{Z}(2)$. Denote by (t,k), where $t \in \mathbb{R}$, $k \in \mathbb{Z}(2)$, elements of $\mathbb{R}\times\mathbb{Z}(2)$. Since $(\mathbb{R}\times\mathbb{Z}(2))^* \cong \mathbb{R}\times\mathbb{Z}(2)$, in order not to complicate the notation, suppose $(\mathbb{R}\times\mathbb{Z}(2))^* = \mathbb{R}\times\mathbb{Z}(2)$. Denote by (s,l), where $s \in \mathbb{R}$, $l \in \mathbb{Z}(2)$, elements of the group $(\mathbb{R}\times\mathbb{Z}(2))^*$. The value of a character $(s,l) \in (\mathbb{R}\times\mathbb{Z}(2))^*$ at an element $(t,k) \in \mathbb{R}\times\mathbb{Z}(2)$ is defined by the formula $((t,k),(s,l)) = e^{its}(-1)^{kl}$.

Let $a \in \mathrm{Aut}(\mathbb{R}\times\mathbb{Z}(2))$. It is obvious that a is of the form $a(t,k) = (c_a t, k)$, where $c_a \in \mathbb{R}$, $c_a \neq 0$. We identify a and c_a, i.e., we write $a(t,k) = (at,k)$ and assume that $a \in \mathbb{R}$, $a \neq 0$. The converse is also true. Any nonzero real number corresponds to a topological automorphism of the group $\mathbb{R}\times\mathbb{Z}(2)$. Observe that if $a \in \mathrm{Aut}(\mathbb{R}\times\mathbb{Z}(2))$, then $\widetilde{a}$ is of the form $\widetilde{a}(s,l) = (as,l)$, i.e., $\widetilde{a} = a$.

Let $\mu \in \Gamma(\mathbb{R}) * \mathrm{M}^1(\mathbb{Z}(2))$. It is easy to see that the characteristic function of the distribution μ is of the form

$$\hat{\mu}(s,l) = \begin{cases} \exp\{-\sigma s^2 + i\beta s\}, & \text{if} \quad s \in \mathbb{R},\ l = 0, \\ \kappa \exp\{-\sigma s^2 + i\beta s\}, & \text{if} \quad s \in \mathbb{R},\ l = 1, \end{cases}$$

where $\sigma \geq 0$, $\beta, \kappa \in \mathbb{R}$, $|\kappa| \leq 1$.

Let us take into consideration a class Θ of distributions on the group $\mathbb{R}\times\mathbb{Z}(2)$ which is wider than the class $\Gamma(\mathbb{R}) * \mathrm{M}^1(\mathbb{Z}(2))$. For this we need the following lemma.

LEMMA 11.1. *Consider the group $\mathbb{R}\times\mathbb{Z}(2)$. Let $f(s,l)$ be a function on the character group of the group $\mathbb{R}\times\mathbb{Z}(2)$ of the form*

$$f(s,l) = \begin{cases} \exp\{-\sigma s^2 + i\beta s\}, & \textit{if} \quad s \in \mathbb{R},\ l = 0, \\ \kappa \exp\{-\sigma' s^2 + i\beta' s\}, & \textit{if} \quad s \in \mathbb{R},\ l = 1, \end{cases}$$

where $\sigma \geq 0$, $\sigma' \geq 0$, $\beta, \beta', \kappa \in \mathbb{R}$. Then $f(s,l)$ is the characteristic function of a signed measure μ on the group $\mathbb{R}\times\mathbb{Z}(2)$. Moreover, μ is a distribution if and only if either

$$0 < \sigma' < \sigma, \quad 0 < |\kappa| \leq \sqrt{\frac{\sigma'}{\sigma}} \exp\left\{-\frac{(\beta-\beta')^2}{4(\sigma-\sigma')}\right\}$$

or

$$\sigma = \sigma', \quad \beta = \beta', \quad |\kappa| \leq 1,$$

In the latter case $\mu \in \Gamma(\mathbb{R}) * \mathrm{M}^1(\mathbb{Z}(2))$.

PROOF. We can prove the lemma assuming that $\kappa \neq 0$. Multiplying if necessary the function $f(s,l)$ by a suitable character of the group $\mathbb{R}\times\mathbb{Z}(2)$, we can assume without loss of generality that $\kappa > 0$.

Let $a > 0$. Denote by γ_a the Gaussian distribution on the group $\mathbb{R}$ with the characteristic function

$$\hat{\gamma}_a(s) = \exp\{-as^2\}, \quad s \in \mathbb{R},$$

and the density

$$f_a(t) = \frac{1}{2\sqrt{\pi a}} \exp\left\{-\frac{t^2}{4a}\right\}, \quad t \in \mathbb{R}. \tag{11.1}$$

Let μ be the signed measure on the group $\mathbb{R} \times \mathbb{Z}(2)$ defined as follows

$$\mu(B \times \{k\}) = \begin{cases} \frac{1}{2}(\gamma_\sigma * E_\beta + \kappa\gamma_{\sigma'} * E_{\beta'})(B), & \text{if } k = 0, \\ \frac{1}{2}(\gamma_\sigma * E_\beta - \kappa\gamma_{\sigma'} * E_{\beta'})(B), & \text{if } k = 1, \end{cases}$$

where B is a Borel subset of $\mathbb{R}$. Put

$$\lambda_0 = \frac{1}{2}(\gamma_\sigma * E_\beta + \kappa\gamma_{\sigma'} * E_{\beta'}), \quad \lambda_1 = \frac{1}{2}(\gamma_\sigma * E_\beta - \kappa\gamma_{\sigma'} * E_{\beta'}).$$

Taking into account that $\hat\lambda_0(s)+\hat\lambda_1(s) = \hat\gamma_\sigma(s)e^{i\beta s}$ and $\hat\lambda_0(s)-\hat\lambda_1(s) = \kappa\hat\gamma_{\sigma'}(s)e^{i\beta' s}$, we have

$$\hat\mu(s,l) = \int\limits_{\mathbb{R}\times\mathbb{Z}(2)} e^{its}(-1)^{kl} d\mu(t,k)$$

$$= \int\limits_{\mathbb{R}\times\{0\}} e^{its} d\mu(t,0) + \int\limits_{\mathbb{R}\times\{1\}} e^{its}(-1)^l d\mu(t,1) = f(s,l).$$

Thus, $f(s,l)$ is the characteristic function of the signed measure μ. The signed measure μ is a measure if and only if the signed measure λ_1 is a measure. It is obvious that if the signed measure λ_1 is a measure, then either $\sigma > 0$ and $\sigma' > 0$ or $\sigma = \sigma' = 0$. It is clear that if $\sigma = \sigma' = 0$, then the signed measure μ is a measure if and only if $\beta = \beta'$ and $\kappa \le 1$. The statement of the lemma is proved in this case.

Let $\sigma > 0$ and $\sigma' > 0$. Taking into account (11.1), the signed measure λ_1 is a measure if and only if the inequality

$$\frac{1}{2\sqrt{\pi\sigma}} \exp\left\{-\frac{(t-\beta)^2}{4\sigma}\right\} - \frac{\kappa}{2\sqrt{\pi\sigma'}} \exp\left\{-\frac{(t-\beta')^2}{4\sigma'}\right\} \ge 0$$

holds for all $t \in \mathbb{R}$. This inequality is equivalent to the following

$$\kappa \le \sqrt{\frac{\sigma'}{\sigma}} \exp\left\{-\frac{(t-\beta)^2}{4\sigma} + \frac{(t-\beta')^2}{4\sigma'}\right\}, \quad t \in \mathbb{R}. \tag{11.2}$$

Let $\sigma \ne \sigma'$. Inasmuch as $\kappa > 0$, we have $\sigma' < \sigma$. The function in the right-hand side of inequality (11.2) reaches its minimum at the point

$$t = \frac{\sigma\beta' - \sigma'\beta}{\sigma - \sigma'}$$

and this minimum is equal to

$$\sqrt{\frac{\sigma'}{\sigma}} \exp\left\{-\frac{(\beta-\beta')^2}{4(\sigma-\sigma')}\right\}.$$

Assume that $\sigma = \sigma'$. Then (11.2) implies that $\beta = \beta'$ and $\kappa \le 1$. Thus, the signed measure λ_1 and hence the signed measure μ, is a measure if and only if either $\sigma' < \sigma$ and

$$\kappa \le \sqrt{\frac{\sigma'}{\sigma}} \exp\left\{-\frac{(\beta-\beta')^2}{4(\sigma-\sigma')}\right\}$$

or $\sigma' = \sigma$, $\beta = \beta'$ and $\kappa \le 1$. It is also obvious that in the latter case $\mu \in \Gamma(\mathbb{R}) * \mathrm{M}^1(\mathbb{Z}(2))$. $\square$

DEFINITION 11.2. Let $\mu \in \mathrm{M}^1(\mathbb{R} \times \mathbb{Z}(2))$. We say that $\mu \in \Theta$ if the characteristic function $\hat{\mu}(s, l)$ is represented in the form

$$\hat{\mu}(s,l) = \begin{cases} \exp\{-\sigma s^2 + i\beta s\}, & \text{if} \quad s \in \mathbb{R}, \ l = 0, \\ \kappa \exp\{-\sigma' s^2 + i\beta' s\}, & \text{if} \quad s \in \mathbb{R}, \ l = 1, \end{cases} \tag{11.3}$$

where either

$$0 < \sigma' < \sigma, \quad 0 < |\kappa| \le \sqrt{\frac{\sigma'}{\sigma}} \exp\left\{-\frac{(\beta - \beta')^2}{4(\sigma - \sigma')}\right\}$$

or

$$\sigma = \sigma', \quad \beta = \beta', \quad |\kappa| \le 1.$$

Take into consideration another class of probability distributions on the group $\mathbb{R} \times \mathbb{Z}(2)$. Let $\mu \in \mathrm{M}^1(\mathbb{R} \times \mathbb{Z}(2))$. Define a distribution $\mu_{\mathbb{R}} \in \mathrm{M}^1(\mathbb{R})$ by the formula

$$\mu_{\mathbb{R}}(E) = \mu(E \times \mathbb{Z}(2)),$$

where E is a Borel subset of $\mathbb{R}$.

DEFINITION 11.3. Let $\mu \in \mathrm{M}^1(\mathbb{R} \times \mathbb{Z}(2))$. We say that $\mu \in \Lambda$ if $\mu_{\mathbb{R}} \in \Gamma(\mathbb{R})$. In other words, $\mu \in \Lambda$ if and only if $\hat{\mu}(s, 0)$ is the characteristic function of a Gaussian distribution on the real line.

Let $\mu \in \Lambda$. Then μ can be obtained as follows. Let γ be the Gaussian distribution on $\mathbb{R}$ with the characteristic function $\hat{\gamma}(s) = e^{-\sigma s^2 + i\beta s}$, $s \in \mathbb{R}$, where $\sigma \ge 0$, $\beta \in \mathbb{R}$. Let $\gamma = \lambda + \omega$, where λ and ω are measures on $\mathbb{R}$. Define a distribution $\mu \in \mathrm{M}^1(\mathbb{R} \times \mathbb{Z}(2))$ as follows:

$$\mu(E \times \{k\}) = \begin{cases} \lambda(E), & \text{if} \quad k = 0, \\ \omega(E), & \text{if} \quad k = 1, \end{cases}$$

where E is a Borel subset of $\mathbb{R}$. Then $\mu \in \Lambda$ and $\mu_{\mathbb{R}} = \gamma$. We note that the classes Θ and Λ are subsemigroups of $\mathrm{M}^1(\mathbb{R} \times \mathbb{Z}(2))$ and $\Theta \subset \Lambda$.

For what follows, we need the following lemmas.

LEMMA 11.4 ([**69**, Lemma 1.5.1]). *Let $\varepsilon > 0$. Consider the equation*

$$\sum_{j=1}^{n} \psi_j(s_1 + c_j s_2) = 0, \quad |s_1| < \varepsilon, \ |s_2| < \varepsilon,$$

where all numbers c_j are pairwise distinct and the complex-valued functions $\psi_j(s)$ of a real variable s are continuous. Then the functions $\psi_j(s)$ are polynomials in some neighborhood of the zero.

LEMMA 11.5. *Suppose $\mu \in \Theta$ and the characteristic function $\hat{\mu}(s, l)$ does not vanish. If $\mu = \mu_1 * \mu_2$, where $\mu_j \in \mathrm{M}^1(\mathbb{R} \times \mathbb{Z}(2))$, then $\mu_j \in \Theta$, $j = 1, 2$.*

PROOF. We have

$$\hat{\mu}(s,l) = \hat{\mu}_1(s,l)\hat{\mu}_2(s,l), \quad s \in \mathbb{R}, \ l \in \mathbb{Z}(2). \tag{11.4}$$

Substituting $l = 0$ in (11.4) and taking into account Definition 11.2, we obtain by Cramér's decomposition theorem for the Gaussian distribution on the real line

$$\hat{\mu}_j(s,0) = \exp\{-\sigma_j s^2 + i\beta_j s\}, \quad s \in \mathbb{R}, \ j = 1, 2, \tag{11.5}$$

where $\sigma_j \ge 0$, $\beta_j \in \mathbb{R}$. By Proposition 2.13, $\hat{\mu}_j(s, 1)$ are entire functions. Moreover, (11.3) and (11.4) imply that they do not vanish. By Proposition 2.13, it follows from (11.5) that the entire functions $\hat{\mu}_j(s,1)$ are of at most order 2 and type σ_j. Taking into account that $\hat{\mu}_j(-s,1) = \overline{\hat{\mu}_j(s,1)}$, by the Hadamard theorem on the representation of an entire function of finite order and Lemma 11.1, we obtain the desired statement. □

The following statement can be viewed as an analogue of Heyde's theorem for the group $\mathbb{R} \times \mathbb{Z}(2)$.

THEOREM 11.6. *Let a_j, b_j, $j = 1, 2, \dots, n$, $n \geq 2$, be topological automorphisms of the group $\mathbb{R} \times \mathbb{Z}(2)$ satisfying the conditions $b_i a_i^{-1} + b_j a_j^{-1} \neq 0$ for all i, j. Let ξ_j be independent random variables with values in $\mathbb{R} \times \mathbb{Z}(2)$ and distributions μ_j. Assume that the conditional distribution of the linear form $L_2 = b_1\xi_1 + \cdots + b_n\xi_n$ given $L_1 = a_1\xi_1 + \cdots + a_n\xi_n$ is symmetric. Then the following alternative holds.*

1. *All distributions μ_j belong to the class Λ, and at least one of the distributions μ_j is represented in the form $\mu_j = \gamma_j * m_{\mathbb{Z}(2)}$, where $\gamma_j \in \Gamma(\mathbb{R})$.*
2. *All distributions μ_j belong to the class Θ and their characteristic functions do not vanish.*

PROOF. Passing to the new random variables $\zeta_j = a_j\xi_j$, we reduce the proof of the theorem to the case when L_1 and L_2 are of the form $L_1 = \xi_1 + \cdots + \xi_n$ and $L_2 = b_1\xi_1 + \cdots + b_n\xi_n$ and b_j satisfy the conditions $b_i + b_j \neq 0$ for all i, j. By Lemma 9.1, the characteristic functions $\hat{\mu}_j(s,l)$ satisfy equation (9.1) which takes the form

$$\prod_{j=1}^{n} \hat{\mu}_j(s_1+b_js_2, l_1+l_2) = \prod_{j=1}^{n} \hat{\mu}_j(s_1-b_js_2, l_1+l_2), \quad s_1, s_2 \in \mathbb{R},\ l_1, l_2 \in \mathbb{Z}(2). \tag{11.6}$$

Substituting $l_1 = l_2 = 0$ in equation (11.6), we get

$$\prod_{j=1}^{n} \hat{\mu}_j(s_1 + b_js_2, 0) = \prod_{j=1}^{n} \hat{\mu}_j(s_1 - b_js_2, 0), \quad s_1, s_2 \in \mathbb{R}. \tag{11.7}$$

Taking into account Lemma 9.1 and Heyde's theorem, equation (11.7) implies that

$$\hat{\mu}_j(s,0) = e^{-\sigma_j s^2 + i\beta_j s}, \quad s \in \mathbb{R},\ j = 1, 2, \dots, n, \tag{11.8}$$

where $\sigma_j \geq 0$, $\beta_j \in \mathbb{R}$. Substituting (11.8) into equation (11.7), we obtain

$$\sum_{j=1}^{n} b_j\sigma_j = 0, \quad \sum_{j=1}^{n} b_j\beta_j = 0. \tag{11.9}$$

It follows from (11.8) that $\mu_j \in \Lambda$, $j = 1, 2, \dots, n$. Denote by γ_j the Gaussian distributions on the group $\mathbb{R} \times \mathbb{Z}(2)$ with the characteristic functions

$$\hat{\gamma}_j(s,l) = \hat{\mu}_j(s,0), \quad s \in \mathbb{R},\ l \in \mathbb{Z}(2),\ j = 1, 2, \dots, n. \tag{11.10}$$

Since $\hat{\gamma}_j(0,l) = 1$ for all $l \in \mathbb{Z}(2)$, by Proposition 2.10, $\sigma(\gamma_j) \subset A(\mathbb{R}\times\mathbb{Z}(2), \mathbb{Z}(2)) = \mathbb{R}$, i.e., $\gamma_j \in \Gamma(\mathbb{R})$. Inasmuch as $A(\mathbb{R} \times \mathbb{Z}(2), \mathbb{Z}(2)) = \mathbb{R}$, it follows from (2.3) that

$$\widehat{m}_{\mathbb{Z}(2)}(s,l) = \begin{cases} 1, & \text{if} \quad s \in \mathbb{R},\ l = 0, \\ 0, & \text{if} \quad s \in \mathbb{R},\ l = 1. \end{cases} \tag{11.11}$$

Substituting $l_1 = 1$, $l_2 = 0$ in equation (11.6), we obtain

$$\prod_{j=1}^{n} \hat{\mu}_j(s_1 + b_js_2, 1) = \prod_{j=1}^{n} \hat{\mu}_j(s_1 - b_js_2, 1), \quad s_1, s_2 \in \mathbb{R}. \tag{11.12}$$

Assume that at least for one $j = j_0$ we have $\hat{\mu}_{j_0}(0,1) = 0$. Substituting $s_1 = b_{j_0}s$, $s_2 = s$ in equation (11.12), we get

$$\prod_{j=1}^{n} \hat{\mu}_j((b_{j_0} + b_j)s, 1) = 0, \quad s \in \mathbb{R}. \tag{11.13}$$

By Proposition 2.13, it follows from (11.8) that $\hat{\mu}_j(s,1)$, $j = 1, 2, \dots, n$, are entire functions. Since by the assumption $b_i + b_j \neq 0$ for all i, j, we find from (11.13) that $\hat{\mu}_j(s,1) = 0$ for all $s \in \mathbb{R}$, for at least one of the characteristic functions $\hat{\mu}_j(s,l)$. Taking into account (11.10) and (11.11), this characteristic function can be represented in the form

$$\hat{\mu}_j(s,l) = \hat{\gamma}_j(s,l)\widehat{m}_{\mathbb{Z}(2)}(s,l), \quad s \in \mathbb{R},\ l \in \mathbb{Z}(2).$$

This implies that $\mu_j = \gamma_j * m_{\mathbb{Z}(2)}$.

Assume now that $\hat{\mu}_j(0,1) \neq 0$ for all $j = 1, 2, \dots, n$. Then there exists $\varepsilon > 0$ such that $\hat{\mu}_j(s,1) \neq 0$, $j = 1, 2, \dots, n$, for all $|s| < \varepsilon$. Put

$$\psi_j(s) = \ln \hat{\mu}_j(s,1), \quad |s| < \varepsilon,\ j = 1, 2, \dots, n$$

(we consider the principal branch of the logarithm). It follows from equation (11.12) that in a neighborhood of the zero the continuous functions $\psi_j(s)$ satisfy the equation

$$\sum_{j=1}^{n} (\psi_j(s_1 + b_j s_2) - \psi_j(s_2 - b_j s_2)) = 0, \quad s_1, s_2 \in \mathbb{R}. \tag{11.14}$$

Assume first that all numbers b_j are pairwise distinct. Since by the assumption $b_i + b_j \neq 0$ for all i, j, we may apply Lemma 11.4. It follows from equation (11.14) that in some neighborhood of the zero the continuous functions $\psi_j(s)$ are polynomials. Inasmuch as $\hat{\mu}_j(s,1)$ are entire functions, we have the representation

$$\hat{\mu}_j(s,1) = \exp\{\psi_j(s)\}, \quad j = 1, 2, \dots, n,$$

in the complex plane. By Proposition 2.13, it follows from (11.8) that the entire functions $\hat{\mu}_j(s,1)$ are of at most order 2. For this reason the degrees of the polynomials $\psi_j(s)$ are at most 2. Taking into account that $\hat{\mu}_j(-s,1) = \overline{\hat{\mu}_j(s,1)}$ and $|\hat{\mu}_j(s,1)| \leq 1$ for all $s \in \mathbb{R}$, we get from (2.5) and (11.8) the representations

$$\hat{\mu}_j(s,1) = \kappa_j \exp\{-\sigma'_j s^2 + i\beta'_j s\}, \quad s \in \mathbb{R},\ j = 1, 2, \dots, n, \tag{11.15}$$

where $0 \leq \sigma'_j \leq \sigma_j$, $\beta'_j, \kappa_j \in \mathbb{R}$, $|\kappa_j| \leq 1$, $\kappa_j \neq 0$. It follows from (11.8) and (11.15) that $\mu_j \in \Theta$, $j = 1, 2, \dots, n$. Moreover, all the characteristic functions $\hat{\mu}_j(s,l)$ do not vanish. Thus, in the case when all numbers b_j are pairwise distinct, the theorem is proved.

Let us get rid of the restriction that all numbers b_j are pairwise distinct. Assume that $b_1 = \dots = b_{j_1}$, $b_{j_1+1} = \dots = b_{j_2}$, $\dots$, $b_{j_{l-1}+1} = \dots = b_{j_l} = b_n$, where the numbers $b_{j_1}, b_{j_2}, \dots, b_{j_l}$ are pairwise distinct. Put $\eta_1 = \xi_1 + \dots + \xi_{j_1}$, $\eta_2 = \xi_{j_1+1} + \dots + \xi_{j_2}$, $\dots$, $\eta_l = \xi_{j_{l-1}+1} + \dots + \xi_{j_l}$. Then $\eta_1, \dots, \eta_l$ are the independent random variables with values in the group $\mathbb{R} \times \mathbb{Z}(2)$ and distributions $\nu_1 = \mu_1 * \dots * \mu_{j_1}$, $\nu_2 = \mu_{j_1+1} * \dots * \mu_{j_2}$, $\dots$, $\nu_l = \mu_{j_{l-1}+1} * \dots * \mu_{j_l}$. The linear forms L_1 and L_2 of the random variables η_j are of the form $L_1 = \eta_1 + \dots + \eta_l$ and $L_2 = b_{j_1}\eta_1 + \dots + b_{j_l}\eta_l$. Since the numbers $b_{j_1}, b_{j_2}, \dots, b_{j_l}$ are pairwise distinct, as proved above, $\nu_j \in \Theta$, $j = 1, 2, \dots, l$. The characteristic functions of the distributions ν_j do not vanish.

By Lemma 11.5, this implies that $\mu_j \in \Theta$, $j = 1, 2, \ldots, n$, and the characteristic functions $\hat{\mu}_j(s, l)$ do not vanish. $\square$

COROLLARY 11.7. *Assume that in Theorem 11.6 $a_j = 1$ and $b_j > 0$, $j = 1, 2, \ldots, n$. Then some shifts of the distributions μ_j are supported in $\mathbb{Z}(2)$.*

PROOF. It follows from (11.9) that $\sigma_1 = \cdots = \sigma_n = 0$. Put $\nu_j = \mu_j * \bar{\mu}_j$. Then $\hat{\nu}_j(s, 0) = 1$ for all $s \in \mathbb{R}$. By Proposition 2.10, this implies that $\sigma(\nu_j) \subset A(\mathbb{R} \times \mathbb{Z}(2), \mathbb{R}) = \mathbb{Z}(2)$, $j = 1, 2, \ldots, n$. The required assertion follows from Proposition 2.1. $\square$

REMARK 11.8. Consider the group $\mathbb{R} \times \mathbb{Z}(2)$ and let $b_j \in \mathrm{Aut}(\mathbb{R} \times \mathbb{Z}(2))$, $j = 1, 2 \ldots, n$. We can construct distributions μ_j such as in items 1 and 2 of Theorem 11.6 with the following property: if ξ_j are independent random variables with values in the group $\mathbb{R} \times \mathbb{Z}(2)$ and distributions μ_j, then the conditional distribution of the linear form $L_2 = b_1\xi_1 + \cdots + b_n\xi_n$ given $L_1 = \xi_1 + \cdots + \xi_n$ is symmetric.

1. Let γ_j be Gaussian distributions on the group $\mathbb{R}$ with the characteristic functions

$$\hat{\gamma}_j(s) = \exp\{-\sigma_j s^2 + i\beta_j s\}, \quad s \in \mathbb{R},\ \sigma_j > 0,\ \beta_j \in \mathbb{R},\ j = 1, 2, \ldots, n.$$

Let μ_1 be a distribution on the group $\mathbb{R} \times \mathbb{Z}(2)$ of the form $\mu_1 = \gamma_1 * m_{\mathbb{Z}(2)}$. Assume that conditions (11.9) hold. Let $\gamma_j = \lambda_j + \omega_j$, where λ_j and ω_j are measures on $\mathbb{R}$. Define distributions $\mu_j \in \mathrm{M}^1(\mathbb{R} \times \mathbb{Z}(2))$, $j = 2, \ldots, n$, as follows

$$\mu_j(E \times \{k\}) = \begin{cases} \lambda_j(E), & \text{if } k = 0, \\ \omega_j(E), & \text{if } k = 1, \end{cases}$$

where E is a Borel subset of $\mathbb{R}$. It is obvious that all $\mu_j \in \Lambda$ and $(\mu_j)_{\mathbb{R}} = \gamma_j$. Taking into account that $\hat{\mu}_j(s, 0) = \hat{\gamma}_j(s)$, $s \in \mathbb{R}$ and (11.11), it is easy to see that the characteristic functions $\hat{\mu}_j(s, l)$ satisfy equation (11.6). By Lemma 9.1, this implies that the conditional distribution of the linear form $L_2 = b_1\xi_1 + \cdots + b_n\xi_n$ given $L_1 = \xi_1 + \cdots + \xi_n$ is symmetric.

2. Let $\mu_j \in \Theta$, $j = 1, 2, \ldots, n$. Assume that the characteristic functions $\hat{\mu}_j(s, l)$ are of the form

$$\hat{\mu}_j(s, l) = \begin{cases} \exp\{-\sigma_j s^2 + i\beta_j s\}, & \text{if } s \in \mathbb{R},\ l = 0, \\ \kappa_j \exp\{-\sigma'_j s^2 + i\beta'_j s\}, & \text{if } s \in \mathbb{R},\ l = 1. \end{cases}$$

Suppose that conditions (11.9) hold and the equalities

$$\sum_{j=1}^{n} b_j\sigma'_j = 0, \quad \sum_{j=1}^{n} b_j\beta'_j = 0 \tag{11.16}$$

are fulfilled. It follows from (11.9) and (11.16) that the characteristic functions $\hat{\mu}_j(s, l)$ satisfy equation (11.6). By Lemma 9.1, this implies that the conditional distribution of the linear form $L_2 = b_1\xi_1 + \cdots + b_n\xi_n$ given $L_1 = \xi_1 + \cdots + \xi_n$ is symmetric.

Thus, we cannot narrow in Theorem 11.6 the class of distributions which is characterized by the symmetry of the conditional distribution of one linear form of n independent random variables given another.

PROPOSITION 11.9. *Let X be a locally compact Abelian group. Put $G = X_{(2)}$ and assume that $G \cong \mathbb{Z}(2)$. Let ξ_1 and ξ_2 be independent random variables with values in the group X and distributions μ_1 and μ_2. The conditional distribution of the linear form $L_2 = \xi_1 - \xi_2$ given $L_1 = \xi_1 + \xi_2$ is symmetric if and only if either $\mu_1 = \mu_2 * \omega_2$ or $\mu_2 = \mu_1 * \omega_1$, where $\omega_j \in \mathrm{M}^1(G)$, $j = 1, 2$.*

PROOF. Necessity. Denote by Y the character group of the group X. By Theorems 1.9 and 1.12, we have $(Y/\overline{Y^{(2)}})^* \cong A(X, \overline{Y^{(2)}}) = X_{(2)}$. It follows from $G = X_{(2)} \cong \mathbb{Z}(2)$ that $Y = \overline{Y^{(2)}} \cup (y_0 + \overline{Y^{(2)}})$ is a $\overline{Y^{(2)}}$-coset decomposition of the group Y, where $y_0 \notin \overline{Y^{(2)}}$. Inasmuch as $A(Y, G) = \overline{Y^{(2)}}$, it follows from (2.3) that

$$\widehat{m}_G(y) = \begin{cases} 1, & \text{if } \ y \in \overline{Y^{(2)}}, \\ 0, & \text{if } \ y \in y_0 + \overline{Y^{(2)}}. \end{cases} \tag{11.17}$$

Assume that the conditional distribution of the linear form L_2 given L_1 is symmetric. By Corollary 9.2, the characteristic functions $\hat{\mu}_j(y)$ satisfy equation (9.3) which takes the form

$$\hat{\mu}_1(u + v)\hat{\mu}_2(u - v) = \hat{\mu}_1(u - v)\hat{\mu}_2(u + v), \quad u, v \in Y. \tag{11.18}$$

Substituting $u = v = y$ in equation (11.18), we get $\hat{\mu}_1(2y) = \hat{\mu}_2(2y)$ for all $y \in Y$. Hence

$$\hat{\mu}_1(y) = \hat{\mu}_2(y), \quad y \in \overline{Y^{(2)}}. \tag{11.19}$$

Substituting $u = y$, $v = y + y_0$ in equation (11.18), we obtain

$$\hat{\mu}_1(2y + y_0)\hat{\mu}_2(y_0) = \hat{\mu}_1(y_0)\hat{\mu}_2(2y + y_0), \quad y \in Y. \tag{11.20}$$

1. Assume that there exists an element $y_0 \notin \overline{Y^{(2)}}$ such that either $\hat{\mu}_1(y_0) \neq 0$ or $\hat{\mu}_2(y_0) \neq 0$. Suppose for definiteness that $\hat{\mu}_1(y_0) \neq 0$. Two cases are possible: either $\hat{\mu}_2(y_0) = 0$ or $\hat{\mu}_2(y_0) \neq 0$.

Suppose $\hat{\mu}_2(y_0) = 0$. Then (11.20) implies that $\hat{\mu}_2(y) = 0$ for all $y \in y_0 + \overline{Y^{(2)}}$. Then it follows from (11.17) and (11.19) that $\hat{\mu}_2(y) = \hat{\mu}_1(y)\widehat{m}_G(y)$, $y \in Y$. This implies that $\mu_2 = \mu_1 * m_G$.

Suppose $\hat{\mu}_2(y_0) \neq 0$. Let $|\hat{\mu}_2(y_0)| \leq |\hat{\mu}_1(y_0)|$. Put $d_1 = \hat{\mu}_2(y_0)/\hat{\mu}_1(y_0)$. It follows from (11.20) that d_1 is a real number and

$$\hat{\mu}_2(y) = d_1\hat{\mu}_1(y), \quad y \in y_0 + \overline{Y^{(2)}}. \tag{11.21}$$

Let ω_1 be the distribution on the group G with the characteristic function

$$\hat{\omega}_1(y) = \begin{cases} 1, & \text{if } \ y \in \overline{Y^{(2)}}, \\ d_1, & \text{if } \ y \in y_0 + \overline{Y^{(2)}}. \end{cases} \tag{11.22}$$

We find from (11.19), (11.21), and (11.22) that $\hat{\mu}_2(y) = \hat{\mu}_1(y)\hat{\omega}_1(y)$, $y \in Y$. This implies that $\mu_2 = \mu_1 * \omega_1$. If $|\hat{\mu}_1(y_0)| \leq |\hat{\mu}_2(y_0)|$, we reason similarly, define d_2, ω_2 and get that $\mu_1 = \mu_2 * \omega_2$.

2. Assume that $\hat{\mu}_1(y) = \hat{\mu}_2(y) = 0$ for all $y \notin \overline{Y^{(2)}}$. Taking into account (11.19), this implies that $\mu_1 = \mu_2$. Moreover, it follows from (11.17) that $\mu_1 = \mu_2 * m_G$ and $\mu_2 = \mu_1 * m_G$.

Sufficiency follows directly from Corollary 9.2. $\square$

REMARK 11.10. Compare Theorem 11.6 with an analogue of the Skitovich–Darmois theorem for the group $\mathbb{R} \times \mathbb{Z}(2)$. It is not difficult to prove that the following assertion holds.

Let a_j, b_j, $j = 1, 2, \ldots, n$, $n \geq 2$, be topological automorphisms of the group $\mathbb{R} \times \mathbb{Z}(2)$. Let ξ_j be independent random variables with values in $\mathbb{R} \times \mathbb{Z}(2)$ and distributions μ_j. Assume that the linear forms $L_1 = a_1\xi_1 + \cdots + a_n\xi_n$ and $L_2 = b_1\xi_1 + \cdots + b_n\xi_n$ are independent. Then all distributions μ_j are Gaussian.

We see that on the one hand, the class of Gaussian distributions on the real line is characterized both by the independence of two linear forms of independent random variables and by the symmetry of the conditional distribution of one linear form given another. On the other hand, on the group $\mathbb{R} \times \mathbb{Z}(2)$ the independence of two linear forms of independent random variables characterizes only the class of Gaussian distributions, whereas the symmetry of the conditional distribution of one linear form given another characterizes a much wider class of distributions (see Theorem 11.6).

Consider the group $X = \mathbb{R} \times \mathbb{T}$ with character group $Y = \mathbb{R} \times \mathbb{Z}$. We recall that each automorphism $\alpha \in \mathrm{Aut}(X)$ is defined by a matrix $\begin{pmatrix} a & b \\ 0 & \pm 1 \end{pmatrix}$, where $a, b \in \mathbb{R}$, $a \neq 0$, and α is given by the formula

$$\alpha(t, z) = (at, e^{ibt}z^{\pm 1}), \quad t \in \mathbb{R},\ z \in \mathbb{T}.$$

The adjoint automorphism $\widetilde{\alpha} \in \mathrm{Aut}(Y)$ is of the form

$$\widetilde{\alpha}(s, n) = (as + bn, \pm n), \quad s \in \mathbb{R},\ n \in \mathbb{Z}.$$

Put $G = \{(0, \pm 1)\}$, i.e., G is the subgroup of X generated by the element of X of order 2. Let $\alpha \in \mathrm{Aut}(X)$. Assume that α is of the form $\alpha = \begin{pmatrix} a & b \\ 0 & 1 \end{pmatrix}$, where $a \neq 1$. Define the continuous monomorphism $\tau\colon \mathbb{R} \times \mathbb{Z}(2) \to X$ by the formula

$$\tau(t, k) = \left(t, (-1)^k e^{\frac{ibt}{a-1}}\right), \quad t \in \mathbb{R},\ k \in \mathbb{Z}(2). \tag{11.23}$$

Put

$$F = \tau(\mathbb{R}) = \left\{\left(t, e^{\frac{ibt}{a-1}}\right) : t \in \mathbb{R}\right\}. \tag{11.24}$$

Then F is a closed subgroup of X topologically isomorphic to $\mathbb{R}$. Note that $G = \tau(\mathbb{Z}(2))$.

We will study an analogue of Heyde's theorem for two independent random variables with values in the group $X = \mathbb{R} \times \mathbb{T}$. Without loss of generality, we can assume that L_1 and L_2 are of the form $L_1 = \xi_1 + \xi_2$ and $L_2 = \xi_1 + \alpha\xi_2$, where $\alpha \in \mathrm{Aut}(X)$. We need the following lemma.

LEMMA 11.11. *Let $X = \mathbb{R} \times \mathbb{T}$ and let $\alpha = \begin{pmatrix} a & b \\ 0 & 1 \end{pmatrix} \in \mathrm{Aut}(X)$. If $a \neq 1$, let $\tau\colon \mathbb{R} \times \mathbb{Z}(2) \to X$ be given by (11.23) and let F be given by (11.24). Let G be the subgroup of X generated by the element of X of order 2. Let ξ_1 and ξ_2 be independent random variables with values in the group X and distributions μ_1 and μ_2. Assume that the conditional distribution of the linear form $L_2 = \xi_1 + \alpha\xi_2$ given $L_1 = \xi_1 + \xi_2$ is symmetric. Then there exist elements $x_j \in X$ such that* *if $a \neq 1$, then the distributions $\lambda_j = \mu_j * E_{-x_j}$ are supported in the subgroup $F \times G$, and if $a = 1$, then λ_j are supported in the subgroup G. Moreover, if η_j are independent random variables with values in the group X and distributions λ_j, then the conditional distribution of the linear form $N_2 = \eta_1 + \alpha\eta_2$ given $N_1 = \eta_1 + \eta_2$ is symmetric.*

PROOF. Put $L = \mathrm{Ker}(I - \widetilde{\alpha})$. It is obvious that $L = \{(s, n) : (1-a)s - bn = 0\}$. Put $K = A\left(X, L^{(2)}\right)$. By Corollary 9.2, the characteristic functions $\hat{\mu}_j(y)$ satisfy equation (9.3). Consider the restriction of equation (9.3) to the subgroup L. Since $\widetilde{\alpha}y = y$ for all $y \in L$, we have

$$\hat{\mu}_1(u+v)\hat{\mu}_2(u+v) = \hat{\mu}_1(u-v)\hat{\mu}_2(u-v), \quad u, v \in L. \tag{11.25}$$

Substituting $u = v = y$ in equation (11.25), we obtain

$$\hat{\mu}_1(y)\hat{\mu}_2(y) = 1, \quad y \in L^{(2)}. \tag{11.26}$$

This implies that $|\hat{\mu}_1(y)| = |\hat{\mu}_2(y)| = 1$ for all $y \in L^{(2)}$. It follows from this that there exist elements $p_j \in X$ such that

$$\hat{\mu}_j(y) = (p_j, y), \quad y \in L^{(2)}, \ j = 1, 2. \tag{11.27}$$

Substituting (11.27) in (11.26), we get that $p_1 + p_2 \in K$. Let $p_j = (t_j, z_j)$, $t_j \in \mathbb{R}$, $z_j \in \mathbb{T}$, $j = 1, 2$.

1. Suppose $a \neq 1$. We have

$$L^{(2)} = \left\{\left(\frac{2bn}{1-a}, 2n\right) : n \in \mathbb{Z}\right\}$$

and hence

$$K = A\left(X, L^{(2)}\right) = \left\{\left(t, \pm e^{\frac{ibt}{a-1}}\right) : t \in \mathbb{R}\right\}.$$

It is clear that $K = F \times G$ and τ is a topological isomorphism of the groups $\mathbb{R} \times \mathbb{Z}(2)$ and K.

Put $\tilde{x} = \left(t_1, e^{\frac{ibt_1}{a-1}}\right) \in K$. Verify that $x_1 = p_1 - \tilde{x}$ and $x_2 = -p_1 + \tilde{x}$ are the required elements. Consider the distributions

$$\lambda_j = \mu_j * E_{-x_j}, \quad j = 1, 2.$$

It follows from $p_1 + p_2, \tilde{x} \in K$ that $\hat{\lambda}_1(y) = \hat{\lambda}_2(y) = 1$ for all $y \in L^{(2)}$. By Proposition 2.10, this implies that $\sigma(\lambda_j) \subset A\left(X, L^{(2)}\right) = K$, $j = 1, 2$. It is obvious that $x_j \in \mathbb{T}$. Observe that the restriction of α to the subgroup $\mathbb{T}$ is the identity automorphism. Hence $x_1 + \alpha x_2 = 0$. This implies that the characteristic functions $f_j(y) = (-x_j, y)$ satisfy equation (9.3). Since the characteristic functions $\hat{\mu}_j(y)$ satisfy equation (9.3), we obtain that the characteristic functions $\hat{\lambda}_j(y)$ also satisfy equation (9.3). Let η_j be independent random variables with values in the group X and distributions λ_j. Then by Corollary 9.2, the conditional distribution of the linear form $N_2 = \eta_1 + \alpha\eta_2$ given $N_1 = \eta_1 + \eta_2$ is symmetric.

2. Suppose $a = 1$, $b = 0$, i.e., $\alpha = I$. Hence $L = Y$ and then $K = A\left(X, L^{(2)}\right) = G$. It is easy to see that we can put $x_j = p_j$, $j = 1, 2$.

3. Suppose $a = 1$, $b \neq 0$. Then $L = \{(s, 0) : s \in \mathbb{R}\} = \mathbb{R}$ and hence $K = A(X, \mathbb{R}) = \mathbb{T}$. This implies that $t_1 + t_2 = 0$. Put $\hat{x} = (0, z_1 z_2 e^{-ibt_1})$ and $x_1 = p_1 - \hat{x}$, $x_2 = p_2$. Consider the distributions $\lambda_j = \mu_j * E_{-x_j}$. Since $\hat{x} \in \mathbb{T}$, we obtain from (11.27) that $\hat{\lambda}_1(s, 0) = 1$ for all $s \in \mathbb{R}$. By Proposition 2.10, it follows

from this that $\sigma(\lambda_j) \subset A(X, \mathbb{R}) = \mathbb{T}$, $j = 1, 2$. It is obvious that $x_1 + \alpha x_2 = 0$. This implies that the characteristic functions $f_j(y) = (-x_j, y)$ satisfy equation (9.3). Taking into account that the characteristic functions $\hat{\mu}_j(y)$ satisfy equation (9.3), we get that the characteristic functions $\hat{\lambda}_j(y)$ also satisfy equation (9.3). Let η_j be independent random variables with values in the group X and distributions λ_j. By Corollary 9.2, the conditional distribution of the linear form $N_2 = \eta_1 + \alpha\eta_2$ given $N_1 = \eta_1 + \eta_2$ is symmetric. Taking into account that the restriction of α to the subgroup $\mathbb{T}$ is the identity automorphism, we reduce the proof to the case when $X = \mathbb{T}$ and $\alpha = I$. To complete the proof we argue in the same way as in case 2. □

THEOREM 11.12. *Let $X = \mathbb{R} \times \mathbb{T}$ and let $\alpha = \begin{pmatrix} a & b \\ 0 & 1 \end{pmatrix} \in \mathrm{Aut}(X)$. If $a \neq 1$, let $\tau\colon \mathbb{R} \times \mathbb{Z}(2) \to X$ be given by (11.23) and let F be given by (11.24). Let G be the subgroup of X generated by the element of X of order 2. Let ξ_1 and ξ_2 be independent random variables with values in the group X and distributions μ_1 and μ_2. If the conditional distribution of the linear form $L_2 = \xi_1 + \alpha\xi_2$ given $L_1 = \xi_1 + \xi_2$ is symmetric, then the following statements hold.*

1. *If $a < 0$ and $a \neq -1$, then there is the following alternative.*

   1a. *The distributions μ_j are represented in the form $\mu_j = \tau(M_j) * E_{x_j}$, where $M_j \in \Lambda$, $x_j \in X$. Moreover, at least one of the distributions μ_j is represented in the form $\mu_j = \gamma_j * m_G * E_{x_j}$, where $\gamma_j \in \Gamma(F)$, $x_j \in X$.*

   1b. *The distributions μ_j are represented in the form $\mu_j = \tau(M_j) * E_{x_j}$, where $M_j \in \Theta$, $x_j \in X$. Moreover, the characteristic functions of μ_j do not vanish.*
2. *If $a = -1$, then the distributions μ_j are of the form $\mu_j = \lambda_j * E_{x_j}$, where $\lambda_j \in \mathrm{M}^1(F \times G)$ and $x_j \in X$. Furthermore, either $\lambda_2 = \lambda_1 * \omega_1$ or $\lambda_1 = \lambda_2 * \omega_2$, where $\omega_j \in \mathrm{M}^1(G)$.*
3. *If $a > 0$, then some shifts of the distributions μ_j are supported in G.*

PROOF. Assume that $a \neq 1$. We apply Lemma 11.11 and get independent random variables η_j with values in the group X and distributions λ_j. Put $K = F \times G$. Since $\sigma(\lambda_j) \subset K$, we may consider η_j as independent random variables with values in the group K. It is easy to see that the restriction of α to the subgroup K is a topological automorphism of the subgroup K and α operates on the elements of K as follows

$$\alpha\left(t, \pm e^{\frac{ibt}{a-1}}\right) = \left(at, \pm e^{\frac{ibat}{a-1}}\right), \quad t \in \mathbb{R}. \tag{11.28}$$

We also note that since τ is a topological isomorphism of the groups $\mathbb{R} \times \mathbb{Z}(2)$ and K, by Corollary 2.4, τ generates an isomorphism of the semigroups $\mathrm{M}^1(\mathbb{R} \times \mathbb{Z}(2))$ and $\mathrm{M}^1(K)$. We keep the notation τ for this isomorphism.

Statement 1. Put $\zeta_j = \tau^{-1}\eta_j$. Then ζ_j are independent random variables with values in the group $\mathbb{R} \times \mathbb{Z}(2)$. Denote by M_j the distribution of the random variable ζ_j. Then $\lambda_j = \tau(M_j)$, $j = 1, 2$. It follows from (11.28) that $\tau^{-1}\alpha = a\tau^{-1}$ and hence $\tau^{-1}\alpha\eta_j = a\zeta_j$. The symmetry of the conditional distribution of the linear form $N_2 = \eta_1 + \alpha\eta_2$ given $N_1 = \eta_1 + \eta_2$ implies the symmetry of the conditional distribution of the linear form $P_2 = \zeta_1 + a\zeta_2$ given $P_1 = \zeta_1 + \zeta_2$. Since $\tau(\Gamma(\mathbb{R})) = \Gamma(F)$ and $\tau(m_{\mathbb{Z}(2)}) = m_G$, statement 1 follows from Theorem 11.6.

Statement 2. Inasmuch as $a = -1$, it follows from (11.28) that the restriction of α to the subgroup K coincides with $-I$. It means that if we consider η_j as independent random variables with values in the group K, then the conditional distribution of the linear form $N_2 = \eta_1 - \eta_2$ given $N_1 = \eta_1 + \eta_2$ is symmetric. Applying Proposition 11.9 to the group K, the independent random variables η_1 and η_2, and the linear forms N_1 and N_2, we prove statement 2.

Statement 3. It follows from Lemma 11.11 that it suffices to consider the case when $a \neq 1$. We retain the notation used in the proof of statement 1. Applying Corollary 11.7 to the random variables ζ_j and the linear forms P_1 and P_2, we obtain that some shifts of the distributions M_j are supported in the subgroup $\mathbb{Z}(2)$. Since the semigroups $\mathrm{M}^1(\mathbb{R} \times \mathbb{Z}(2))$ and $\mathrm{M}^1(K)$ are isomorphic and $G = \tau(\mathbb{Z}(2))$, some shifts of the distributions λ_j are supported in the subgroup G. Hence some shifts of the distributions μ_j are also supported in G. □

REMARK 11.13. Let $X = \mathbb{R} \times \mathbb{T}$. Suppose $\alpha = \begin{pmatrix} a & b \\ 0 & 1 \end{pmatrix} \in \mathrm{Aut}(X)$, where a is as in each of statements 1–3 of Theorem 11.12. Then we can construct the distributions μ_1 and μ_2 as in the corresponding statements of Theorem 11.12 with the following property: if ξ_1 and ξ_2 are independent random variables with values in the group X and distributions μ_1 and μ_2, then the conditional distribution of the linear form $L_2 = \xi_1 + \alpha\xi_2$ given $L_1 = \xi_1 + \xi_2$ is symmetric. Thus, we cannot narrow the class of distributions in Theorem 11.12 which is characterized by the symmetry of the conditional distribution of one linear form of two independent random variables given another.

REMARK 11.14. Let $X = \mathbb{R} \times \mathbb{T}$ and let G be the subgroup of X generated by the element of X of order 2. Let $\alpha = \begin{pmatrix} a & b \\ 0 & 1 \end{pmatrix} \in \mathrm{Aut}(X)$. Theorem 11.12 gives us a complete description of possible distributions of independent random variables ξ_1 and ξ_2 with values in the group X provided that the conditional distribution of the linear form $L_2 = \xi_1 + \alpha\xi_2$ given $L_1 = \xi_1 + \xi_2$ is symmetric. Now we will discuss what happens when $\alpha = \begin{pmatrix} a & b \\ 0 & -1 \end{pmatrix} \in \mathrm{Aut}(X)$.

Let either $a > 0$ or $a = -1$ and $b = 0$. Assume that ξ_1 and ξ_2 are independent random variables with values in the group X and distributions μ_1 and μ_2. Suppose that the conditional distribution of the linear form $L_2 = \xi_1 + \alpha\xi_2$ given $L_1 = \xi_1 + \xi_2$ is symmetric. What can we say about the distributions μ_j? By Corollary 9.2, the characteristic functions $\hat{\mu}_j(y)$ satisfy equation (9.3), which takes the form

$$\begin{aligned}(11.29)\quad &\hat{\mu}_1(s_1 + s_2, n_1 + n_2)\hat{\mu}_2(s_1 + as_2 + bn_2, n_1 - n_2) \\ &= \hat{\mu}_1(s_1 - s_2, n_1 - n_2)\hat{\mu}_2(s_1 - as_2 - bn_2, n_1 + n_2), \quad s_1, s_2 \in \mathbb{R},\ n_1, n_2 \in \mathbb{Z}.\end{aligned}$$

1. Let $a > 0$. Substitute $n_1 = n_2 = 0$ in equation (11.29). It follows from Corollary 9.2 and Heyde's theorem that

$$\hat{\mu}_j(s, 0) = \exp\{-\sigma_j s^2 + i\beta_j s\}, \quad s \in \mathbb{R},\ j = 1, 2,$$

where $\sigma_j \geq 0$, $\beta_j \in \mathbb{R}$. Furthermore, $\sigma_1 + a\sigma_2 = 0$ and $\beta_1 + a\beta_2 = 0$. Since $a > 0$, we have $\sigma_1 = \sigma_2 = 0$. Thus,

$$\hat{\mu}_j(s, 0) = \exp\{i\beta_j s\}, \quad s \in \mathbb{R},\ j = 1, 2.$$

Put $\lambda_j = \mu_j * E_{(-\beta_j,1)}$. Since the characteristic functions $\hat{\mu}_j(s,n)$ satisfy equation (11.29) and $\beta_1 + a\beta_2 = 0$, the characteristic functions $\hat{\lambda}_j(s,n)$ also satisfy equation (11.29). Let η_j be independent random variables with values in the group X and distributions λ_j. By Corollary 9.2, the conditional distribution of the linear form $N_2 = \eta_1 + \alpha\eta_2$ given $N_1 = \eta_1 + \eta_2$ is symmetric. It is obvious that $\hat{\lambda}_1(s,0) = \hat{\lambda}_2(s,0) = 1$ for all $s \in \mathbb{R}$. By Proposition 2.10, it follows from this that $\sigma(\lambda_j) \subset A(X,\mathbb{R}) = \mathbb{T}$, $j = 1,2$. Note that the restriction of α to the subgroup $\mathbb{T}$ coincides with $-I$. This implies that if we consider η_j as independent random variables with values in the circle group $\mathbb{T}$ and distributions λ_j, then the conditional distribution of the linear form $N_2 = \eta_1 - \eta_2$ given $N_1 = \eta_1 + \eta_2$ is symmetric. Applying Proposition 11.9 to the circle group $\mathbb{T}$, the independent random variables η_1 and η_2, and the linear forms N_1 and N_2, we obtain that either $\lambda_2 = \lambda_1 * \omega_1$ or $\lambda_1 = \lambda_2 * \omega_2$, where $\omega_j \in \mathrm{M}^1(G)$.

2. Let $a = -1$ and $b = 0$. Then $\alpha = -I$ and the description of the distributions μ_j follows from Proposition 11.9.

We see that in cases 1 and 2 the description of possible distributions of independent random variables ξ_j which are characterized by the symmetry of the conditional distribution of the linear form $L_2 = \xi_1 + \alpha\xi_2$ given $L_1 = \xi_1 + \xi_2$ is reduced to the case when $L_2 = \xi_1 - \xi_2$, i.e., to Proposition 11.9.

Now suppose that either $a = -1$ and $b \neq 0$ or $a < 0$ and $a \neq -1$. In each of these cases we construct distributions μ_1 and μ_2 on the group X with the following property: if ξ_1 and ξ_2 are independent random variables with values in the group X and distributions μ_1 and μ_2, then the conditional distribution of the linear form $L_2 = \xi_1 + \alpha\xi_2$ given $L_1 = \xi_1 + \xi_2$ is symmetric.

3. Let $a = -1$ and $b \neq 0$. In this case we construct two different types of distributions μ_j.

3a. Consider the subgroup $H = \mathrm{Ker}(I + \alpha)$. Then

$$H = \left\{ \left(\frac{2\pi n}{b}, z \right) : n \in \mathbb{Z}, z \in \mathbb{T} \right\}.$$

It is obvious that H is topologically isomorphic to the group $\mathbb{Z} \times \mathbb{T}$. Let either $\mu_2 = \mu_1 * \omega_1$, where $\mu_1 \in \mathrm{M}^1(H)$, $\omega_1 \in \mathrm{M}^1(G)$ or $\mu_1 = \mu_2 * \omega_2$, where $\mu_2 \in \mathrm{M}^1(H)$, $\omega_2 \in \mathrm{M}^1(G)$. By Proposition 11.9, if ξ_1 and ξ_2 are independent random variables with values in the group H and distributions μ_1 and μ_2, then the conditional distribution of the linear form $L_2 = \xi_1 - \xi_2$ given $L_1 = \xi_1 + \xi_2$ is symmetric. Observe that the restriction of α to the subgroup H coincides with $-I$. Hence if we consider ξ_1 and ξ_2 as independent random variables with values in the group X, then the conditional distribution of the linear form $L_2 = \xi_1 + \alpha\xi_2$ given $L_1 = \xi_1 + \xi_2$ is also symmetric.

3b. Let $\{c_n\}_{n \in \mathbb{Z}}$ be a sequence of complex numbers satisfying the conditions:

(a) $c_0 = 1$,
(b) $c_{-n} = \bar{c}_n$ for all $n \in \mathbb{Z}$,
(c) $\sum_{n \in \mathbb{Z}} |c_n| < 2$.

It follows from this that

$$f(t) = \sum_{n \in \mathbb{Z}} c_n \exp\{-int\} > 0, \quad t \in \mathbb{R}.$$

It is also obvious that

$$\frac{1}{2\pi}\int_{-\pi}^{\pi} f(t)dt = 1.$$

Assume that $c_{2n} = 0$ for all $n \in \mathbb{Z}$, $n \neq 0$. Denote by λ the distribution on the circle group $\mathbb{T}$ with the density $r(e^{it}) = f(t)$ with respect to $m_{\mathbb{T}}$. Then $\hat{\lambda}(n) = c_n$, $n \in \mathbb{Z}$, in particular, $\hat{\lambda}(2n) = 0$ for all $n \in \mathbb{Z}$, $n \neq 0$. Let μ be an arbitrary distribution on the group $\mathbb{R}$. Consider the groups $\mathbb{R}$ and $\mathbb{T}$ as subgroups of X and put $\mu_1 = \mu \times m_{\mathbb{T}} = \mu * m_{\mathbb{T}}$, $\mu_2 = \mu \times \lambda = \mu * \lambda$. We then have

$$\hat{\mu}_1(s,0) = \hat{\mu}_2(s,0) = \hat{\mu}(s), \quad s \in \mathbb{R}, \tag{11.30}$$

$$\hat{\mu}_1(s,n) = \hat{\mu}_2(s,2n) = 0, \quad s \in \mathbb{R},\ n \in \mathbb{Z},\ n \neq 0. \tag{11.31}$$

It is not difficult to check that the characteristic functions $\hat{\mu}_j(s,n)$ satisfy the equation

$$\begin{aligned} &\hat{\mu}_1(s_1+s_2, n_1+n_2)\hat{\mu}_2(s_1-s_2+bn_2, n_1-n_2) \\ &\quad = \hat{\mu}_1(s_1-s_2, n_1-n_2)\hat{\mu}_2(s_1+s_2-bn_2, n_1+n_2), \quad s_1, s_2 \in \mathbb{R},\ n_1, n_2 \in \mathbb{Z}. \end{aligned} \tag{11.32}$$

Indeed, if $n_1 = n_2 = 0$, then taking into account (11.30), equation (11.32) turns into the equality

$$\hat{\mu}(s_1+s_2)\hat{\mu}(s_1-s_2) = \hat{\mu}(s_1-s_2)\hat{\mu}(s_1+s_2), \quad s_1, s_2 \in \mathbb{R}.$$

If either $n_1 \neq 0$ or $n_2 \neq 0$, then (11.31) implies that both parts of equation (11.32) are equal to zero. Let ξ_1 and ξ_2 be independent random variables with values in the group X and distributions μ_1 and μ_2. By Corollary 9.2, it follows from (11.32) that the conditional distribution of the linear form $L_2 = \xi_1 + \alpha\xi_2$ given $L_1 = \xi_1 + \xi_2$ is symmetric.

4. Let $a < 0$ and $a \neq -1$. In this case we construct four different types of distributions μ_j.

4a. Consider the subgroup $H = \mathrm{Ker}(I+\alpha)$. Then $H = \{(0,z) : z \in \mathbb{T}\} = \mathbb{T}$. In this case we argue as in case 3a.

4b. Let γ_j be Gaussian distributions on the group $\mathbb{R}$ with the characteristic functions $\hat{\gamma}_j(s) = \exp\{-\sigma_j s^2\}$, $j = 1, 2$. Assume that

$$\sigma_1 + a\sigma_2 = 0. \tag{11.33}$$

Consider the groups $\mathbb{R}$ and $\mathbb{T}$ as subgroups of X and put $\mu_1 = \gamma_1 \times m_{\mathbb{T}} = \gamma_1 * m_{\mathbb{T}}$, $\mu_2 = \gamma_2 \times \lambda = \gamma_2 * \lambda$, where λ is the distribution on the circle group $\mathbb{T}$ is constructed as in case 3b. Then the characteristic functions $\hat{\mu}_j(s,n)$ satisfy equality (11.29). Indeed, taking into account that $\hat{\mu}_j(s,0) = \hat{\gamma}_j(s)$, $j = 1, 2$, (11.33) implies that equality (11.29) holds for $n_1 = n_2 = 0$. If either $n_1 \neq 0$ or $n_2 \neq 0$, then (11.31) implies that both parts of equality (11.29) are equal to zero.

4c. Let $d \geq \frac{b^2}{(a+1)^2}$ and $\sigma > 0$. Consider the matrices

$$A = \sigma\begin{pmatrix} -a & -\frac{ab}{a+1} \\ -\frac{ab}{a+1} & d - \frac{b^2}{a+1} \end{pmatrix}, \quad B = \sigma\begin{pmatrix} 1 & \frac{b}{a+1} \\ \frac{b}{a+1} & d \end{pmatrix}.$$

It is easy to check that the matrices A and B are positive semidefinite and the equality

$$A + B\widetilde{\alpha} = 0 \tag{11.34}$$

holds. Consider Gaussian distributions μ_1 and μ_2 on the group X with the characteristic functions

$$\hat{\mu}_1(s,n) = \exp\{-\langle A(s,n),(s,n)\rangle\}, \quad s\in\mathbb{R},\ n\in\mathbb{Z},$$

$$\hat{\mu}_2(s,n) = \exp\{-\langle B(s,n),(s,n)\rangle\}, \quad s\in\mathbb{R},\ n\in\mathbb{Z}.$$

It follows from (11.34) that the characteristic functions $\hat{\mu}_j(s,n)$ satisfy equation (11.29).

4d. Define the continuous monomorphism $\theta\colon \mathbb{R}\times\mathbb{Z}(2)\to X$ by the formula

$$\theta(t,k) = \left(t, (-1)^k e^{\frac{ibt}{a+1}}\right), \quad t\in\mathbb{R},\ k\in\mathbb{Z}(2).$$

Put $S = \theta(\mathbb{R}\times\mathbb{Z}(2))$. Then S is a closed subgroup of the group X and S is topologically isomorphic to the group $\mathbb{R}\times\mathbb{Z}(2)$. It is easy to see that the restriction of α to the subgroup S is a topological automorphism of the group S. Let $\lambda_j\in\Theta$, $j=1,2$. Assume that the characteristic functions $\hat{\lambda}_j(s,l)$ are of the form

$$\hat{\lambda}_j(s,l) = \begin{cases} \exp\{-\sigma_j s^2\}, & \text{if } s\in\mathbb{R},\ l=0,\\ \kappa_j\exp\{-\sigma_j' s^2\}, & \text{if } s\in\mathbb{R},\ l=1.\end{cases}$$

Assume that the equalities $\sigma_1 + a\sigma_2 = 0$ and $\sigma_1' + a\sigma_2' = 0$ hold. Put $\mu_j = \theta(\lambda_j)$. It is not difficult to verify that the characteristic functions $\hat{\mu}_j(s,n)$ satisfy equation (11.29).

Let ξ_1 and ξ_2 be independent random variables with values in the group X and distributions μ_1 and μ_2, where μ_j are constructed as in cases 4b–4d. Since the characteristic functions $\hat{\mu}_j(s,n)$ satisfy equation (11.29), by Corollary 9.2, the conditional distribution of the linear form $L_2 = \xi_1+\alpha\xi_2$ given $L_1 = \xi_1+\xi_2$ is symmetric.

Cases 1–4 exhaust all possibilities for a and b, when $\alpha = \begin{pmatrix} a & b\\ 0 & -1\end{pmatrix}\in \mathrm{Aut}(X)$. The examples of distributions μ_j constructed in cases 3 and 4 show that when either $a=-1$ and $b\neq 0$ or $a<0$ and $a\neq -1$ we can hardly expect to get a reasonable description of distributions which are characterized by the symmetry of the conditional distribution of the linear form $L_2 = \xi_1+\alpha\xi_2$ given $L_1=\xi_1+\xi_2$.

From what has been said above, when we consider cases 1–4 in Remark 11.14, and Theorem 11.12 we get the following assertion.

PROPOSITION 11.15. *Let $X=\mathbb{R}\times\mathbb{T}$ and let G be the subgroup of X generated by the element of X of order 2. There is no topological automorphism α of the group X such that the following statements hold.*

1. *If ξ_1 and ξ_2 are independent random variables with values in the group X and distributions μ_1 and μ_2 with nonvanishing characteristic functions, then the symmetry of the conditional distribution of the linear form $L_2 = \xi_1+\alpha\xi_2$ given $L_1=\xi_1+\xi_2$ implies that $\mu_j=\gamma_j*\rho_j$, where γ_j are Gaussian distributions on X and $\rho_j\in \mathrm{M}^1(G)$.*
2. *There are independent random variables ξ_1 and ξ_2 with values in the group X and distributions $\mu_j=\gamma_j*\rho_j$, where γ_j are nondegenerate Gaussian distributions on X and $\rho_j\in\mathrm{M}^1(G)$, such that the conditional distribution of the linear form $L_2=\xi_1+\alpha\xi_2$ given $L_1=\xi_1+\xi_2$ is symmetric.*

It is interesting to note that if we consider the 2-dimensional torus $\mathbb{T}^2$ and G is the subgroup of $\mathbb{T}^2$ generated by all elements of $\mathbb{T}^2$ of order 2, then there exist topological automorphisms α of the group $\mathbb{T}^2$ such that statements 1 and 2 of Proposition 11.15 are fulfilled [**34**, Theorem 2], see also [**36**, Theorem 16.8].

Consider the group $X = \Sigma_{\boldsymbol{a}} \times \mathbb{T}$ with character group $Y = H_{\boldsymbol{a}} \times \mathbb{Z}$. Denote by $x = (g, z)$, where $g \in \Sigma_{\boldsymbol{a}}$, $z \in \mathbb{T}$, elements of the group X and by $y = (r, n)$, where $r \in H_{\boldsymbol{a}}$, $n \in \mathbb{Z}$, elements of the group Y. It is easy to verify that each automorphism $\alpha \in \mathrm{Aut}(X)$ is defined by a matrix $\begin{pmatrix} a & b \\ 0 & \pm 1 \end{pmatrix}$, where $a \in \mathrm{Aut}(\Sigma_{\boldsymbol{a}})$, $b \in H_{\boldsymbol{a}}$, and α is given by the formula

$$\alpha(g, z) = (ag, (g, b)z^{\pm 1}), \quad g \in \Sigma_{\boldsymbol{a}}, \ z \in \mathbb{T}.$$

Then the adjoint automorphism $\widetilde{\alpha} \in \mathrm{Aut}(Y)$ is of the form

$$\widetilde{\alpha}(r, n) = (ar + bn, \pm n), \quad r \in H_{\boldsymbol{a}}, \ n \in \mathbb{Z}.$$

We identify the automorphisms α and $\widetilde{\alpha}$ with the corresponding matrix $\begin{pmatrix} a & b \\ 0 & \pm 1 \end{pmatrix}$.

Assume that an $\boldsymbol{a}$-adic solenoid $\Sigma_{\boldsymbol{a}}$ contains no elements of order 2. Based on Theorem 11.12, we prove a statement which can be viewed as an analogue of Heyde's theorem for the group $X = \Sigma_{\boldsymbol{a}} \times \mathbb{T}$ for two independent random variables ξ_1 and ξ_2 with values in X. We assume that $L_1 = \xi_1 + \xi_2$, $L_2 = \xi_1 + \alpha\xi_2$, where $\alpha \in \mathrm{Aut}(X)$, and α is of the form $\alpha = \begin{pmatrix} a & b \\ 0 & 1 \end{pmatrix}$.

THEOREM 11.16. *Let $X = \Sigma_{\boldsymbol{a}} \times \mathbb{T}$, where the $\boldsymbol{a}$-adic solenoid $\Sigma_{\boldsymbol{a}}$ contains no elements of order 2. Let G be the subgroup of X generated by the element of X of order 2. Let $\alpha \in \mathrm{Aut}(X)$ and let $\alpha = \begin{pmatrix} a & b \\ 0 & 1 \end{pmatrix}$, where $\mathrm{Ker}(I + a) = \{0\}$. Let ξ_1 and ξ_2 be independent random variables with values in the group X and distributions μ_1 and μ_2 with nonvanishing characteristic functions. Assume that the conditional distribution of the linear form $L_2 = \xi_1 + \alpha\xi_2$ given $L_1 = \xi_1 + \xi_2$ is symmetric. If $a < 0$, then there is a continuous monomorphism $\phi\colon \mathbb{R} \times \mathbb{Z}(2) \to X$ such that the distributions μ_j are represented in the form $\mu_j = \phi(M_j) * E_{x_j}$, where $M_j \in \Theta$, $x_j \in X$. If $a > 0$, then some shifts of the distributions μ_j are supported in G.*

PROOF. Put $\nu_j = \mu_j * \bar{\mu}_j$. This implies that $\hat{\nu}_j(y) = |\hat{\mu}_j(y)|^2 > 0$ for all $y \in Y$, $j = 1, 2$. By Corollary 9.2, the characteristic functions $\hat{\mu}_j(y)$ satisfy equation (9.3). The characteristic functions $\hat{\nu}_j(y)$ also satisfy equation (9.3) which takes the form

$$\begin{aligned} (11.35) \quad & \hat{\nu}_1(r_1 + r_2, n_1 + n_2)\hat{\nu}_2(r_1 + ar_2 + bn_2, n_1 + n_2) \\ & = \hat{\nu}_1(r_1 - r_2, n_1 - n_2)\hat{\nu}_2(r_1 - ar_2 - bn_2, n_1 - n_2), \quad r_1, r_2 \in H_{\boldsymbol{a}}, \ n_1, n_2 \in \mathbb{Z}. \end{aligned}$$

Substituting $n_1 = n_2 = 0$ in equation (11.35), we get

$$(11.36) \quad \hat{\nu}_1(r_1 + r_2, 0)\hat{\nu}_2(r_1 + ar_2, 0) = \hat{\nu}_1(r_1 - r_2, 0)\hat{\nu}_2(r_1 - ar_2, 0), \quad r_1, r_2 \in H_{\boldsymbol{a}}.$$

Applying Corollary 9.2 and Theorem 9.9 to the $\boldsymbol{a}$-adic solenoid $\Sigma_{\boldsymbol{a}}$, we obtain from (11.36) that $\hat{\nu}_j(r, 0)$ are the characteristic functions of some Gaussian distributions on the group $\Sigma_{\boldsymbol{a}}$. In view of (3.14), we have

$$(11.37) \qquad \hat{\nu}_j(r, 0) = e^{-\sigma_j r^2}, \quad r \in H_{\boldsymbol{a}},$$

where $\sigma_j \geq 0$, $j = 1, 2$.

Taking into account statement (vii) of Theorem 2.5, it follows from (11.37) that the characteristic functions $\hat{\nu}_j(r,n)$ are uniformly continuous on Y in the topology induced on Y by the topology of the group $\mathbb{R}\times\mathbb{Z}$. This implies that the characteristic functions $\hat{\nu}_j(r,n)$ can be extended by continuity to some continuous positive definite functions $g_j(s,n)$ on the group $\mathbb{R}\times\mathbb{Z}$. Since the group $\mathbb{R}\times\mathbb{Z}$ is topologically isomorphic to the character group of the group $\mathbb{R}\times\mathbb{T}$, by Bochner's theorem, there are distributions λ_j on the group $\mathbb{R}\times\mathbb{T}$ such that $\hat{\lambda}_j(s,n)=g_j(s,n)$ for all $s\in\mathbb{R}$, $n\in\mathbb{Z}$, $j=1,2$. Denote by ι the natural embedding of the group $Y=H_{\boldsymbol{a}}\times\mathbb{Z}$ into the group $\mathbb{R}\times\mathbb{Z}$, $\iota(r,n)=(r,n)$, $r\in H_{\boldsymbol{a}}$, $n\in\mathbb{Z}$. Put

$$\upsilon=\widetilde{\iota},\quad \upsilon\colon \mathbb{R}\times\mathbb{T}\to\Sigma_{\boldsymbol{a}}\times\mathbb{T}.$$

Then υ is a continuous homomorphism and $\nu_j=\upsilon(\lambda_j)$, $j=1,2$. Since the subgroup $\iota(Y)$ is dense in the group $\mathbb{R}\times\mathbb{Z}$, by statement (b) of Theorem 1.20, υ is a monomorphism.

It follows from (11.35) that the characteristic functions $\hat{\lambda}_j(s,n)$ satisfy the equation

$$\begin{aligned}(11.38)\quad &\hat{\lambda}_1(s_1+s_2,n_1+n_2)\hat{\lambda}_2(s_1+as_2+bn_2,n_1+n_2)\\ &=\hat{\lambda}_1(s_1-s_2,n_1-n_2)\hat{\lambda}_2(s_1-as_2-bn_2,n_1-n_2),\quad s_1,s_2\in\mathbb{R},\ n_1,n_2\in\mathbb{Z}.\end{aligned}$$

We retain the notation G for the subgroup of $\mathbb{R}\times\mathbb{T}$ generated by the element of $\mathbb{R}\times\mathbb{T}$ of order 2 and the notation α for the topological automorphism of the group $\mathbb{R}\times\mathbb{T}$ which is defined by the matrix $\begin{pmatrix}a & b\\ 0 & 1\end{pmatrix}$. Let η_1 and η_2 be independent random variables with values in the group $\mathbb{R}\times\mathbb{T}$ and distributions λ_1 and λ_2. By Corollary 9.2, (11.38) implies that the conditional distribution of the linear form $P_2=\eta_1+\alpha\eta_2$ given $P_1=\eta_1+\eta_2$ is symmetric.

Let $a<0$. It follows from $\mathrm{Ker}(I+a)=\{0\}$ that $a\neq-1$. Since the characteristic functions $\hat{\nu}_j(y)$ do not vanish, by Theorem 11.12, $\lambda_j=\tau(M_j)$, where a continuous monomorphism $\tau\colon\mathbb{R}\times\mathbb{Z}(2)\to\mathbb{R}\times\mathbb{T}$ is defined by formula (11.23) and $M_j\in\Theta$. Put $\phi=\upsilon\tau$. Then $\phi\colon\mathbb{R}\times\mathbb{Z}(2)\to\Sigma_{\boldsymbol{a}}\times\mathbb{T}$ is a continuous monomorphism. By Corollary 2.4, ϕ induces an isomorphism of the semigroups $\mathrm{M}^1(\mathbb{R}\times\mathbb{Z}(2))$ and $\mathrm{M}^1(\phi(\mathbb{R}\times\mathbb{Z}(2))$. We keep the notation ϕ for this isomorphism. We have $\nu_j=\phi(M_j)$. This implies that the distributions ν_j are concentrated on the subgroup $\phi(\mathbb{R}\times\mathbb{Z}(2))$. By Proposition 2.1, we can substitute the distributions μ_j by their shifts and assume, without loss of generality, that μ_j are also concentrated on the subgroup $\phi(\mathbb{R}\times\mathbb{Z}(2))$. Since ϕ is an isomorphism of the semigroups $\mathrm{M}^1(\mathbb{R}\times\mathbb{Z}(2))$ and $\mathrm{M}^1(\phi(\mathbb{R}\times\mathbb{Z}(2)))$, we have

$$M_j=\phi^{-1}(\nu_j)=\phi^{-1}(\mu_j)*\phi^{-1}(\bar{\mu}_j)=\phi^{-1}(\mu_j)*\overline{\phi^{-1}(\mu_j)},\quad j=1,2.$$

Put $N_j=\phi^{-1}(\mu_j)$. By Lemma 11.5, $N_j\in\Theta$. We proved that $\mu_j=\phi(N_j)$, $j=1,2$. Thus, in the case when $a<0$, the theorem is proved.

Let $a>0$. Substituting (11.37) into equation (11.36) we get $\sigma_1+a\sigma_2=0$. Hence $\sigma_1=\sigma_2=0$ and (11.37) implies that $\hat{\nu}_j(r,0)=1$ for all $r\in H_{\boldsymbol{a}}$, $j=1,2$. Applying Proposition 2.10, we get that $\sigma(\nu_j)\subset A(X,H_{\boldsymbol{a}})=\mathbb{T}$. Let ζ_j be independent random variables with values in the group X and distributions ν_j. By Corollary 9.2, it follows from (11.35) that the conditional distribution of the linear form $Q_2=\zeta_1+\alpha\zeta_2$ given $Q_1=\zeta_1+\zeta_2$ is symmetric. Note that the restriction of the automorphism α to the subgroup $\mathbb{T}$ is the identity automorphism. Thus, if we consider ζ_j as independent random variables with values in the circle group $\mathbb{T}$, then the conditional distribution of the linear form $Q_2 = \zeta_1 + \zeta_2$ given $Q_1 = \zeta_1 + \zeta_2$ is symmetric. By Corollary 9.2 and Proposition 2.10, this implies that $\sigma(\nu_j) \subset G$. By Proposition 2.1, some shifts of the distributions μ_j are supported in G. □

Let $X = \Sigma_{\boldsymbol{a}} \times \mathbb{T}$, where the $\boldsymbol{a}$-adic solenoid $\Sigma_{\boldsymbol{a}}$ contains no elements of order 2. We note that since $\Sigma_{\boldsymbol{a}}$ is a compact connected Abelian group, by Theorem 1.13, multiplication by 2 is an epimorphism of the group $\Sigma_{\boldsymbol{a}}$. It follows from this that if the group $\Sigma_{\boldsymbol{a}}$ has no elements of order 2, then $\Sigma_{\boldsymbol{a}}$ and $H_{\boldsymbol{a}}$ are groups with unique division by 2. Let $\alpha \in \mathrm{Aut}(X)$ and let $\alpha = \begin{pmatrix} a & b \\ 0 & 1 \end{pmatrix}$. Omit in Theorem 11.16 the condition $\mathrm{Ker}(I + a) = \{0\}$ and consider the case when $\mathrm{Ker}(I + a) = \Sigma_{\boldsymbol{a}}$. The following statement holds.

PROPOSITION 11.17. *Let $X = \Sigma_{\boldsymbol{a}} \times \mathbb{T}$, where the $\boldsymbol{a}$-adic solenoid $\Sigma_{\boldsymbol{a}}$ contains no elements of order 2. Let G be the subgroup of X generated by the element of X of order 2. Let $\alpha = \begin{pmatrix} -1 & b \\ 0 & 1 \end{pmatrix} \in \mathrm{Aut}(X)$. Let K be the subgroup of X of the form*

$$K = \left\{ \left(g, \pm\left(-g, \frac{b}{2}\right)\right) : g \in \Sigma_{\boldsymbol{a}} \right\}.$$

Let ξ_1 and ξ_2 be independent random variables with values in the group X and distributions μ_1 and μ_2. Assume that the conditional distribution of the linear form $L_2 = \xi_1 + \alpha\xi_2$ given $L_1 = \xi_1 + \xi_2$ is symmetric. Then $\mu_j = \lambda_j * E_{x_j}$, where $\lambda_j \in \mathrm{M}^1(K)$, $x_j \in X$. Moreover, either $\lambda_2 = \lambda_1 * \omega_1$ or $\lambda_1 = \lambda_2 * \omega_2$, where $\omega_j \in \mathrm{M}^1(G)$.

PROOF. We observe that $G = \{(0, \pm 1)\} \subset K$. Put $L = \mathrm{Ker}(I - \widetilde{\alpha})$. Then

$$L = \{(r, n) : 2r - bn = 0\} = \left\{ \left(\frac{bn}{2}, n\right) : n \in \mathbb{Z} \right\}$$

and hence $L^{(2)} = \{(bn, 2n) : n \in \mathbb{Z}\}$. It follows from this that $K = A\left(X, L^{(2)}\right)$. Arguing as in the proof of Lemma 11.11, we obtain that $\hat{\mu}_j(y) = (p_j, y)$, $y \in L^{(2)}$, $p_j \in X$, $j = 1, 2$. Moreover, $p_1 + p_2 \in K$. Let $p_j = (g_j, z_j)$, $j = 1, 2$. Put $\tilde{x} = \left(g_1, \left(-g_1, \frac{b}{2}\right)\right)$, $x_1 = p_1 - \tilde{x}$, $x_2 = -p_1 + \tilde{x}$, $\lambda_j = \mu_j * E_{-x_j}$, $j = 1, 2$. Denote by η_j independent random variables with values in the group X and distributions λ_j. As in the proof of Lemma 11.11, we verify that the conditional distribution of the linear form $N_2 = \eta_1 + \alpha\eta_2$ given $N_1 = \eta_1 + \eta_2$ is symmetric and $\sigma(\lambda_j) \subset K$, $j = 1, 2$. It is easy to see that α acts on K as $-I$. Thus, if we consider η_j as independent random variables with values in the group K, then the conditional distribution of the linear form $N_2 = \eta_1 - \eta_2$ given $N_1 = \eta_1 + \eta_2$ is symmetric. Applying Proposition 11.9 to the group K and the independent random variables η_j, we obtain that either $\lambda_2 = \lambda_1 * \omega_1$ or $\lambda_1 = \lambda_2 * \omega_2$, where $\omega_j \in \mathrm{M}^1(G)$, $j = 1, 2$. The statement of the proposition follows from this. □

Theorems 11.12 and 11.16 imply the following statement.

PROPOSITION 11.18. *Let either $X = \mathbb{R} \times \mathbb{T}$ or $X = \Sigma_{\boldsymbol{a}} \times \mathbb{T}$, where the $\boldsymbol{a}$-adic solenoid $\Sigma_{\boldsymbol{a}}$ contains no elements of order 2. Let $\alpha \in \mathrm{Aut}(X)$. Denote by G the subgroup of X generated by the element of X of order 2. Assume that*

(i) $$\mathrm{Ker}(I + \alpha) = G.$$

Let ξ_1 and ξ_2 be independent random variables with values in the group X and distributions μ_1 and μ_2 with nonvanishing characteristic functions. If the conditional distribution of the linear form $L_2 = \xi_1 + \alpha\xi_2$ given $L_1 = \xi_1 + \xi_2$ is symmetric, then there is a continuous monomorphism $\pi\colon \mathbb{R} \times \mathbb{Z}(2) \to X$ such that the distributions μ_j are represented in the form $\mu_j = \pi(M_j) * E_{x_j}$, where $M_j \in \Theta$, $x_j \in X$, $j = 1, 2$.

PROOF. It follows from (i) that $\alpha = \begin{pmatrix} a & b \\ 0 & 1 \end{pmatrix}$. Indeed, suppose $\alpha = \begin{pmatrix} a & b \\ 0 & -1 \end{pmatrix}$. Then $\mathrm{Ker}(I + \alpha) \supset \mathbb{T}$, but this contradicts (i). Let the continuous monomorphisms τ and ϕ be the same as in Theorems 11.12 and 11.16 respectively.

Suppose $X = \mathbb{R} \times \mathbb{T}$. Then (i) holds if and only if $a \neq -1$. Indeed, if $a = -1$, then $\mathrm{Ker}(I + \alpha) = F \times G$, but this contradicts (i). The converse is obvious. Thus, $a \neq -1$. Taking into account that the characteristic functions $\hat{\mu}_j(y)$ do not vanish, in Theorem 11.12 only cases 1b and 3 are possible and we can put $\pi = \tau$.

Suppose $X = \Sigma_{\boldsymbol{a}} \times \mathbb{T}$. Then (i) holds if and only if $\mathrm{Ker}(I + a) = \{0\}$. Indeed, if $\mathrm{Ker}(I + a) \neq \{0\}$, then G is a proper subgroup of $\mathrm{Ker}(I + \alpha)$, but this contradicts (i). The converse is obvious. Thus, $\mathrm{Ker}(I + a) = \{0\}$. The conditions of Theorem 11.16 hold and we can put $\pi = \phi$. □

We now prove that Theorem 9.18 is false if an $\boldsymbol{a}$-adic solenoid $\Sigma_{\boldsymbol{a}}$ contains an element of order 2.

PROPOSITION 11.19. *Assume that an $\boldsymbol{a}$-adic solenoid $\Sigma_{\boldsymbol{a}}$ contains an element of order 2. Let α be a topological automorphism of the group $\Sigma_{\boldsymbol{a}}$ such that $\alpha < 0$, $\alpha \neq -I$. Put $K = \mathrm{Ker}(I + \alpha)$. Then there exist independent random variables ξ_1 and ξ_2 with values in $\Sigma_{\boldsymbol{a}}$ and distributions μ_1 and μ_2 with nonvanishing characteristic functions such that the conditional distribution of the linear form $L_2 = \xi_1 + \alpha\xi_2$ given $L_1 = \xi_1 + \xi_2$ is symmetric, whereas $\mu_j \notin \Gamma(\Sigma_{\boldsymbol{a}}) * \mathrm{M}^1(K)$, $j = 1, 2$.*

PROOF. Put $G = (\Sigma_{\boldsymbol{a}})_{(2)}$. Then $G \cong \mathbb{Z}(2)$. Choose numbers σ_j, σ'_j and κ in such a way that they satisfy the conditions

$$0 < \sigma'_j < \sigma_j, \quad 0 < \kappa \leq \sqrt{\frac{\sigma'_j}{\sigma_j}}, \quad j = 1, 2.$$

Consider on the group $\mathbb{R} \times \mathbb{Z}(2)$ the functions

$$f_j(s, l) = \begin{cases} \exp\{-\sigma_j s^2\}, & \text{if } \ s \in \mathbb{R},\ l = 0, \\ \kappa\exp\{-\sigma'_j s^2\}, & \text{if } \ s \in \mathbb{R},\ l = 1, \quad j = 1, 2. \end{cases} \tag{11.39}$$

By Lemma 11.1, there exist distributions $\lambda_j \in \Theta$ such that $\hat{\lambda}_j(s, l) = f_j(s, l)$. It is obvious that $\lambda_j \notin \Gamma(\mathbb{R}) * \mathrm{M}^1(\mathbb{Z}(2))$, $j = 1, 2$. We have $\alpha = f_p f_q^{-1}$ for some mutually prime p and q, where $f_p, f_q \in \mathrm{Aut}(\Sigma_{\boldsymbol{a}})$. It follows from $G \cong \mathbb{Z}(2)$, that p and q are odd. Hence f_p and f_q are topological automorphisms of the group $\mathbb{R} \times \mathbb{Z}(2)$. Inasmuch as $\alpha < 0$, we can assume that the conditions

$$\sigma_1 + pq^{-1}\sigma_2 = 0, \tag{11.40}$$

$$\sigma'_1 + pq^{-1}\sigma'_2 = 0, \tag{11.41}$$

hold. Verify that the characteristic functions $\hat{\lambda}_j(s, l)$ satisfy equation (9.3). Put $u = (s_1, l_1)$, $v = (s_2, l_2)$. Let either $u, v \in \mathbb{R} \times \{0\}$ or $u, v \in \mathbb{R} \times \{1\}$. Then

$u \pm v, u \pm \alpha v \in \mathbb{R} \times \{0\}$. Taking into account (11.39), equation (9.3) for these u, v is satisfied if the equality

$$\sigma_1(s_1+s_2)^2 + \sigma_2(s_1+pq^{-1}s_2)^2 = \sigma_1(s_1-s_2)^2 + \sigma_2(s_1-pq^{-1}s_2)^2, \quad s_1, s_2 \in \mathbb{R},$$

holds true. It is obvious that this equality follows from (11.40). Let either $u \in \mathbb{R} \times \{0\}$, $v \in \mathbb{R} \times \{1\}$ or $u \in \mathbb{R} \times \{1\}$, $v \in \mathbb{R} \times \{0\}$. Then $u \pm v, u \pm \alpha v \in \mathbb{R} \times \{1\}$. Taking into account (11.39), equation (9.3) for these u, v is satisfied if the equality

$$\sigma_1'(s_1+s_2)^2 + \sigma_2'(s_1+pq^{-1}s_2)^2 = \sigma_1'(s_1-s_2)^2 + \sigma_2'(s_1-pq^{-1}s_2)^2, \quad s_1, s_2 \in \mathbb{R},$$

is valid. It is clear that this equality follows from (11.41). We exhausted all possibilities for u and v. Thus, we verified that the characteristic functions $\hat\lambda_j(l, s)$ satisfy equation (9.3).

Let ζ_1 and ζ_2 be independent random variables with values in the group $\mathbb{R} \times \mathbb{Z}(2)$ and distributions λ_1 and λ_2. Since the characteristic functions $\hat\lambda_j(l, s)$ satisfy equation (9.3), by Corollary 9.2, the conditional distribution of the linear form $T_2 = \zeta_1 + \alpha\zeta_2$ given $T_1 = \zeta_1 + \zeta_2$ is symmetric.

Let $H_{\boldsymbol{a}}$ be the character group of the group $\Sigma_{\boldsymbol{a}}$. Let $y_0 \notin H_{\boldsymbol{a}}^{(2)}$ and let

$$H_{\boldsymbol{a}} = H_{\boldsymbol{a}}^{(2)} \cup (y_0 + H_{\boldsymbol{a}}^{(2)})$$

be an $H_{\boldsymbol{a}}^{(2)}$-coset decomposition of the group $H_{\boldsymbol{a}}$. Define the mapping

$$\tau : H_{\boldsymbol{a}} \to \mathbb{R} \times \mathbb{Z}(2), \quad \tau(y) = \begin{cases} (y, 0), & \text{if} \quad y \in H_{\boldsymbol{a}}^{(2)}, \\ (y, 1), & \text{if} \quad y \in y_0 + H_{\boldsymbol{a}}^{(2)}. \end{cases} \tag{11.42}$$

It is obvious that τ is a homomorphism.

Put

$$\pi = \widetilde{\tau}, \quad \pi : \mathbb{R} \times \mathbb{Z}(2) \to \Sigma_{\boldsymbol{a}}. \tag{11.43}$$

Inasmuch as the subgroup $\tau(H_{\boldsymbol{a}})$ is dense in $\mathbb{R} \times \mathbb{Z}(2)$, it follows from statement (b) of Theorem 1.20 that π is a continuous monomorphism. It is easy to see that

$$\alpha\pi(t, k) = \pi f_p f_q^{-1}(t, k), \quad t \in \mathbb{R},\ k \in \mathbb{Z}(2). \tag{11.44}$$

Put $\xi_j = \pi(\zeta_j)$, $j = 1, 2$. Then ξ_j are independent random variables with values in the group $\Sigma_{\boldsymbol{a}}$ and distributions $\mu_j = \pi(\lambda_j)$. By Proposition 2.7, we have $\hat\mu_j(y) = \hat\lambda_j(\tau(y))$, $y \in H_{\boldsymbol{a}}$. Then (11.39) and (11.42) imply that

$$\hat\mu_j(y) = \begin{cases} \exp\{-\sigma_j y^2\}, & \text{if} \quad y \in H_{\boldsymbol{a}}^{(2)}, \\ \kappa \exp\{-\sigma_j' y^2\}, & \text{if} \quad y \in y_0 + H_{\boldsymbol{a}}^{(2)}, \quad j = 1, 2. \end{cases}$$

Inasmuch as the conditional distribution of the linear form T_2 given T_1 is symmetric, the conditional distribution of the linear form $L_2 = \pi(T_2)$ given $L_1 = \pi(T_1)$ is also symmetric. We have $L_1 = \xi_1 + \xi_2$, and it follows from (11.44) that $L_2 = \xi_1 + \alpha\xi_2$. Since $\sigma_j \neq \sigma_j'$ and $\alpha \neq -I$, it is obvious that $\mu_j \notin \Gamma(\Sigma_{\boldsymbol{a}}) * \mathrm{M}^1(K)$, $j = 1, 2$. □

Consider an $\boldsymbol{a}$-adic solenoid $\Sigma_{\boldsymbol{a}}$. Assume that the group $\Sigma_{\boldsymbol{a}}$ contains an element of order 2. Then $(\Sigma_{\boldsymbol{a}})_{(2)} \cong \mathbb{Z}(2)$. Let α be a topological automorphism of $\Sigma_{\boldsymbol{a}}$. It is clear that $(\Sigma_{\boldsymbol{a}})_{(2)} \subset \mathrm{Ker}(I + \alpha)$ for all $\alpha \in \mathrm{Aut}(\Sigma_{\boldsymbol{a}})$. Let ξ_1 and ξ_2 be independent random variables with values in the group $\Sigma_{\boldsymbol{a}}$ and distributions μ_1 and μ_2 with nonvanishing characteristic functions. Assume that the conditional distribution of the linear form $L_2 = \xi_1 + \alpha\xi_2$ given $L_1 = \xi_1 + \xi_2$ is symmetric. It turns out that in the case when $(\Sigma_{\boldsymbol{a}})_{(2)} = \mathrm{Ker}(I + \alpha)$ we can completely describe distributions μ_j.

THEOREM 11.20. *Assume that an $\boldsymbol{a}$-adic solenoid $\Sigma_{\boldsymbol{a}}$ contains an element of order 2. Let α be a topological automorphism of $\Sigma_{\boldsymbol{a}}$ such that $(\Sigma_{\boldsymbol{a}})_{(2)} = \mathrm{Ker}(I+\alpha)$. Let $\pi : \mathbb{R} \times \mathbb{Z}(2) \to \Sigma_{\boldsymbol{a}}$ be given by (11.43). Let ξ_1 and ξ_2 be independent random variables with values in the group $\Sigma_{\boldsymbol{a}}$ and distributions μ_1 and μ_2 with nonvanishing characteristic functions. Assume that the conditional distribution of the linear form $L_2 = \xi_1 + \alpha\xi_2$ given $L_1 = \xi_1 + \xi_2$ is symmetric. Then $\mu_j = \pi(N_j) * E_{x_j}$, where $N_j \in \Theta$, $x_j \in \Sigma_{\boldsymbol{a}}$, $j = 1, 2$. Moreover, if $\alpha > 0$, then some sifts of the distributions μ_j are supported in $(\Sigma_{\boldsymbol{a}})_{(2)}$.*

PROOF. Let $H_{\boldsymbol{a}}$ be the character group of $\Sigma_{\boldsymbol{a}}$. We consider $H_{\boldsymbol{a}}$ as a subset of $\mathbb{R}$. Put $G = (\Sigma_{\boldsymbol{a}})_{(2)}$. Obviously, $G \cong \mathbb{Z}(2)$. Let $\alpha = f_p f_q^{-1}$ for some mutually prime p and q, where $f_p, f_q \in \mathrm{Aut}(\Sigma_{\boldsymbol{a}})$. It follows from $G \cong \mathbb{Z}(2)$ that p and q are odd.

Reasoning as in the proof of Theorem 9.18, we consider the distributions $\nu_j = \mu_j * \bar{\mu}_j$ and get that the functions $\varphi_j(y) = \ln \hat{\nu}_j(y)$, $j = 1, 2$, satisfy equation (9.69). Moreover, each of the functions $\varphi_j(y)$ satisfies equation (9.73), where H is of the form (9.72). Since $G \cong \mathbb{Z}(2)$ and $\mathrm{Ker}(I + \alpha) = G$, it is easy to see that

$$I + \alpha = 2\beta, \tag{11.45}$$

where $\beta \in \mathrm{Aut}(\Sigma_{\boldsymbol{a}})$. Inasmuch as p and q are odd, (9.72) and (11.45) imply that $H = (I - \tilde{\alpha})(H_{\boldsymbol{a}}) = H_{\boldsymbol{a}}^{(n)}$, where n is even. We can assume that n is the minimum possible. Consider the factor-group $H_{\boldsymbol{a}}/H$. It is obvious that $H_{\boldsymbol{a}}/H \cong \mathbb{Z}(n)$. Take an element $y_0 \in H_{\boldsymbol{a}}$ in such a way that the coset $y_0 + H$ is a generator of the factor-group $H_{\boldsymbol{a}}/H$. Let

$$H_{\boldsymbol{a}} = H \cup (y_0 + H) \cup (2y_0 + H) \cup \cdots \cup ((n-1)y_0 + H)$$

be an H-coset decomposition of the group $H_{\boldsymbol{a}}$. It follows from (9.73) that for any coset $ly_0 + H$ there exist polynomials $P_{lj}(s)$ on the real line with real coefficients of the form (9.75) such that

$$P_{lj}(y) = \varphi_j(y), \quad y \in ly_0 + H, \ l = 0, 1, \dots, n-1, \ j = 1, 2. \tag{11.46}$$

Denote by ζ_1 and ζ_2 independent random variables with values in the group $\Sigma_{\boldsymbol{a}}$ and distributions ν_1 and ν_2. Since the characteristic functions $\hat{\nu}_j(y)$ also satisfy equation (9.3), by Corollary 9.2, the conditional distribution of the linear form $L_2 = \zeta_1 + \alpha\zeta_2$ given $L_1 = \zeta_1 + \zeta_2$ is symmetric. By Corollary 9.7, the linear forms $P_1 = (I+\alpha)\zeta_1 + 2\alpha\zeta_2$ and $P_2 = 2\zeta_1 + (I+\alpha)\zeta_2$ are independent. Consider the new independent random variables $\eta_j = 2\zeta_j$, $j = 1, 2$. Taking into account (11.45), the linear forms L_1' and L_2' can be written as follows: $P_1 = \beta\eta_1 + \alpha\eta_2$ and $P_2 = \eta_1 + \beta\eta_2$. Inasmuch as $\alpha, \beta \in \mathrm{Aut}(\Sigma_{\boldsymbol{a}})$ and the group $\Sigma_{\boldsymbol{a}}$ contains no subgroups topologically isomorphic to the circle group $\mathbb{T}$, by Theorem 6.3, the random variables η_j have Gaussian distributions. In view of (3.14), this implies that

$$\varphi_j(y) = -\sigma_j y^2, \quad y \in H_{\boldsymbol{a}}^{(2)}, \quad j = 1, 2, \tag{11.47}$$

where $\sigma_j \geq 0$.

It is obvious that $H_{\boldsymbol{a}} = H_{\boldsymbol{a}}^{(2)} \cup (y_0 + H_{\boldsymbol{a}}^{(2)})$ is an $H_{\boldsymbol{a}}^{(2)}$-coset decomposition of the group $H_{\boldsymbol{a}}$. It follows from statement (vii) of Theorem 2.5 and (11.47) that each of the functions $\hat{\nu}_j(y)$ is uniformly continuous on the coset $y_0 + H_{\boldsymbol{a}}^{(2)}$ in the topology induced on $y_0 + H_{\boldsymbol{a}}^{(2)}$ by the topology of the group $\mathbb{R}$. Observe that

$$y_0 + H_{\boldsymbol{a}}^{(2)} = (y_0 + H) \cup (3y_0 + H) \cup \cdots \cup ((n-1)y_0 + H)$$

and each coset $ly_0 + H$ is a dense subset of $\mathbb{R}$. Taking into account (11.46) this implies that there exist polynomials $a_js^2 + b_js + c_j$, $j = 1, 2$, on the real line with real coefficients such that

$$P_{lj}(y) = a_jy^2 + b_jy + c_j, \quad y \in ly_0 + H, \quad l = 1, 3, \dots, n-1, \quad j = 1, 2.$$

and hence the representations

$$\varphi_j(y) = a_jy^2 + b_jy + c_j, \quad y \in y_0 + H_{\boldsymbol{a}}^{(2)}, \quad j = 1, 2, \tag{11.48}$$

hold. Since $\varphi_j(-y) = \varphi_j(y)$, $y \in H_{\boldsymbol{a}}$, we have $b_j = 0$, $j = 1, 2$. Put $a_j = -\sigma_j'$, $\kappa_j = e^{c_j}$. As a result, (11.47) and (11.48) imply the representations

$$\hat{\nu}_j(y) = \begin{cases} \exp\{-\sigma_jy^2\}, & \text{if} \quad y \in H_{\boldsymbol{a}}^{(2)}, \\ \kappa_j \exp\{-\sigma_j'y^2\}, & \text{if} \quad y \in y_0 + H_{\boldsymbol{a}}^{(2)}, \end{cases} \tag{11.49}$$

where $0 < \kappa_j \le 1$, $\sigma_j \ge 0$, $\sigma_j' \ge 0$, $j = 1, 2$.

By Lemma 11.1, there exist signed measures M_j on the group $\mathbb{R} \times \mathbb{Z}(2)$ with the characteristic functions of the form

$$\widehat{M}_j(s, l) = \begin{cases} \exp\{-\sigma_js^2\}, & \text{if} \quad s \in \mathbb{R},\ l = 0, \\ \kappa_j \exp\{-\sigma_j's^2\}, & \text{if} \quad s \in \mathbb{R},\ l = 1, \quad j = 1, 2. \end{cases} \tag{11.50}$$

Let τ be the homomorphism $\tau : Y \to \mathbb{R} \times \mathbb{Z}(2)$ defined by formula (11.42). Taking into account statement (ii) of Theorem 2.5 and Proposition 2.7, it follows from (11.42), (11.49), and (11.50) that $\nu_j = \pi(M_j)$, $j = 1, 2$. Since π is a continuous monomorphism, the signed measures M_j are measures. Obviously, $M_j \in \Theta$, $j = 1, 2$. The distributions ν_j are concentrated on the Borel subgroup $F = \pi(\mathbb{R} \times \mathbb{Z}(2))$. By Proposition 2.1, we can substitute the distributions μ_j by their shifts and suppose, without loss of generality, that μ_j are also concentrated on F. Inasmuch as π is a continuous monomorphism, by Corollary 2.4, π induces an isomorphism of the semigroups $\mathrm{M}^1(\mathbb{R} \times \mathbb{Z}(2))$ and $\mathrm{M}^1(\pi(\mathbb{R} \times \mathbb{Z}(2))$. We keep the notation π for this isomorphism. Put $N_j = \pi^{-1}(\mu_j) \in \mathrm{M}^1(\mathbb{R} \times \mathbb{Z}(2))$. It is obvious that $M_j = N_j * \overline{N}_j$. In view of $M_j \in \Theta$, by Lemma 11.5, $N_j \in \Theta$. Hence $\mu_j = \pi(N_j) * E_{x_j}$, where $x_j \in \Sigma_{\boldsymbol{a}}$, $j = 1, 2$.

Let $\alpha > 0$. Assuming that $u, v \in H_{\boldsymbol{a}}^{(2)}$, substitute (11.47) in equation (9.69). We obtain $\sigma_1 + pq^{-1}\sigma_2 = 0$. This implies that $\sigma_1 = \sigma_2 = 0$. Hence $\hat{\nu}_j(y) = 1$ for all $y \in H_{\boldsymbol{a}}^{(2)}$. It follows from Proposition 2.10 that $\sigma(\nu_j) \subset A(\Sigma_{\boldsymbol{a}}, H_{\boldsymbol{a}}^{(2)}) = (\Sigma_{\boldsymbol{a}})_{(2)} = G$, $j = 1, 2$. By Proposition 2.1, some shifts of the distributions μ_j are supported in G. □

REMARK 11.21. Let $\boldsymbol{a} = (3, 3, 3, \dots)$. Consider the corresponding the $\boldsymbol{a}$-adic solenoid $\Sigma_{\boldsymbol{a}}$ and let α be the multiplication by -3. Then $\alpha \in \mathrm{Aut}(\Sigma_{\boldsymbol{a}})$, $\Sigma_{\boldsymbol{a}}$ contains an element of order 2 and $(\Sigma_{\boldsymbol{a}})_{(2)} = \mathrm{Ker}(I + \alpha)$, i.e., the conditions of Theorem 11.20 are satisfied.

At the same time, if $\boldsymbol{a} = (5, 5, 5, \dots)$, then the corresponding the $\boldsymbol{a}$-adic solenoid $\Sigma_{\boldsymbol{a}}$ give us an example of an $\boldsymbol{a}$-adic solenoid containing an element of order 2 and for which the only $\alpha \in \mathrm{Aut}(\Sigma_{\boldsymbol{a}})$ such that $(\Sigma_{\boldsymbol{a}})_{(2)} = \mathrm{Ker}(I + \alpha)$ is $\alpha = I$.

REMARK 11.22. Consider an $\boldsymbol{a}$-adic solenoid $\Sigma_{\boldsymbol{a}}$, and assume that the group $\Sigma_{\boldsymbol{a}}$ contains an element of order 2. Put $G = (\Sigma_{\boldsymbol{a}})_{(2)}$ and let $\pi : \mathbb{R} \times \mathbb{Z}(2) \to \Sigma_{\boldsymbol{a}}$ be given by (11.43). Then the class of distributions $\Gamma(\Sigma_{\boldsymbol{a}}) * \mathrm{M}^1(G)$ is a subclass of distributions μ which are represented in the form $\mu = \pi(N) * E_x$, where $N \in \Theta$, $x \in \Sigma_{\boldsymbol{a}}$. If $G = (\Sigma_{\boldsymbol{a}})_{(2)} = \mathrm{Ker}(I+\alpha)$, then Proposition 11.19 implies, in particular, that we cannot assert in Theorem 11.20 that $\mu_j \in \Gamma(\Sigma_{\boldsymbol{a}}) * \mathrm{M}^1(G)$ for at least one j.

REMARK 11.23. Consider an $\boldsymbol{a}$-adic solenoid $\Sigma_{\boldsymbol{a}}$. Let α be a topological automorphism of $\Sigma_{\boldsymbol{a}}$. Let ξ_1 and ξ_2 be independent random variables with values in the group $\Sigma_{\boldsymbol{a}}$ and distributions μ_1 and μ_2 with nonvanishing characteristic functions. Assume that the linear forms $L_1 = \xi_1 + \xi_2$ and $L_2 = \xi_1 + \alpha\xi_2$ are independent. Since the group $\Sigma_{\boldsymbol{a}}$ has no subgroups topologically isomorphic to the circle group $\mathbb{T}$, by Theorem 6.3, $\mu_j \in \Gamma(X)$. Thus, in contrast to a group analogue of Heyde's theorem, a group analogue of the Skitovich–Darmois theorem depends neither on elements of order 2 in $\Sigma_{\boldsymbol{a}}$ nor on a topological automorphism α. We always get Gaussian distributions.

Now we will study a generalization of Theorem 10.8 to the case when a discrete Abelian group X contains an element of order 2. To do this, we need the following lemma.

LEMMA 11.24. *Let X be a discrete torsion Abelian group. Let α be an automorphism of X satisfying condition* (i) *of Theorem* 10.2. *Let ξ_1 and ξ_2 be independent random variables with values in the group X and distributions μ_1 and μ_2. If the conditional distribution of the linear form $L_2 = \xi_1 + \alpha\xi_2$ given $L_1 = \xi_1 + \xi_2$ is symmetric, then μ_j are supported in a subgroup $X_{(n)}$ for some n.*

PROOF. First we prove the lemma supposing that $\hat{\mu}_j(y) \geq 0$, $j = 1, 2$. Reasoning as in the proof of Lemma 10.3 and retaining the same notation, we obtain that

$$\hat{\mu}_1(y) = \hat{\mu}_1(y) = 1, \quad y \in B, \tag{11.51}$$

where $B = (I+\tilde{\alpha})(W)$, $W = V^{(2)} \cap (I+\tilde{\alpha})(V) \cap (\tilde{\alpha}(V))^{(2)}$. In contrast to Lemma 10.3, if the group X contains an element of order 2, then the subgroup B does not need to be open.

Put $L = V \cap (I+\tilde{\alpha})(V) \cap \tilde{\alpha}(V)$, $M = (I+\tilde{\alpha})(L)$. Inasmuch as $(I+\tilde{\alpha})(Y) = Y$, the continuous endomorphism $I+\tilde{\alpha}$ is open. This implies that L and hence M are open subgroups of Y. It is obvious that $W \supset L^{(2)}$. Hence $B \supset M^{(2)}$. Put $G = A(X, M^{(2)})$. By Proposition 2.10, it follows from (11.51) that $\sigma(\mu_j) \subset A(X, B) \subset G$, $j = 1, 2$.

We have $G = \{x \in X : (x, 2y) = 1 \text{ for all } y \in M\} = \{x \in X : (2x, y) = 1 \text{ for all } y \in M\} = \{x \in X : 2x \in A(X, M)\} = f_2^{-1}(A(X, M))$. Note that $\mathrm{Ker}\, f_2 = X_{(2)}$. Since M is an open subgroup of Y, by Theorem 1.11, the annihilator $A(X, M)$ is a compact subgroup of X. Hence $A(X, M)$ is a finite subgroup of X. This implies that the group $G = f_2^{-1}(A(X, M))$ is generated by $X_{(2)}$ and a finite subgroup of X.

We proved the lemma assuming that $\hat{\mu}_j(y) \geq 0$, $j = 1, 2$. Get rid of this restriction. Put $\nu_j = \mu_j * \bar{\mu}_j$. Then $\hat{\nu}_j(y) = |\hat{\mu}_j(y)|^2 \geq 0$, $y \in Y$. The characteristic functions $\hat{\nu}_j(y)$ also satisfy equation (9.3). Denote by η_j independent random variables with values in the group X and distributions ν_j, $j = 1, 2$. By Corollary 9.2, the conditional distribution of the linear form $M_2 = \eta_1 + \alpha\eta_2$ given $M_1 = \eta_1 + \eta_2$ is symmetric. As has been proved above, the distributions ν_j are supported in a subgroup N, generated by $X_{(2)}$ and a finite subgroup. By Proposition 2.1, it follows from this that each of the distributions μ_j is supported in a set $x_j + N$ for some $x_j \in X$. Since X is a torsion group, each of the elements x_j has finite order. It is obvious that the subgroup generated by N and the elements x_j is contained in some subgroup of the form $X_{(n)}$. □

THEOREM 11.25. *Let X be a discrete Abelian group, let F be the 2-component of X, and let G be the subgroup of X generated by all elements of X of odd order. Let α be an automorphism of X satisfying condition* (i) *of Theorem* 10.2. *Let ξ_1 and ξ_2 be independent random variables with values in the group X and distributions μ_1 and μ_2. If the conditional distribution of the linear form $L_2 = \xi_1 + \alpha\xi_2$ given $L_1 = \xi_1 + \xi_2$ is symmetric, then $\mu_j = \rho_j * m_K * E_{x_j}$, where $\rho_j \in \mathrm{M}^1(F)$, K is a finite subgroup of G, $x_j \in X$, $j = 1, 2$.*

PROOF. Inasmuch as $\alpha(b_X) = b_X$, we can apply Lemma 10.7 and assume that X is a torsion group. This implies by Theorem 1.24, that X is a weak direct product of its p-components. Hence $X = F \times G$. Denote by Y the character group of the group X. The group Y is topologically isomorphic to the group $L \times H$, where $L = F^*$ and $H = G^*$. In order not to complicate the notation we assume that $Y = L \times H$. Since $\alpha(F) = F$ and $\alpha(G) = G$, the automorphism α can be written in the form $\alpha = (\alpha_F, \alpha_G)$. Denote by $y = (l, h)$, where $l \in L$, $h \in H$, elements of the group Y. We shall write $\widetilde{\alpha} = (\widetilde{\alpha}_F, \widetilde{\alpha}_G)$. By Corollary 9.2, the characteristic functions $\hat{\mu}_j(y)$ satisfy equation (9.3) which takes the form

$$\begin{aligned}(11.52)\quad &\hat{\mu}_1(l_1 + l_2, h_1 + h_2)\hat{\mu}_2(l_1 + \widetilde{\alpha}_F l_2, h_1 + \widetilde{\alpha}_G h_2)\\ &\qquad = \hat{\mu}_1(l_1 - l_2, h_1 - h_2)\hat{\mu}_2(l_1 - \widetilde{\alpha}_F l_2, h_1 - \widetilde{\alpha}_G h_2), \quad l_j \in L,\ h_j \in H.\end{aligned}$$

Reasoning as in the proof of Lemma 10.13, but applying Theorem 10.8 instead of Theorem 10.12, and considering the group F instead of $\mathbb{R}^n$, we reduce the proof of the theorem to the case when G is a finite group and

$$(11.53)\qquad \hat{\mu}_1(0, h) = \hat{\mu}_2(0, h) = \widehat{m}_G(h), \quad h \in H.$$

Substitute in (11.52) $l_1 = l_2 = l$, $h_1 = h_2 = h$. We get

$$\begin{aligned}(11.54)\quad &\hat{\mu}_1(2l, 2h)\hat{\mu}_2((I + \widetilde{\alpha}_F)l, (I + \widetilde{\alpha}_G)h))\\ &\qquad = \hat{\mu}_2((I - \widetilde{\alpha}_F)l, (I - \widetilde{\alpha}_G)h)), \quad l \in L,\ h \in H.\end{aligned}$$

Substitute in (11.54) $l = 0$, $h \neq 0$. Inasmuch as $2h \neq 0$, it follows from (2.3) and (11.53) that the left-hand side of the resulting equation is equal to zero. Therefore, $\hat{\mu}_2(0, (I - \widetilde{\alpha}_G)h) = 0$. Hence if $h \neq 0$, then $(I - \widetilde{\alpha}_G)h \neq 0$, i.e., $\mathrm{Ker}(I - \widetilde{\alpha}_G) = \{0\}$. Since H is a finite group, this implies that

$$(11.55)\qquad I - \widetilde{\alpha}_G \in \mathrm{Aut}(H).$$

Obviously, the automorphism α_F satisfies the condition

$$(11.56)\qquad \mathrm{Ker}(I + \alpha_F) = \{0\}.$$

It is easy to see that (11.56) is equivalent to the condition

$$(11.57)\qquad \mathrm{Ker}(I - \alpha_F) = \{0\}.$$

Indeed, assume that (11.56) is fulfilled. Take $x \in \operatorname{Ker}(I-\alpha_F)$, $x \neq 0$. We can assume without loss of generality that $2x = 0$. Then the equality $(I+\alpha_F)x = (I-\alpha_F)x + 2\alpha_F x = 0$ contradicts (11.56). Reasoning similarly, we get that (11.57) implies (11.56).

Consider the factor group $X/X_{(2)}$. By Theorem 1.9, its character group is topologically isomorphic to the annihilator $A(Y, X_{(2)})$. By Theorem 1.12, $A(Y, X_{(2)}) = Y^{(2)}$. It follows from $\alpha(X_{(2)}) = X_{(2)}$ that α induces an automorphism $\hat{\alpha}$ of the factor-group $X/X_{(2)}$ by the formula $\hat{\alpha}[x] = [\alpha x]$. In so doing, a topological automorphism of the group $Y^{(2)}$ which is adjoint to $\hat{\alpha}$ is the restriction of $\widetilde{\alpha}$ to $Y^{(2)}$.

Obviously, $\widetilde{\alpha}(Y^{(2)}) = Y^{(2)}$. Verify that $\operatorname{Ker}(I+\hat{\alpha}) = \{0\}$. Take $x_0 \in X$ such that $[x_0] \in \operatorname{Ker}(I+\hat{\alpha})$. Then $[(I+\alpha)x_0] = 0$. Hence $(I+\alpha)x_0 \in X_{(2)}$, i.e., $2(I+\alpha)x_0 = 0$. Since α satisfies condition (i) of Theorem 10.2, we have $2x_0 = 0$, i.e., $[x_0] = 0$.

Inasmuch as $\alpha(X_{(n)}) = X_{(n)}$, applying Lemma 11.24 we can suppose that $X = X_{(n)}$ for some n. Hence the group X is bounded. Therefore, the group F is also bounded. Denote by k_X the least nonnegative integer such that $F_{(2^{k_X})} = \{0\}$. If $k_X = 0$, then $X = G$. In this case the statement of the theorem follows from Theorem 10.8. We will prove by induction that if the theorem holds true for the groups X satisfying the condition $k_X = m-1$, then it is valid for the groups X satisfying the condition $k_X = m$.

Suppose $k_X = m$. Put $\hat{\mu}_1(l,0) = a_1(l)$, $\hat{\mu}_2(l,0) = a_2(l)$. We will prove that

$$(11.58)\qquad \hat{\mu}_1(l,h) = \begin{cases} a_1(l), & \text{if } h = 0,\\ 0, & \text{if } h \neq 0, \end{cases} \qquad \hat{\mu}_2(l,h) = \begin{cases} a_2(l), & \text{if } h = 0,\\ 0, & \text{if } h \neq 0. \end{cases}$$

The statement of the theorem easily follows from this.

The restrictions of the characteristic functions $\hat{\mu}_j(l,h)$ to the subgroup $Y^{(2)}$ are the characteristic functions of some independent random variables η_1 and η_2 with values in the factor-group $X/X_{(2)}$. By Corollary 9.2, the conditional distribution of the linear form $M_2 = \eta_1 + \hat{\alpha}\eta_2$ given $M_1 = \eta_1 + \eta_2$ is symmetric. As has been noted above, $\operatorname{Ker}(I+\hat{\alpha}) = \{0\}$. Obviously, $k_{X/X_{(2)}} = m-1$. Then by the induction hypothesis, (11.58) holds true for $(l,h) \in Y^{(2)}$.

Substitute in (11.54) $l \in L$, $h \in H$, $h \neq 0$. Since $(2l, 2h) \in Y^{(2)}$ and $2h \neq 0$, the left-hand side of equation (11.54) is equal to zero by the induction hypothesis. Hence $\hat{\mu}_2((I-\widetilde{\alpha}_F)l, (I-\widetilde{\alpha}_G)h)) = 0$ for all $l \in L$, $h \in H$, $h \neq 0$. By statement (b) of Theorem 1.20, it follows from (11.57) that $(I-\widetilde{\alpha}_F)(L) = L$. Taking into account (11.55), we get that representation (11.58) for the function $\hat{\mu}_2(l,h)$ holds true for all $(l,h) \in Y$. Substituting in (11.52) $l_1 = l$, $h_1 = h$, $l_2 = \widetilde{\alpha}_F^{-1} l$, $h_2 = \widetilde{\alpha}_G^{-1} h$ and reasoning similarly, we obtain the required representation for the function $\hat{\mu}_1(l,h)$. □

It is easy to see that the condition

$$(11.59)\qquad \operatorname{Ker}(I+\alpha) \subset X_2$$

is necessary if we want Theorem 11.25 to be valid. Below we consider the case when condition (11.59) holds, but condition (i) of Theorem 10.2 is not satisfied, i.e.,

$$(11.60)\qquad \operatorname{Ker}(I+\alpha) \neq \{0\}.$$

We complement Theorem 11.25 by the following statement which shows that, generally speaking, Theorem 11.25 is not valid if we change condition (i) of Theorem 10.2 for condition (11.59).

PROPOSITION 11.26. *Let X be a discrete Abelian group. Denote by F the 2-component of X and by G the subgroup of X generated by all elements of X of odd order. Let α be an automorphism of the group X satisfying condition* (11.59). *Then the following statements hold.*

1. *Assume that the only finite subgroup K of the group G satisfying the condition $(I-\alpha)(K) = K$, is $K = \{0\}$. Let ξ_1 and ξ_2 be independent random variables with values in the group X and distributions μ_1 and μ_2. If the conditional distribution of the linear form $L_2 = \xi_1+\alpha\xi_2$ given $L_1 = \xi_1+\xi_2$ is symmetric, then $\mu_j = \rho_j * E_{x_j}$, where $\rho_j \in \mathrm{M}^1(F)$, $x_j \in X$, $j = 1, 2$.*
2. *Assume that there exists a nonzero finite subgroup K_0 of the group G, satisfying the condition $(I-\alpha)(K_0) = K_0$. Assume that the automorphism α satisfies the condition* (11.60). *Then there exist independent identically distributed random variables ξ_1 and ξ_2 with values in the group X and distribution μ such that the conditional distribution of the linear form $L_2 = \xi_1 + \alpha\xi_2$ given $L_1 = \xi_1 + \xi_2$ is symmetric, whereas μ cannot be represented in the form $\mu = \rho * m_K * E_x$, where $\rho \in \mathrm{M}^1(F)$, K is a finite subgroup of the group G, $x \in X$.*

PROOF. In view of (11.59), condition (i) of Lemma 10.7 is satisfied. Moreover, it follows from (11.59) that $\mathrm{Ker}(I + \alpha_G) = \{0\}$. Reasoning as in the proof of Theorem 11.25 and retaining the same notation, we can suppose $X = F \times G$ and G is a finite group.

Statement 1. By Corollary 9.2 the characteristic functions $\hat\mu_j(l, h)$ satisfy equation (11.52). Put $\nu_j = \mu_j * \bar\mu_j$. Then $\hat\nu_j(y) = |\hat\mu_j(y)|^2 \geq 0$ for all $y \in Y$, $j = 1, 2$. The characteristic functions $\hat\nu_j(y)$ also satisfy equation (11.52) which takes the form

$$\begin{aligned}(11.61)\quad & \hat\nu_1(l_1 + l_2, h_1 + h_2)\hat\nu_2(l_1 + \tilde\alpha_F l_2, h_1 + \tilde\alpha_G h_2) \\ &= \hat\nu_1(l_1 - l_2, h_1 - h_2)\hat\nu_2(l_1 - \tilde\alpha_F l_2, h_1 - \tilde\alpha_G h_2), \quad l_j \in L,\ h_j \in H.\end{aligned}$$

Putting $l_1 = l_2 = 0$ in (11.61) we obtain

$$\begin{aligned}(11.62)\quad & \hat\nu_1(0, h_1 + h_2)\hat\nu_2(0, h_1 + \tilde\alpha_G h_2) \\ &= \hat\nu_1(0, h_1 - h_2)\hat\nu_2(0, h_1 - \tilde\alpha_G h_2), \quad h_1, h_2 \in H.\end{aligned}$$

Taking this into account and applying Corollary 9.2 and Theorem 10.2 to the group G, we get

$$(11.63)\qquad \hat\nu_1(0, h) = \hat\nu_2(0, h) = \hat m_K(0, h), \quad h \in H,$$

where K is a finite subgroup of G. In view of (2.3), as easily follows from (11.62) and (11.63), $(I - \alpha)(K) = K$. In this case, by the condition of the proposition, $K = \{0\}$. Hence $\hat\nu_1(0, h) = \hat\nu_2(0, h) = 1$ for all $h \in H$. By Proposition 2.10, this implies that $\sigma(\nu_j) \subset A(X, H) = F$ and by Proposition 2.1, $\mu_j = \rho_j * E_{x_j}$, where $\rho_j \in \mathrm{M}^1(F)$, $x_j \in X$, $j = 1, 2$. Thus, we proved statement 1.

Statement 2. It is easy to see that $\alpha(K_0) = K_0$. Consider the subgroup $T = F_{(2)} \times K_0$ of the group X. Obviously, $\alpha(T) = T$, i.e., the restriction of α to T is an automorphism of the group T. Then the group T and the automorphism

α_T satisfy the conditions of statement 2. Taking this into account we can prove statement 2 assuming that $X = F \times G$, where

$$F = F_{(2)} \tag{11.64}$$

and G is a finite group such that $(I - \alpha_G)(G) = G$. Since G is a finite group, it means that

$$I - \alpha_G \in \mathrm{Aut}(G). \tag{11.65}$$

It follows from (11.59) and (11.60) that $\mathrm{Ker}(I + \alpha_F) \neq \{0\}$. By Theorems 1.8, 1.19 and statement (a) of Theorem 1.20, we have $(I + \widetilde{\alpha}_F)(L) = A(L, \mathrm{Ker}(I + \alpha_F))$. Hence

$$(I + \widetilde{\alpha}_F)(L) \neq L.$$

Take $l_0 \notin (I + \widetilde{\alpha}_F)(L)$. Since F is a discrete torsion group, by Theorems 1.2 and 1.5, L is a compact totally disconnected group. For this reason there exists an open subgroup U of L such that $U \cap (l_0 + U) = \emptyset$ and $(l_0 + U) \cap (I + \widetilde{\alpha}_F)(L) = \emptyset$. Consider the factor-group $A = Y/U \cong L/U \times H$. Denote by $[y] = ([l], h)$, where $[l] \in L/U$, $h \in H$, its elements. Let n be the number of elements of the group H. Consider on the group A the function

$$f([y]) = f([l], h) = \begin{cases} 1, & \text{if } [l] = 0,\ h = 0, \\ 0, & \text{if } [l] = 0,\ h \neq 0, \\ \frac{1}{n}, & \text{if } [l] = [l_0],\ h \in H, \\ 0, & \text{if } [l] \notin \{0, [l_0]\},\ h \in H. \end{cases}$$

Put $B = A(F, U) \times G$. By Theorem 1.9, the character group of the group A is topologically isomorphic to the group B and the character group of the group B is topologically isomorphic to the group A. Inasmuch as L is a compact group and U is an open subgroup of L, the factor-group L/U is discrete and hence finite. Thus, A and B are finite Abelian groups. Put

$$r(x) = \sum_{[y] \in A} f([y]) \overline{(x, [y])}, \quad x \in B.$$

It is obvious that $r(x) \geq 0$ and

$$\int_B r(x) dm_B(x) = 1.$$

Let μ be the distribution on the group B with the density $r(x)$ with respect to m_B. We have $\hat{\mu}([y]) = f([y])$, $[y] \in A$. If we consider μ as a distribution on the group X, then the characteristic function $\hat{\mu}(y)$ is of the form

$$\hat{\mu}(y) = \hat{\mu}(l, h) = \begin{cases} 1, & \text{if } l \in U,\ h = 0, \\ 0, & \text{if } l \in U,\ h \neq 0, \\ \frac{1}{n}, & \text{if } l \in l_0 + U,\ h \in H, \\ 0, & \text{if } l \notin U \cup (l_0 + U),\ h \in H. \end{cases} \tag{11.66}$$

Obviously, μ cannot be represented in the form $\mu = \rho * m_K * E_x$, where $\rho \in \mathrm{M}^1(F)$, K is a finite subgroup of G, $x \in X$. Let ξ_1 and ξ_2 be independent identically distributed random variables with values in the group X and distribution μ. It follows from (11.64) that $L_{(2)} = L$. Hence $l = -l$ for all $l \in L$. We shall verify that the characteristic function $f(l, h) = \hat{\mu}(l, h)$ satisfies the equation

$$\begin{aligned}(11.67) \quad & f(l_1 + l_2, h_1 + h_2) f(l_1 + \widetilde{\alpha}_F l_2, h_1 + \widetilde{\alpha}_G h_2) \\ &= f(l_1 - l_2, h_1 - h_2) f(l_1 - \widetilde{\alpha}_F l_2, h_1 - \widetilde{\alpha}_G h_2), \quad l_j \in L, \ h_j \in H.\end{aligned}$$

Then, by Corollary 9.2, the conditional distribution of the linear form $L_2 = \xi_1 + \alpha\xi_2$ given $L_1 = \xi_1 + \xi_2$ is symmetric. The following cases exhaust all possibilities for l_j.

1. $l_1 + l_2 \in U$ and $l_1 + \widetilde{\alpha}_F l_2 \in U$. It follows from (11.66) that the left-hand side and the right-hand side of equation (11.67) can take values either 1 or 0. Assume that the left-hand side of equation (11.67) is equal to 1. This implies that $h_1 + h_2 = 0$ and $h_1 + \widetilde{\alpha}_G h_2 = 0$. Hence $(I - \widetilde{\alpha}_G)h_2 = 0$. It follows from (11.65) that $I - \widetilde{\alpha}_G \in \mathrm{Aut}(H)$. Therefore, $h_2 = 0$. Hence $h_1 = 0$. It means that the right-hand side of equation (11.67) is also equal to 1. Reasoning similarly, we verify that if the right-hand side of equation (11.67) is equal to 1, then the left-hand side of equation (11.67) is equal to 1. Thus, in this case both sides of equation (11.67) are equal.

2. $l_1 + l_2 \in l_0 + U$ and $l_1 + \widetilde{\alpha}_F l_2 \in l_0 + U$. It follows from (11.66) that both sides of equation (11.67) are equal $\frac{1}{n^2}$.

3. $l_1 + l_2 \in U$ and $l_1 + \widetilde{\alpha}_F l_2 \in l_0 + U$. This implies that $(I + \widetilde{\alpha}_F)l_2 \in l_0 + U$. But this contradicts the fact that $(l_0 + U) \cap (I + \widetilde{\alpha}_F)(L) = \emptyset$. Thus, this case is impossible.

4. $l_1 + l_2 \in l_0 + U$ and $l_1 + \widetilde{\alpha}_F l_2 \in U$. This case is impossible. The reasoning is similar to the case 3.

5. $l_1 + l_2 \notin l_0 + U$ and $l_1 + l_2 \notin U$. Then in view of (11.66), we have $f(l_1 + l_2, h_1 + h_2) = f(l_1 + l_2, h_1 - h_2) = 0$ and both sides of equation (11.67) are equal to zero.

6. $l_1 + \widetilde{\alpha}_F l_2 \notin l_0 + U$ and $l_1 + \widetilde{\alpha}_F l_2 \notin U$. Then in view of (11.66), we have $f(l_1 + \widetilde{\alpha}_F l_2, h_1 + \widetilde{\alpha}_G h_2) = f(l_1 + \widetilde{\alpha}_F l_2, h_1 - \widetilde{\alpha}_G h_2) = 0$ and both sides of equation (11.67) are equal to zero.

We see that for all $u, v \in Y$ equation (11.67) becomes the equality. □

Notes

A theorem where the Gaussian distribution on the real line is characterized by the symmetry of the conditional distribution of one linear form of independent random variables given another was proved by C.C. Heyde in [**65**].

A group analogue of Heyde's theorem in the case when independent random variables take values in a locally compact Abelian group and coefficients of linear forms are topological automorphisms of the group was considered first by G.M. Feldman in [**33**] for finite Abelian groups. Theorems 9.3 and 9.9 were proved by G.M. Feldman in [**34**] and [**44**] respectively. Lemma 9.6 belongs to M.V. Myronyuk [**78**]. Lemmas 9.10, 9.13 and 9.20 and Theorems 9.11, 9.15, and 9.21 belong to G.M. Feldman [**51**]. Theorem 9.18 was proved by G.M. Feldman in [**47**].

The results of Section 10 belong to G.M. Feldman. Theorem 10.2 was proved in [**33**]. Lemmas 10.1, 10.3 and 10.7 and Theorem 10.8 was proved in [**46**]. Observe that when an automorphism α of a discrete Abelian group X satisfies the conditions $I \pm \alpha \in \mathrm{Aut}(X)$, Theorem 10.8 was proved in [**35**]. Lemmas 10.4 and 10.5 were proved in [**35**] and [**44**] respectively. Lemma 10.10, Theorems 10.12 and 10.16 and Propositions 10.17 and 10.18 were proved in [**48**]. The results of Section 11 belong to G.M. Feldman. Lemmas 11.1 and 11.5, Proposition 11.19 and Theorem 11.20 were proved in [**47**]. Lemmas 11.24, Theorem 11.25 and Proposition 11.26 were proved in [**46**]. The remaining results of Section 11 were obtained in [**50**].

Observe that Theorem 11.25 has been generalized by M.V. Myronyuk in [**80**] on the group of the form $\mathbb{R} \times D$, where D is a discrete Abelian group.

We note that if a locally compact Abelian group X contains no subgroups topologically isomorphic to the circle group $\mathbb{T}$ and a topological automorphism α satisfies the conditions $I \pm \alpha \in \mathrm{Aut}(X)$, then the symmetry of the conditional distribution of the linear form $L_2 = \xi_1 + \alpha\xi_2$ given $L_1 = \xi_1 + \xi_2$ implies that the distributions of the random variables ξ_j are convolutions of Gaussian distributions on X and distributions supported in the subgroup of X generated by all elements of X of order 2 [**38**]. A similar result holds true for the 2-dimensional torus $\mathbb{T}^2$ [**34**].

CHAPTER V

Characterization theorems on the field of p-adic numbers

12. Skitovich–Darmois theorem

In this section for two independent random variables we prove an analogue of the Skitovich–Darmois theorem for the additive group of the field of p-adic numbers Ω_p.

Let p be a prime number. We need some properties of the group of p-adic numbers Ω_p (see [**63**, §10]). As a set Ω_p coincides with the set of sequences of integers of the form $x = (\dots, x_{-n}, x_{-n+1}, \dots, x_{-1}, x_0, x_1, \dots, x_n, \dots)$, where $x_n \in \{0, 1, \dots, p-1\}$, such that $x_n = 0$ for all $n < n_0$, where the number n_0 depends on x. We correspond to each element $x \in \Omega_p$ the series $\sum\limits_{k=-\infty}^{\infty} x_k p^k$. Addition and multiplication of the series are defined in a natural way and they define the operations of addition and multiplication in Ω_p. With respect to these operations Ω_p is a field. Elements of the group $\Delta_p = \{x \in \Omega_p : x_n = 0 \text{ for all } n < 0\}$ we write in the form $x = (x_0, x_1, \dots, x_n, \dots)$. The family of the subgroups $\{p^l \Delta_p\}_{l=-\infty}^{\infty}$ forms an open basis at the zero of the group Ω_p and defines a topology on Ω_p. With respect to this topology the group Ω_p is locally compact noncompact and totally disconnected. We note that the group Ω_p is represented as the union

$$\Omega_p = \bigcup_{l=-\infty}^{\infty} p^l \Delta_p.$$

Note also that each nonzero proper closed subgroup of the group Ω_p is of the form $p^l \Delta_p$, $l = 0, \pm 1, \dots$. The character group Ω_p^* of the group Ω_p is topologically isomorphic to Ω_p, and the value of a character $y \in \Omega_p^*$ at an element $x \in \Omega_p$ is defined by the formula

$$(x, y) = \exp\Big[2\pi i\Big(\sum_{n=-\infty}^{\infty} x_n\Big(\sum_{s=n}^{\infty} y_{-s} p^{-s+n-1}\Big)\Big)\Big], \tag{12.1}$$

where for a given x and y the sums in (12.1) actually are finite [**63**, (25.1)]. Each automorphism $\alpha \in \mathrm{Aut}(\Omega_p)$ is of the form $\alpha g = x_\alpha g$, $g \in \Omega_p$, where $x_\alpha \in \Omega_p$, $x_\alpha \neq 0$. For $\alpha \in \mathrm{Aut}(\Omega_p)$ we identify the automorphism $\alpha \in \mathrm{Aut}(\Omega_p)$ with the corresponding element $x_\alpha \in \Omega_p$, i.e., when we write αg, we suppose $\alpha \in \Omega_p$. We note that $\widetilde{\alpha} = \alpha$. We recall that $\Delta_p^\times$ is the subset of Δ_p consisting of all invertible elements of Δ_p and

$$\Delta_p^\times = \{x = (x_0, x_1, \dots, x_n, \dots) \in \Delta_p : x_0 \neq 0\}.$$

We note that each element $g \in \Omega_p$ is represented in the form $g = p^k c$, where $k \in \mathbb{Z}$ and $c \in \Delta_p^{\times}$. Multiplication by c is a topological automorphism of the group Δ_p.

We recall that $\mathbb{Z}(p^\infty)$ is the set of rational numbers of the form $\{k/p^n : k = 0, 1, \dots, p^n - 1,\ n = 0, 1, \dots\}$. If we define the operation in $\mathbb{Z}(p^\infty)$ as addition modulo 1, then $\mathbb{Z}(p^\infty)$ is transformed into an Abelian group which we consider in the discrete topology. The groups $\mathbb{Z}(p^\infty)$ and Δ_p are the character groups of one another. For a fixed n consider a subgroup of $\mathbb{Z}(p^\infty)$ consisting of all elements of the form $\{k/p^n : k = 0, 1, \dots, p^n - 1\}$. This subgroup is isomorphic to $\mathbb{Z}(p^n)$ and we keep the notation $\mathbb{Z}(p^n)$ for it.

To prove the main theorem of this section we need the following lemma.

LEMMA 12.1. *Consider the group of p-adic numbers Ω_p. Let $\alpha = p^k c$, where $k \in \mathbb{Z}$ and $c \in \Delta_p^{\times}$, be a topological automorphism of Ω_p. Let ξ_1 and ξ_2 be independent random variables with values in the group Ω_p and distributions μ_1 and μ_2 such that $\hat\mu_j(y) \geq 0$, $j = 1, 2$. If the linear forms $L_1 = \xi_1 + \xi_2$ and $L_2 = \xi_1 + \alpha\xi_2$ are independent, then there exists a closed subgroup B of Ω_p such that $\hat\mu_j(y) = 1$ for all $y \in B$, $j = 1, 2$.*

PROOF. We use the fact that the family of the subgroups $\{p^l\Delta_p\}_{l=-\infty}^{\infty}$ forms an open basis at the zero of the group Ω_p. Inasmuch as $\hat\mu_1(0) = \hat\mu_2(0) = 1$, we can choose m in such a way that $\hat\mu_j(y) > 0$ for all $y \in L = p^m\Delta_p$, $j = 1, 2$. Put $M = L$ if $k \geq 0$ and $M = p^{-k}L$ if $k < 0$. Then M is a subgroup of L and $\alpha(M) \subset L$. Put $\psi_j(y) = -\ln \hat\mu_j(y)$, $y \in L$, $j = 1, 2$.

By Lemma 6.1, the characteristic functions $\hat\mu_j(y)$ satisfy equation (6.1) which takes the form

$$\hat\mu_1(u + v)\hat\mu_2(u + \tilde\alpha v) = \hat\mu_1(u)\hat\mu_1(v)\hat\mu_2(u)\hat\mu_2(\tilde\alpha v), \quad u, v \in \Omega_p. \tag{12.2}$$

Taking into account that $\tilde\alpha = \alpha$, we get from (12.2) that the functions $\psi_j(y)$ satisfy the equation

$$\psi_1(u+v) + \psi_2(u+\alpha v) = \psi_1(u) + \psi_2(u) + \psi_1(v) + \psi_2(\alpha v), \quad u \in L,\ v \in M. \tag{12.3}$$

Integrating equation (12.3) over the group L with respect to the Haar distribution m_L and using that the Haar distribution m_L is L-invariant, we obtain

$$\psi_1(v) + \psi_2(\alpha v) = 0, \quad v \in M.$$

It follows from this that $\psi_1(v) = \psi_2(\alpha v) = 0$ for all $v \in M$. Hence $\hat\mu_1(y) = \hat\mu_2(\alpha y) = 1$ for all $y \in M$. Put $B = M \cap \alpha(M)$. Then B is the required subgroup. □

THEOREM 12.2. *Consider the group of p-adic numbers Ω_p. Let $\alpha = p^k c$, where $k \in \mathbb{Z}$ and $c = (c_0, c_1, \dots, c_n, \dots) \in \Delta_p^{\times}$, be a topological automorphism of Ω_p. Then the following statements hold.*

1. *Assume that either $k = 0$ or $|k| = 1$. Let ξ_1 and ξ_2 be independent random variables with values in the group Ω_p and distributions μ_1 and μ_2. Assume that the linear forms $L_1 = \xi_1 + \xi_2$ and $L_2 = \xi_1 + \alpha\xi_2$ are independent. Then the following statements are valid.*

   1(i) *If $k = 0$, then $\mu_j = m_K * E_{x_j}$, where K is a compact subgroup of Ω_p and $x_j \in \Omega_p$, $j = 1, 2$. Moreover, if $c_0 = 1$, then μ_1 and μ_2 are degenerate distributions.*

   1(ii) *If $|k| = 1$, then either $\mu_1 \in I(\Omega_p)$ or $\mu_2 \in I(\Omega_p)$.*

2. *If $|k| \geq 2$, then there exist independent random variables ξ_1 and ξ_2 with values in the group Ω_p and distributions μ_1 and μ_2 such that the linear forms $L_1 = \xi_1 + \xi_2$ and $L_2 = \xi_1 + \alpha\xi_2$ are independent, whereas $\mu_1, \mu_2 \notin I(\Omega_p)$.*

PROOF. Statement 1. By Lemma 6.1, the characteristic functions $\hat{\mu}_j(y)$ satisfy equation (6.1) which takes the form (12.2). Put $\nu_j = \mu_j * \bar{\mu}_j$. This implies that $\hat{\nu}_j(y) = |\hat{\mu}_j(y)|^2 \geq 0$ for all $y \in \Omega_p$. The characteristic functions $\hat{\nu}_j(y)$ also satisfy equation (12.2). Hence when we prove statements 1(i) and 1(ii) we may assume without loss of generality that $\mu_j(y) \geq 0$, $j = 1, 2$, because μ_j and ν_j are either degenerate distributions or belong to the class $I(\Omega_p)$ simultaneously. Moreover, if it is necessary we can consider the new independent random variables $\eta_1 = \xi_1$ and $\eta_2 = \alpha\xi_2$ and for this reason suppose $k \geq 0$.

Statement 1(i). We can assume that $\alpha \neq I$. In the opposite case μ_1 and μ_2 are degenerate distributions. Since by the condition $k = 0$, we have $\alpha = c$, where $c \in \Delta_p^{\times}$. Hence the restriction of α to every subgroup $p^m\Delta_p$ is a topological automorphism of $p^m\Delta_p$. By Lemma 12.1, there exists a subgroup $B = p^l\Delta_p$ such that $\hat{\mu}_j(y) = 1$ for all $y \in B$, $j = 1, 2$. It follows from Proposition 2.10 that $\sigma(\mu_j) \subset A(\Omega_p, B)$, $j = 1, 2$. Put $G = A(\Omega_p, B)$. It is easy to see that $G = p^{-l+1}\Delta_p$. We have $G \cong \Delta_p$ and the restriction of α to the subgroup G is a topological automorphism of G. Thus, we get that the independent random variables ξ_1 and ξ_2 take values in a group $G \cong \Delta_p$, they have distributions μ_1 and μ_2, and the linear forms $L_1 = \xi_1 + \xi_2$ and $L_2 = \xi_1 + \alpha\xi_2$, where $\alpha \in \mathrm{Aut}(G)$, are independent. Applying Corollary 7.23 and taking into account that $\mu_j(y) \geq 0$ for all $y \in \Omega_p$, $j = 1, 2$, we obtain that $\mu_1 = \mu_2 = m_K$, where K is a compact subgroup of G. Thus, we proved the first part of statement 1(i). Note that we have independent identically distributed random variables ξ_1 and ξ_2 with values in the group Ω_p and distribution m_K such that the linear forms $L_1 = \xi_1 + \xi_2$ and $L_2 = \xi_1 + \alpha\xi_2$ are independent. It follows from Lemma 7.11 that $(I - \alpha)(K) \supset K$. Suppose $c = (1, c_1, \ldots, c_n, \ldots)$ and $K \neq \{0\}$. It is obvious that in this case $(I - \alpha)(K)$ is a proper subgroup of K. The obtained contradiction shows that $K = \{0\}$, i.e., μ_1 and μ_2 are degenerate distributions. Thus, we proved the second part of statement 1(i).

In particular, it follows from this reasoning that in the case of the group Ω_2, if $k = 0$, then μ_1 and μ_2 are degenerate distributions, because if $c \in \Delta_2^{\times}$, then $c_0 = 1$.

Statement 1(ii). Put $f(y) = \hat{\mu}_1(y)$, $g(y) = \hat{\mu}_2(y)$, $\beta = I - \alpha$. Taking into account that $\alpha = \widetilde{\alpha}$, we rewrite equation (12.2) in the form

$$f(u + v)g(u + \alpha v) = f(u)g(u)f(v)g(\alpha v), \quad u, v \in \Omega_p. \tag{12.4}$$

Put

$$E = \{y \in \Omega_p : f(y) = g(y) = 1\}. \tag{12.5}$$

We can assume that μ_j are nondegenerate distributions. Hence $E \neq \Omega_p$. It follows from Lemma 12.1 that $E \neq \{0\}$. By Proposition 2.10, E is a closed subgroup of Ω_p. This implies that E, as a nonzero proper closed subgroup of Ω_p, is of the form $E = p^l\Delta_p$. Inasmuch as $k \geq 1$, we have $\alpha(E) \subset E$. Hence α induces a continuous endomorphism $\hat{\alpha}$ on the factor-group $L = \Omega_p/E$. Taking into account that by Proposition 2.10,

$$f(y + h) = f(y), \quad g(y + h) = g(y), \quad y \in \Omega_p, \ h \in E,$$

we can consider the functions $\hat{f}(y)$ and $\hat{g}(y)$ which are induced by the functions $f(y)$ and $g(y)$ on L. It follows from (12.5) that

$$\{y \in L : \hat{f}(y) = \hat{g}(y) = 1\} = \{0\}. \tag{12.6}$$

Passing from equation (12.4) on the group Ω_p to the induced equation on the factor-group L, we obtain

$$\hat{f}(u+v)\hat{g}(u+\hat{\alpha}v) = \hat{f}(u)\hat{g}(u)\hat{f}(v)\hat{g}(\hat{\alpha}v), \quad u, v \in L. \tag{12.7}$$

Substituting first $u = -\hat{\alpha}y$, $v = y$ and then $u = y$, $v = -y$ in equation (12.7) and taking into account that $\hat{f}(-y) = \hat{f}(y)$ and $\hat{g}(-y) = \hat{g}(y)$ for all $y \in L$, we get

$$\hat{f}((I - \hat{\alpha})y) = \hat{f}(\hat{\alpha}y)\hat{g}^2(\hat{\alpha}y)\hat{f}(y), \quad y \in L, \tag{12.8}$$

$$\hat{g}((I - \hat{\alpha})y) = \hat{f}^2(y)\hat{g}(y)\hat{g}(\hat{\alpha}y), \quad y \in L. \tag{12.9}$$

It is easy to see that $L \cong \mathbb{Z}(p^\infty)$ and $\hat{\beta} = (I - \hat{\alpha}) \in \mathrm{Aut}(L)$. Since $0 \le \hat{f}(y) \le 1$ and $0 \le \hat{g}(y) \le 1$, $y \in L$, it follows from equation (12.8) that

$$\hat{f}(\hat{\beta}y) \le \hat{f}(y), \quad y \in L. \tag{12.10}$$

We note now that each element of the group L belongs to some subgroup H such that $H \cong \mathbb{Z}(p^m)$. Moreover, $\hat{\beta}(H) = H$. Inasmuch as H is a finite subgroup, $\hat{\beta}^n y = y$ for all $y \in H$, where n depends generally on y. Then (12.10) implies that

$$\hat{f}(y) = \hat{f}(\hat{\beta}^n y) \le \cdots \le \hat{f}(\hat{\beta}y) \le \hat{f}(y), \quad y \in L.$$

Thus, on each orbit $\{y, \hat{\beta}y, \dots, \hat{\beta}^{n-1}y\}$ the function $\hat{f}(y)$ takes a constant value. The similar statement for the function $\hat{g}(y)$ follows from equation (12.9).

Assume $\hat{f}(y_1) \ne 0$ at an element $y_1 \in L$, $y_1 \ne 0$. Then $\hat{f}(\hat{\beta}y_1) = \hat{f}(y_1) \ne 0$, and equation (12.8) implies that

$$\hat{f}(\hat{\alpha}y_1) = \hat{g}(\hat{\alpha}y_1) = 1. \tag{12.11}$$

It follows from (12.6) and (12.11) that $\hat{\alpha}y_1 = 0$. By the condition, $\alpha = pc$, where $c \in \Delta_p^\times$. This implies that $\hat{\alpha} = p\hat{c}$, where $\hat{c}$ is an automorphism of the group L induced by the automorphism c. Hence y_1 is an element of order p. Reasoning similarly we get from equation (12.9) that if $\hat{g}(y_2) \ne 0$, $y_2 \in L$, $y_2 \ne 0$, then $\hat{f}(y_2) = \hat{g}(\hat{\alpha}y_2) = 1$.

Let w be an arbitrary element of L. Denote by $\langle w \rangle$ the subgroup of L generated by w. By Proposition 2.10, it follows from $\hat{f}(y_2) = 1$ that $\hat{f}(y) = 1$ for all $y \in \langle y_2 \rangle$. Since $L \cong \mathbb{Z}(p^\infty)$ and $\langle y_2 \rangle$ is a subgroup of L, we have $\langle y_2 \rangle \cong \mathbb{Z}(p^m)$ for some m, and hence $\hat{\alpha}(\langle y_2 \rangle) \subset \langle y_2 \rangle$. Moreover, $\hat{f}(\hat{\alpha}y_2) = 1$. Thus, the equalities

$$\hat{f}(\hat{\alpha}y_2) = \hat{g}(\hat{\alpha}y_2) = 1 \tag{12.12}$$

hold true. It follows from (12.6) and (12.12) that $\hat{\alpha}y_2 = 0$. Hence y_2 is also an element of order p. Inasmuch as $L \cong \mathbb{Z}(p^\infty)$, the group L contains the only subgroup A isomorphic to $\mathbb{Z}(p)$. So, we proved that the functions $\hat{f}(y)$ and $\hat{g}(y)$ vanish for all $y \notin A$.

Consider the restriction of equation (12.7) to the subgroup A. Taking into account that $\hat{\alpha}y = 0$ for all $y \in A$, we obtain

$$\hat{f}(u+v)\hat{g}(u) = \hat{f}(u)\hat{g}(u)\hat{f}(v), \quad u, v \in A. \tag{12.13}$$

If $\hat{g}(u_0) \neq 0$ at an element $u_0 \in A$, $u_0 \neq 0$, then we conclude from (12.13) that

$$\hat{f}(u_0 + v) = \hat{f}(u_0)\hat{f}(v), \quad v \in A.$$

Putting here $v = (p-1)u_0$, we get $\hat{f}(u_0) = 1$. Since p is a prime number, we have $A = \langle u_0 \rangle$. Hence by Proposition 2.10, $\hat{f}(y) = 1$ for all $y \in A$. If $\hat{g}(y) = 0$ for all $y \in A$, $y \neq 0$, then $\hat{f}(y)$ may be an arbitrary positive definite function on A. So we proved that either

$$\hat{g}(y) = \begin{cases} 1, & \text{if } y \in A, \\ 0, & \text{if } y \notin A, \end{cases} \tag{12.14}$$

or

$$\hat{g}(y) = \begin{cases} 1, & \text{if } y = 0, \\ 0, & \text{if } y \neq 0. \end{cases} \tag{12.15}$$

Returning from the induced functions $\hat{f}(y)$ and $\hat{g}(y)$ on L to the functions $f(y)$ and $g(y)$ on the group Ω_p and taking into account (2.3), we obtain from (12.14) and (12.15) that either $\mu_1 \in I(\Omega_p)$ or $\mu_2 \in I(\Omega_p)$. Statement 1(ii) is proved.

Statement 2. It is easy to see that without loss of generality we can assume that $k \geq 2$. Take $0 < a < 1$ and consider on the group Ω_p the distributions

$$\mu_1 = am_{p\Delta_p} + (1-a)m_{p^{-k+2}\Delta_p}, \quad \mu_2 = am_{p^{-k+2}\Delta_p} + (1-a)m_{p^{-k+1}\Delta_p}.$$

As has been noted earlier, $A(\Omega_p, p^m\Delta_p) = p^{-m+1}\Delta_p$. Therefore, (2.3) implies that the characteristic functions $f(y) = \hat{\mu}_1(y)$ and $g(y) = \hat{\mu}_2(y)$ are of the form

$$f(y) = \begin{cases} 1 & \text{if } \ y \in p^{k-1}\Delta_p, \\ a & \text{if } \ y \in \Delta_p \backslash p^{k-1}\Delta_p, \\ 0 & \text{if } \ y \notin \Delta_p, \end{cases} \qquad g(y) = \begin{cases} 1 & \text{if } \ y \in p^{k}\Delta_p, \\ a & \text{if } \ y \in p^{k-1}\Delta_p \backslash p^k\Delta_p, \\ 0 & \text{if } \ y \notin p^{k-1}\Delta_p. \end{cases}$$

Let ξ_1 and ξ_2 be independent random variables with values in the group Ω_p and distributions μ_1 and μ_2. It is obvious that $\mu_1, \mu_2 \notin I(\Omega_p)$. We will check that the characteristic functions $f(y)$ and $g(y)$ satisfy equation (12.4). Then by Lemma 6.1, the linear forms $L_1 = \xi_1 + \xi_2$ and $L_2 = \xi_1 + \alpha\xi_2$ are independent and statement 2 will be proved.

Consider three cases:

1. $u, v \in \Delta_p$. Note that since $k \geq 2$, we have $p^{k-1}\Delta_p \subset \Delta_p$. Consider three subcases:

1a. $u \in p^{k-1}\Delta_p$, $v \in \Delta_p$. Inasmuch as $u \in p^{k-1}\Delta_p$, we have $f(u) = 1$. Hence by Proposition 2.10, $f(u+v) = f(v)$. Since $\alpha v \in p^k\Delta_p$, we have $g(\alpha v) = 1$. Hence $g(u + \alpha v) = g(u)$. Equation (12.4) takes the form $f(v)g(u) = f(v)g(u)$ and it is true.

1b. $u \in \Delta_p \backslash p^{k-1}\Delta_p$, $v \in p^{k-1}\Delta_p$. Inasmuch as $v \in p^{k-1}\Delta_p$, we have $\alpha v \in p^{2k-1}\Delta_p \subset p^k\Delta_p$. This implies that $g(\alpha v) = 1$. Hence by Proposition 2.10, $g(u + \alpha v) = g(u)$. Since $v \in p^{k-1}\Delta_p$, we have $f(v) = 1$. Hence $f(u+v) = f(u)$. Equation (12.4) takes the form $f(u)g(u) = f(u)g(u)$ and it is true.

1c. $u \in \Delta_p \backslash p^{k-1}\Delta_p$, $v \in \Delta_p \backslash p^{k-1}\Delta_p$. Inasmuch as $v \in \Delta_p$, we have $\alpha v \in p^k\Delta_p$. This implies that $g(\alpha v) = 1$. Hence by Proposition 2.10, $g(u + \alpha v) = g(u)$. Since $u \notin p^{k-1}\Delta_p$, we have $g(u + \alpha v) = g(u) = 0$. Thus, both sides of equation (12.4) vanish.

2. $u \notin \Delta_p$, $v \in \Delta_p$. This implies that $u+v \notin \Delta_p$. Hence $f(u) = 0$ and $f(u+v) = 0$. Thus both sides of equation (12.4) vanish.

3. $v \notin \Delta_p$. This implies that $f(v) = 0$. Hence the right-hand side of equation (12.4) vanishes. If the left-hand side of equation (12.4) does not vanish, then the following inclusions

$$\begin{cases} u+v \in \Delta_p, \\ u+\alpha v \in p^{k-1}\Delta_p \end{cases} \tag{12.16}$$

hold true. Since $k \geq 2$, it follows from (12.16) that

$$(I-\alpha)v \in \Delta_p. \tag{12.17}$$

Inasmuch as $k \geq 2$, we have $(I-\alpha) \in \mathrm{Aut}(\Delta_p)$. It follows from (12.17) that $v \in \Delta_p$. The obtained contradiction shows that the left-hand side of equation (12.4) vanishes too.

We showed that the characteristic functions $f(y)$ and $g(y)$ satisfy equation (12.4). Thus, we proved statement 2. $\square$

REMARK 12.3. Consider the group of p-adic numbers Ω_p, where $p > 2$. Let ξ_1 and ξ_2 be independent identically distributed random variables with values in Ω_p and distribution m_{Δ_p}. Consider the topological automorphism α of the group Ω_p of the form $\alpha = (x_0, x_1, \ldots, x_n, \ldots)$, where $x_0 \neq 1$. It is easy to verify that the characteristic functions $\hat{\mu}_1(y) = \hat{\mu}_2(y) = \widehat{m}_{\Delta_p}(y)$ satisfy equation (12.2). By Lemma 6.1, this implies that the linear forms $L_1 = \xi_1 + \xi_2$ and $L_2 = \xi_1 + \alpha\xi_2$ are independent. Thus, for the group Ω_p, where $p > 2$, statement 1(i) in Theorem 12.2 cannot be strengthened to the statement that μ_j are degenerate distributions.

Statement 1(ii) in Theorem 12.2 cannot be strengthened to the statement that both μ_1 and μ_2 are shifts of an idempotent distribution. Namely, if $k = 1$, then there exist independent random variables ξ_1 and ξ_2 with values in the group Ω_p and distributions μ_1 and μ_2 such that the linear forms $L_1 = \xi_1+\xi_2$ and $L_2 = \xi_1+\alpha\xi_2$ are independent, whereas one of the distributions μ_j, say μ_2, such that $\mu_2 \notin I(\Omega_p)$. We get the corresponding example if put $\mu_1 = m_{p\Delta_p}$ and $\mu_2 = am_{p\Delta_p} + (1-a)m_{\Delta_p}$, where $0 < a < 1$. The proof is similar to the reasoning given in the proof of statement 2 in Theorem 12.2.

13. Heyde theorem

In this section for two independent random variables we prove an analogue of Heyde's theorem for the additive group of the field of p-adic numbers Ω_p.

We need some additional information on the field of p-adic numbers Ω_p. Let $\alpha \in \mathrm{Aut}(\Omega_p)$, $\alpha = p^k c$, where $k \in \mathbb{Z}$ and $c \in \Delta_p^{\times}$. Suppose $k \geq 0$. Let l be an integer. Consider the subgroup $p^l\Delta_p$ of the group Ω_p. It is easy to see that α induces an epimorphism $\bar{\alpha}$ and c induces an automorphism $\bar{c}$ on the factor-group $\Omega_p/p^l\Delta_p$. Let

$$x = (\ldots, x_{-n}, x_{-n+1}, \ldots, x_{-1}, x_0, x_1, \ldots, x_n, \ldots) \in \Omega_p.$$

Define the mapping $\tau : \Omega_p/p^l\Delta_p \to \mathbb{Z}(p^\infty)$ by the formula

$$\tau(x + p^l\Delta_p) = \sum_{n=-\infty}^{l-1} x_n p^{n-l}, \quad x + p^l\Delta_p \in \Omega_p/p^l\Delta_p. \tag{13.1}$$

Then τ is a topological isomorphism of the groups $\Omega_p/p^l\Delta_p$ and $\mathbb{Z}(p^\infty)$. Put

$$\hat{\alpha} = \tau\bar{\alpha}\tau^{-1}, \quad \hat{c} = \tau\bar{c}\tau^{-1}. \tag{13.2}$$

Observe that $\hat{\alpha} = p^k\hat{c}$, $\hat{c} \in \mathrm{Aut}(\mathbb{Z}(p^\infty))$, and $\hat{\alpha}$ is an epimorphism. If $c = (c_0, c_1, \dots, c_n, \dots) \in \Delta_p^\times$, then the automorphism $\hat{c}$ acts in the following way. Put $s_n = c_0 + c_1p + c_2p^2 + \dots + c_{n-1}p^{n-1}$. The restriction of the automorphism $\hat{c}$ to the subgroup $\mathbb{Z}(p^n) \subset \mathbb{Z}(p^\infty)$ is of the form $\hat{c}y = s_ny$, $y \in \mathbb{Z}(p^n)$, i.e., $\hat{c}$ acts in $\mathbb{Z}(p^n)$ as the multiplication by s_n.

To prove the main theorem of this section we need the following lemma.

LEMMA 13.1. *Consider the group of p-adic numbers Ω_p. Let $\alpha = p^kc$, where $k \in \mathbb{Z}$ and $c \in \Delta_p^\times$, be a topological automorphism of Ω_p. Assume that $\alpha \neq -I$. Let ξ_1 and ξ_2 be independent random variables with values in the group Ω_p and distributions μ_1 and μ_2 such that $\hat{\mu}_j(y) \geq 0$, $j = 1, 2$. If the conditional distribution of the linear form $L_2 = \xi_1 + \alpha\xi_2$ given $L_1 = \xi_1 + \xi_2$ is symmetric, then there exists a closed subgroup H of Ω_p such that $\hat{\mu}_j(y) = 1$ for all $y \in H$, $j = 1, 2$.*

PROOF. Taking into account that the conditional distribution of the linear form L_2 given L_1 is symmetric if and only if the conditional distribution of the linear form $\alpha^{-1}L_2$ given L_1 is symmetric, we may assume without loss of generality that $k \geq 0$. We have $\widetilde{\alpha} = \alpha$. By Corollary 9.2, the characteristic functions $\hat{\mu}_j(y)$ satisfy the equation

$$\hat{\mu}_1(u+v)\hat{\mu}_2(u+\alpha v) = \hat{\mu}_1(u-v)\hat{\mu}_2(u-\alpha v), \quad u, v \in \Omega_p. \tag{13.3}$$

Assume that $\alpha = I$. Substituting $u = v = y$ in equation (13.3), we get that $\hat{\mu}_1(y) = \hat{\mu}_1(y) = 1$ for all $y \in \Omega_p$, i.e., $H = \Omega_p$ and μ_j are degenerate distributions. So, we will assume that $\alpha \neq I$. Inasmuch as $\hat{\mu}_1(0) = \hat{\mu}_2(0) = 1$, we can choose a neighborhood of the zero V in Ω_p such that $\hat{\mu}_j(y) > 0$ for all $y \in V$, $j = 1, 2$. We can assume that $V = p^l\Delta_p$ for some l. Since $k \geq 0$, we have $\alpha(V) \subset V$. Put $\psi_j(y) = -\ln \hat{\mu}_j(y)$, $y \in V$, $j = 1, 2$. It follows from (13.3) that the functions $\varphi_j(y)$ satisfy the equation

$$\psi_1(u+v) + \psi_2(u+\alpha v) - \psi_1(u-v) - \psi_2(u-\alpha v) = 0, \quad u, v \in V.$$

Applying Lemma 9.17, we come to the conclusion that there exists a subgroup $p^m\Delta_p \subset V$, where the function $\psi_1(y)$ satisfies the equation

$$\Delta_h^3\psi_1(y) = 0, \quad h, y \in p^m\Delta_p. \tag{13.4}$$

Taking into account that $p^m\Delta_p$ is a compact group, by Proposition 1.30, we conclude from (13.4) that $\psi_1(y) = \text{const}$ for all $y \in p^m\Delta_p$. Since $\psi_1(0) = 0$, we have $\psi_1(y) = 0$ for all $y \in p^m\Delta_p$. This implies that $\hat{\mu}_1(y) = 1$ for all $y \in p^m\Delta_p$. For the distribution μ_2 we argue similarly and find a subgroup $p^n\Delta_p$ such that $\hat{\mu}_2(y) = 1$ for all $y \in p^n\Delta_p$. Put $H = p^m\Delta_p \cap p^n\Delta_p$. □

THEOREM 13.2. *Consider the group of p-adic numbers Ω_p. Let $\alpha = p^kc$, where $k \in \mathbb{Z}$ and $c = (c_0, c_1, \dots, c_n, \dots) \in \Delta_p^\times$, be a topological automorphism of Ω_p. Then the following statements hold.*

1. *Let ξ_1 and ξ_2 be independent random variables with values in the group Ω_p and distributions μ_1 and μ_2. Assume that the conditional distribution of the linear form $L_2 = \xi_1 + \alpha\xi_2$ given $L_1 = \xi_1 + \xi_2$ is symmetric. Then the following statements are valid.*

1(i) *If* $p > 2$, $k = 0$, *and* $c_0 \neq p-1$, *then* $\mu_j = m_K * E_{x_j}$, *where* K *is a compact subgroup of* Ω_p *and* $x_j \in \Omega_p$, $j = 1, 2$. *Moreover, if* $c_0 = 1$, *then* μ_1 *and* μ_2 *are degenerate distributions.*

1(ii) *If* $p = 2$, $k = 0$, $c_0 = 1$, *and* $c_1 = 0$, *then* μ_1 *and* μ_2 *are degenerate distributions.*

1(iii) *If* $p > 2$ *and* $|k| = 1$, *then either* $\mu_1 \in I(\Omega_p)$ *or* $\mu_2 \in I(\Omega_p)$.

2. *If one of the following conditions holds:*
   2(i) $p > 2$, $k = 0$, *and* $c_0 = p - 1$;
   2(ii) $p = 2$, $k = 0$, *and* $c_0 = c_1 = 1$;
   2(iii) $p = 2$ *and* $|k| = 1$;
   2(iv) $p \geq 2$ *and* $|k| \geq 2$,

   then there exist independent random variables ξ_1 *and* ξ_2 *with values in the group* Ω_p *and distributions* μ_1 *and* μ_2 *such that the conditional distribution of the linear form* $L_2 = \xi_1 + \alpha\xi_2$ *given* $L_1 = \xi_1 + \xi_2$ *is symmetric, whereas* $\mu_1, \mu_2 \notin I(\Omega_p)$.

PROOF. Statement 1. By Corollary 9.2, the characteristic functions $\hat{\mu}_j(y)$ satisfy equation (13.3). Put $\nu_j = \mu_j * \bar{\mu}_j$. This implies that $\hat{\nu}_j(y) = |\hat{\mu}_j(y)|^2 \geq 0$ for all $y \in \Omega_p$. The characteristic functions $\hat{\nu}_j(y)$ also satisfy equation (13.3). Hence when we prove statements 1(i)–1(iii) we may assume, without loss of generality, that $\mu_j(y) \geq 0$, $j = 1, 2$, because μ_j and ν_j are either degenerate distributions or belong to the class $I(\Omega_p)$ simultaneously. Moreover, if necessary we can consider the new independent random variables $\eta_1 = \xi_1$ and $\eta_2 = \alpha\xi_2$ and assume, without loss of generality, that $k \geq 0$.

Put $f(y) = \hat{\mu}_1(y)$, $g(y) = \hat{\mu}_2(y)$ and write equation (13.3) in the form

$$f(u+v)g(u+\alpha v) = f(u-v)g(u-\alpha v), \quad u, v \in \Omega_p, \tag{13.5}$$

where $\alpha \in \mathrm{Aut}(\Omega_p)$, $\alpha = p^k c$, $k \geq 0$, $c \in \Delta_p^{\times}$, and $f(y) \geq 0$, $g(y) \geq 0$ for all $y \in \Omega_p$.

Put

$$E = \{y \in \Omega_p : f(y) = g(y) = 1\}. \tag{13.6}$$

If either μ_1 or μ_2 is a nondegenerate distribution, then $E \neq \Omega_p$. Observe that $\alpha \neq -I$ in cases 1(i) and 1(ii). Then by Lemma 13.1, $E \neq \{0\}$. Thus, E is a nonzero proper closed subgroup of Ω_p. Hence $E = p^l\Delta_p$ for some l. By Proposition 2.10,

$$f(y+h) = f(y), \quad g(y+h) = g(y), \quad y \in \Omega_p, \ h \in E = p^l\Delta_p. \tag{13.7}$$

Taking into account (13.7), denote by $\bar{f}(y)$ and $\bar{g}(y)$ the functions induced by the functions $f(y)$ and $g(y)$ on the factor-group $\Omega_p/p^l\Delta_p$. Let τ be defined by formula (13.1). Since τ is a topological isomorphism of the groups $\Omega_p/p^l\Delta_p$ and $\mathbb{Z}(p^\infty)$, we can put $\hat{f} = \bar{f}\tau^{-1}$ and get from (13.5) that the functions $\hat{f}(y)$ and $\hat{g}(y)$ satisfy the equation

$$\hat{f}(u+v)\hat{g}(u+\hat{\alpha}v) = \hat{f}(u-v)\hat{g}(u-\hat{\alpha}v), \quad u, v \in \mathbb{Z}(p^\infty), \tag{13.8}$$

where $\hat{\alpha}$ is defined by formula (13.2). It follows from (13.6) that

$$\{y \in \mathbb{Z}(p^\infty) : \hat{f}(y) = \hat{g}(y) = 1\} = \{0\}. \tag{13.9}$$

Statement 1(i). It is obvious that $\hat{\alpha}$ is an automorphism of the group $\mathbb{Z}(p^\infty)$. First assume that $c_0 \neq 1$. Inasmuch as $p > 2$, $k = 0$, and $c_0 \neq p - 1$, we have

$$\mathrm{Ker}(I + \hat{\alpha}) = \{0\}. \tag{13.10}$$

We note that for every natural n the restriction of each automorphism of the group $\mathbb{Z}(p^\infty)$ to the subgroup $\mathbb{Z}(p^n) \subset \mathbb{Z}(p^\infty)$ is an automorphism of the subgroup $\mathbb{Z}(p^n)$. Consider the restriction of equation (13.8) to the subgroup $\mathbb{Z}(p^n)$. Observe that $(\mathbb{Z}(p^n))^* \cong \mathbb{Z}(p^n)$ and the group $\mathbb{Z}(p^n)$ contains no elements of order 2. Taking into account that (13.10) holds, we can apply Corollary 9.2 and Theorem 10.2 to the group $\mathbb{Z}(p^n)$ and get that the restrictions of the characteristic functions $\hat{f}(y)$ and $\hat{g}(y)$ to the subgroup $\mathbb{Z}(p^n)$ take only two values, 0 and 1. Moreover, $\hat{f}(y) = \hat{g}(y)$ for all $y \in \mathbb{Z}(p^n)$. This implies that the characteristic functions $\hat{f}(y)$ and $\hat{g}(y)$ on the group $\mathbb{Z}(p^\infty)$ also take only two values, 0 and 1. Furthermore, $\hat{f}(y) = \hat{g}(y)$ for all $y \in \mathbb{Z}(p^\infty)$. Then standard reasoning shows that $\mu_j = m_K * E_{x_j}$, where K is a compact subgroup of Ω_p and $x_j \in \Omega_p$, $j = 1, 2$.

Assume now that $c_0 = 1$. Inasmuch as $k = 0$, the restriction of the automorphism $\hat{\alpha}$ to the subgroup $\mathbb{Z}(p) \subset \mathbb{Z}(p^\infty)$ is the identity automorphism. Hence the restriction of equation (13.8) to $\mathbb{Z}(p)$ takes the form

$$\hat{f}(u+v)\hat{g}(u+v) = \hat{f}(u-v)\hat{g}(u-v), \quad u, v \in \mathbb{Z}(p). \tag{13.11}$$

Substituting here $u = v = y$, we get $\hat{f}(2y)\hat{g}(2y) = 1$ for all $y \in \mathbb{Z}(p)$. Since $p > 2$, this implies that

$$\hat{f}(y) = \hat{g}(y) = 1, \quad y \in \mathbb{Z}(p),$$

but this contradicts (13.9). Thus, μ_1 and μ_2 are degenerate distributions. We proved statement 1(i).

Statement 1(ii). It is obvious that $\hat{\alpha}$ is an automorphism of the group $\mathbb{Z}(2^\infty)$. Since $k = 0$, $c_0 = 1$, and $c_1 = 0$, the restriction of the automorphism $\hat{\alpha}$ to the subgroup $\mathbb{Z}(4) \subset \mathbb{Z}(2^\infty)$ is the identity automorphism. Hence the restriction of equation (13.8) to the subgroup $\mathbb{Z}(4)$ takes the form (13.11), where $u, v \in \mathbb{Z}(4)$. Substituting $u = v = y$ in equation (13.11), we get

$$\hat{f}(2y)\hat{g}(2y) = 1, \quad y \in \mathbb{Z}(4).$$

This implies that $\hat{f}(y) = \hat{g}(y) = 1$ for all $y \in \mathbb{Z}(2)$, but this contradicts (13.9). Thus, μ_1 and μ_2 are degenerate distributions. We proved statement 1(ii).

Statement 1(iii). Set $\beta = I - \alpha$, $\gamma = I + \alpha$. Since $k \geq 1$, we have $\hat{\beta}, \hat{\gamma} \in \mathrm{Aut}(\mathbb{Z}(p^\infty))$. Put $\kappa = \beta\gamma^{-1}$.

Substituting $u = v = y$ into equation (13.8), we obtain

$$\hat{f}(2y)\hat{g}(\hat{\gamma}y) = \hat{g}(\hat{\beta}y), \quad y \in \mathbb{Z}(p^\infty). \tag{13.12}$$

Substituting $u = \hat{\alpha}y$, $v = y$ into equation (13.8), we get

$$\hat{f}(\hat{\gamma}y)\hat{g}(2\hat{\alpha}y) = \hat{f}(\hat{\beta}y), \quad y \in \mathbb{Z}(p^\infty). \tag{13.13}$$

Equations (13.12) and (13.13) imply

$$\hat{g}(\hat{\kappa}y) = \hat{f}(2\hat{\gamma}^{-1}y)\hat{g}(y), \quad y \in \mathbb{Z}(p^\infty), \tag{13.14}$$

and

$$\hat{f}(\hat{\kappa}y) = \hat{f}(y)\hat{g}(2\hat{\alpha}\hat{\gamma}^{-1}y), \quad y \in \mathbb{Z}(p^\infty). \tag{13.15}$$

Inasmuch as $0 \leq \hat{f}(y) \leq 1$, it follows from equation (13.14) that

$$\hat{g}(\hat{\kappa}y) \leq \hat{g}(y), \quad y \in \mathbb{Z}(p^\infty).$$

This implies that for every natural n the inequalities

$$\hat{g}(\hat{\kappa}^n y) \leq \cdots \leq \hat{g}(\hat{\kappa}y) \leq \hat{g}(y), \quad y \in \mathbb{Z}(p^\infty), \tag{13.16}$$

hold. Let $y \in \mathbb{Z}(p^\infty)$. Then $y \in \mathbb{Z}(p^l)$ for some l. Hence $\hat{\kappa}y \in \mathbb{Z}(p^l)$. It follows from this that $\hat{\kappa}^m y = y$ for some m, depending generally speaking on y. Substituting $n = m$ in (13.16) we get

$$\hat{g}(\hat{\kappa}y) = \hat{g}(y), \quad y \in \mathbb{Z}(p^\infty). \tag{13.17}$$

Arguing similarly, we obtain from (13.15) that

$$\hat{f}(\hat{\kappa}y) = \hat{f}(y), \quad y \in \mathbb{Z}(p^\infty). \tag{13.18}$$

Note that in proving (13.17) and (13.18) we used only the fact that $k \geq 1$. Put

$$E_{\hat{f}} = \{y \in \mathbb{Z}(p^\infty) : \hat{f}(y) \neq 0\}, \quad E_{\hat{g}} = \{y \in \mathbb{Z}(p^\infty) : \hat{g}(y) \neq 0\},$$

$$B_{\hat{f}} = \{y \in \mathbb{Z}(p^\infty) : \hat{f}(y) = 1\}, \quad B_{\hat{g}} = \{y \in \mathbb{Z}(p^\infty) : \hat{g}(y) = 1\}.$$

Since $p > 2$ and $k = 1$, we have $2\hat{\gamma}^{-1} \in \mathrm{Aut}(\mathbb{Z}(p^\infty))$. Moreover, (13.14), (13.15), (13.17), and (13.18) hold. Taking this into account, we find from (13.14) and (13.17) that $E_{\hat{g}} \subset B_{\hat{f}}$. Analogously, we find from (13.15) and (13.18) that $pE_{\hat{f}} \subset B_{\hat{g}}$. If μ_2 is a nondegenerate distribution, then $B_{\hat{g}}$ is a proper subgroup of $\mathbb{Z}(p^\infty)$. Hence $B_{\hat{g}} = \mathbb{Z}(p^n)$ for some n. Since $B_{\hat{g}} \subset E_{\hat{g}} \subset B_{\hat{f}}$, it follows from (13.9) that $B_{\hat{g}} = \{0\}$. If $B_{\hat{f}} = \{0\}$, then $E_{\hat{g}} \subset B_{\hat{f}} = \{0\}$. Hence μ_2 is an idempotent distribution. If $B_{\hat{f}} \neq \{0\}$, then $pB_{\hat{f}} \subset pE_{\hat{f}} \subset B_{\hat{g}} = \{0\}$. This implies that $B_{\hat{f}} = E_{\hat{f}} = \mathbb{Z}(p)$. Hence μ_1 is an idempotent distribution. We proved statement 1(iii).

Statement 2. Taking into account that the conditional distribution of the linear form L_2 given L_1 is symmetric if and only if the conditional distribution of the linear form $\alpha^{-1}L_2$ given L_1 is symmetric, we may assume without loss of generality that $k \geq 0$. This implies that the restriction of $\alpha \in \mathrm{Aut}(\Omega_p)$ to the subgroup Δ_p is a continuous endomorphism of Δ_p. We retain the notation α for this restriction. We will construct in cases 2(i)–2(iv) independent random variables ξ_1 and ξ_2 with values in the group Δ_p and distributions μ_1 and μ_2 such that the conditional distribution of the linear form $L_2 = \xi_1 + \alpha\xi_2$ given $L_1 = \xi_1 + \xi_2$ is symmetric, whereas $\mu_1, \mu_2 \notin I(\Delta_p)$. Considering ξ_j as independent random variables with values in the group Ω_p, we prove statements 2(i)–2(iv). Taking into account Corollary 9.2, it suffices to construct distributions $\mu_j \notin I(\Delta_p)$ such that their characteristic functions satisfy equation (9.3). We recall that the groups $\mathbb{Z}(p^\infty)$ and Δ_p are the character groups of one another. Moreover, any topological automorphism of the group Δ_p is the multiplication by an element of $\Delta_p^\times$. For $c = (c_0, c_1, \dots, c_n, \dots) \in \Delta_p^\times$ the restriction of the automorphism $\widetilde{c} \in \mathrm{Aut}(\mathbb{Z}(p^\infty))$ to the subgroup $\mathbb{Z}(p^n) \subset \mathbb{Z}(p^\infty)$ is of the form $\widetilde{c}y = s_n y$, $y \in \mathbb{Z}(p^n)$, where $s_n = c_0 + c_1 p + c_2 p^2 + \dots + c_{n-1}p^{n-1}$. Observe also that $\widetilde{\alpha} = p^k\widetilde{c}$, and for any natural m we have

$$A(\mathbb{Z}(p^\infty), p^m\Delta_p) = \mathbb{Z}(p^m). \tag{13.19}$$

Put $f(y) = \hat{\mu}_1(y)$, $g(y) = \hat{\mu}_2(y)$. In this notation equation (9.3) takes the form

$$f(u+v)g(u+\widetilde{\alpha}v) = f(u-v)g(u-\widetilde{\alpha}v), \quad u, v \in \mathbb{Z}(p^\infty). \tag{13.20}$$

Statement 2(i). Consider on the group Δ_p the distribution $\mu = am_{\Delta_p} + (1-a)m_{p\Delta_p}$, where $0 < a < 1$. It follows from (2.3) and (13.19) that the characteristic function $\hat{\mu}(y)$ is of the form

$$\hat{\mu}(y) = \begin{cases} 1 & \text{if} \quad y = 0, \\ 1 - a & \text{if} \quad y \in \mathbb{Z}(p), \\ 0 & \text{if} \quad y \notin \mathbb{Z}(p). \end{cases} \tag{13.21}$$

Let us check that the characteristic functions $f(y) = g(y) = \hat{\mu}(y)$ satisfy equation (13.20). Consider three cases:

1. $u, v \in \mathbb{Z}(p)$. Since $c_0 = p - 1$, we have $\widetilde{\alpha} y = -y$ for all $y \in \mathbb{Z}(p)$ and the restriction of equation (13.20) to the subgroup $\mathbb{Z}(p)$ takes the form

$$f(u+v)g(u-v) = f(u-v)g(u+v), \quad u, v \in \mathbb{Z}(p). \tag{13.22}$$

It follows from $f(y) = g(y)$ that (13.22) holds.

2. Either $u \in \mathbb{Z}(p)$, $v \notin \mathbb{Z}(p)$ or $u \notin \mathbb{Z}(p)$, $v \in \mathbb{Z}(p)$. Then $u \pm v \notin \mathbb{Z}(p)$. This implies that $f(u \pm v) = 0$, and hence both sides of equation (13.20) are equal to zero.

3. $u, v \notin \mathbb{Z}(p)$. If $u + v, u + \widetilde{\alpha} v \in \mathbb{Z}(p)$, then

$$(I - \widetilde{\alpha})v \in \mathbb{Z}(p). \tag{13.23}$$

Inasmuch as $p > 2$, $k = 0$, and $c_0 = p - 1$, we have $I - \alpha \in \mathrm{Aut}(\Delta_p)$ and hence $I - \widetilde{\alpha} \in \mathrm{Aut}(\mathbb{Z}(p^\infty))$. Then (13.23) implies that $v \in \mathbb{Z}(p)$, contrary to the assumption. Thus, either $u + v \notin \mathbb{Z}(p)$ or $u + \widetilde{\alpha} v \notin \mathbb{Z}(p)$ and the left-hand side of equation (13.20) is equal to zero. Similarly, we check that the right-hand side of equation (13.20) is also equal to zero. Thus, (13.20) holds.

Statement 2(ii). Consider on the group Δ_2 the distribution $\mu = am_{\Delta_2} + (1 - a)m_{2\Delta_2}$, where $0 < a < 1$. Then the characteristic function $\hat{\mu}(y)$ is represented by formula (13.21) for $p = 2$. Let us check that the characteristic functions $f(y) = g(y) = \hat{\mu}(y)$ satisfy equation (13.20). Since $k = 0$ and $c_0 = c_1 = 1$, the restriction of the automorphism $\widetilde{\alpha} \in \mathrm{Aut}(\mathbb{Z}(2^\infty))$ to the subgroup $\mathbb{Z}(2^n) \subset \mathbb{Z}(2^\infty)$ is of the form $\widetilde{\alpha} y = my$, $y \in \mathbb{Z}(2^n)$, where $m = 1 + 2 + c_2 2^2 + \cdots + c_{n-1} 2^{n-1} = 4l - 1$. Consider three cases:

1. $u, v \in \mathbb{Z}(2)$. We argue as in case 1 of statement 2(i).

2. Either $u \in \mathbb{Z}(2)$, $v \notin \mathbb{Z}(2)$ or $u \notin \mathbb{Z}(2)$, $v \in \mathbb{Z}(2)$. We argue as in case 2 of statement 2(i).

3. $u, v \notin \mathbb{Z}(2)$. First prove that if $u+v, u+\widetilde{\alpha} v \in \mathbb{Z}(2)$, then $u-v, u-\widetilde{\alpha} v \in \mathbb{Z}(2)$, and (13.20) holds. For the proof we note that the inclusions $u + v, u + \widetilde{\alpha} v \in \mathbb{Z}(2)$ are possible only in the following subcases:

(a) $u + v = 0$ and $u + \widetilde{\alpha} v = 0$. This implies that $(\widetilde{\alpha} - I)v = (m-1)v = 2(2l-1)v = 0$. Hence $v \in \mathbb{Z}(2)$, but this contradicts the assumption.

(b) $u + v = \frac{1}{2}$ and $u + \widetilde{\alpha} v = \frac{1}{2}$. The reasoning is the same as in case (a).

(c) $u + v = 0$ and $u + \widetilde{\alpha} v = \frac{1}{2}$. This implies that $(\widetilde{\alpha} - I)v = (m-1)v = 2(2l-1)v = \frac{1}{2}$. It follows from this that either $v = \frac{1}{4}$, $u = \frac{3}{4}$ or $v = \frac{3}{4}$, $u = \frac{1}{4}$. In both cases $u - v = \frac{1}{2}$, $u - \widetilde{\alpha} v = u - mv = 0$. Hence (13.20) holds.

(d) $u + v = \frac{1}{2}$ and $u + \widetilde{\alpha} v = 0$. The reasoning is the same as in case (c).

Reasoning similarly, we verify that if $u - v, u - \widetilde{\alpha} v \in \mathbb{Z}(2)$, then $u + v, u + \widetilde{\alpha} v \in \mathbb{Z}(2)$ and (13.20) holds.

Statement 2(iii). Consider on the group Δ_2 the distributions $\mu_1 = am_{2\Delta_2} + (1-a)m_{4\Delta_2}$ and $\mu_2 = am_{\Delta_2} + (1-a)m_{2\Delta_2}$, where $0 < a < 1$. It follows from (2.3) and (13.19) that the characteristic function $\hat{\mu}_1(y)$ is of the form

$$\hat{\mu}_1(y) = \begin{cases} 1 & \text{if } \ y \in \mathbb{Z}(2), \\ 1-a & \text{if } \ y \in \mathbb{Z}(4)\backslash\mathbb{Z}(2), \\ 0 & \text{if } \ y \notin \mathbb{Z}(4). \end{cases}$$

The characteristic function $\hat{\mu}_2(y)$ is represented by formula (13.21) for $p = 2$. Let us check that the characteristic functions $f(y) = \hat{\mu}_1(y)$ and $g(y) = \hat{\mu}_2(y)$ satisfy equation (13.20). Consider three cases:

1. $u, v \in \mathbb{Z}(4)$. We can assume that $u \neq 0$, $v \neq 0$. Since either $c_0 = 1$, $c_1 = 0$ or $c_0 = c_1 = 1$, the restriction of the automorphism $\widetilde{c}$ to the subgroup $\mathbb{Z}(4)$ is of the form: either $\widetilde{y} = y$ for all $y \in \mathbb{Z}(4)$ or $\widetilde{y} = -y$ for all $y \in \mathbb{Z}(4)$. Thus, the restriction of equation (13.20) to the subgroup $\mathbb{Z}(4)$ either takes the form

$$f(u+v)g(u+2v) = f(u-v)g(u-2v), \quad u, v \in \mathbb{Z}(4), \tag{13.24}$$

or

$$f(u+v)g(u-2v) = f(u-v)g(u+2v), \quad u, v \in \mathbb{Z}(4). \tag{13.25}$$

Consider equation (13.24). Equation (13.25) can be considered analogously. Consider four subcases:

(a) $u = v = \frac{1}{2}$. Then $u \pm v = 0$, $2v = 0$. Hence $f(u \pm v) = 1$, $g(u \pm 2v) = g(u)$ and (13.24) holds.

(b) $u = \frac{1}{2}$, $v \in \{\frac{1}{4}, \frac{3}{4}\}$. Then $f(u \pm v) = f(v)$. Since $u \pm 2v = 0$, we have $g(u \pm 2v) = 1$ and (13.24) holds.

(c) $u \in \{\frac{1}{4}, \frac{3}{4}\}$, $v = \frac{1}{2}$. Then $f(u \pm v) = f(u)$. Inasmuch as $2v = 0$, we have $g(u \pm 2v) = g(u)$ and (13.24) holds.

(d) $u, v \in \{\frac{1}{4}, \frac{3}{4}\}$. This implies that $u \pm 2v \notin \mathbb{Z}(2)$. Hence $g(u \pm 2v) = 0$ and both sides of equation (13.24) are equal to zero.

2. Either $u \in \mathbb{Z}(4)$, $v \notin \mathbb{Z}(4)$ or $u \notin \mathbb{Z}(4)$, $v \in \mathbb{Z}(4)$. Then $u \pm v \notin \mathbb{Z}(4)$. Hence $f(u \pm v) = 0$ and both sides of equation (13.20) are equal to zero.

3. $u, v \notin \mathbb{Z}(4)$. If $u + v \in \mathbb{Z}(4)$ and $u + \widetilde{\alpha} v \in \mathbb{Z}(2)$, then

$$(I - \widetilde{\alpha})v \in \mathbb{Z}(4). \tag{13.26}$$

Since $k = 1$, we have $I - \alpha \in \mathrm{Aut}(\Delta_2)$ and hence $I - \widetilde{\alpha} \in \mathrm{Aut}(\mathbb{Z}(2^\infty))$. Then (13.26) implies that $v \in \mathbb{Z}(4)$, contrary to the assumption. Hence either $u + v \notin \mathbb{Z}(4)$ or $u + \widetilde{\alpha} v \notin \mathbb{Z}(2)$. This implies that the left-hand side of equation (13.20) is equal to zero. Reasoning analogously, we verify that the right-hand side of equation (13.20) is also equal to zero. Thus, (13.20) holds.

Statement 2(iv). Consider on the group Δ_p the distributions $\mu_1 = a m_{p^{k-1}\Delta_p} + (1-a) m_{p^k\Delta_p}$ and $\mu_2 = a m_{\Delta_p} + (1-a) m_{p^{k-1}\Delta_p}$, where $0 < a < 1$. It follows from (2.3) and (13.19) that the characteristic functions $\hat{\mu}_1(y)$ and $\hat{\mu}_2(y)$ are of the form

$$\hat{\mu}_1(y) = \begin{cases} 1 & \text{if } y \in \mathbb{Z}(p^{k-1}), \\ 1-a & \text{if } y \in \mathbb{Z}(p^k)\backslash\mathbb{Z}(p^{k-1}), \\ 0 & \text{if } y \notin \mathbb{Z}(p^k), \end{cases} \quad \hat{\mu}_2(y) = \begin{cases} 1 & \text{if } y = 0, \\ 1-a & \text{if } y \in \mathbb{Z}(p^{k-1})\backslash\{0\}, \\ 0 & \text{if } y \notin \mathbb{Z}(p^{k-1}). \end{cases}$$

Let us check that the characteristic functions $f(y) = \hat{\mu}_1(y)$ and $g(y) = \hat{\mu}_2(y)$ satisfy equation (13.20). Consider three cases:

1. $u, v \in \mathbb{Z}(p^k)$. Since $\alpha = p^k c$, we have $\widetilde{\alpha} v = 0$ and hence the restriction of equation (13.20) to the subgroup $\mathbb{Z}(p^k)$ takes the form

$$f(u+v)g(u) = f(u-v)g(u), \quad u, v \in \mathbb{Z}(p^k). \tag{13.27}$$

Consider two subcases:

(a) $u \in \mathbb{Z}(p^{k-1})$. Then $f(u \pm v) = f(v)$ and equation (13.27) holds.

(b) $u \in \mathbb{Z}(p^k) \backslash \mathbb{Z}(p^{k-1})$. Then $g(u) = 0$ and both sides of equation (13.27) are equal to zero.

2. Either $u \in \mathbb{Z}(p^k)$, $v \notin \mathbb{Z}(p^k)$ or $u \notin \mathbb{Z}(p^k)$, $v \in \mathbb{Z}(p^k)$. Then $u \pm v \notin \mathbb{Z}(p^k)$. This implies that $f(u \pm v) = 0$, and both sides of equation (13.20) are equal to zero.

3. $u, v \notin \mathbb{Z}(p^k)$. If $u + v \in \mathbb{Z}(p^k)$, $u + \widetilde{\alpha} v \in \mathbb{Z}(p^{k-1})$, then

$$(I - \widetilde{\alpha})v \in \mathbb{Z}(p^k). \tag{13.28}$$

Since $k \geq 1$, we have $I - \alpha \in \mathrm{Aut}(\Delta_p)$ and hence $I - \widetilde{\alpha} \in \mathrm{Aut}(\mathbb{Z}(p^\infty))$. Then (13.28) implies that $v \in \mathbb{Z}(p^k)$, contrary to the assumption. Hence either $u + v \notin \mathbb{Z}(p^k)$ or $u + \widetilde{\alpha} v \notin \mathbb{Z}(p^{k-1})$. This implies that the left-hand side of equation (13.20) is equal to zero. Reasoning analogously, we verify that the right-hand side of equation (13.20) is also equal to zero. Thus, (13.20) holds. □

THEOREM 13.3. *Consider the group of p-adic integers Δ_p. Let $\alpha = c$, where $c = (c_0, c_1, \ldots, c_n, \ldots) \in \Delta_p^\times$, be a topological automorphism of Δ_p. Then the following statements hold.*

1. *Let ξ_1 and ξ_2 be independent random variables with values in the group Δ_p and distributions μ_1 and μ_2. Assume that the conditional distribution of the linear form $L_2 = \xi_1 + \alpha\xi_2$ given $L_1 = \xi_1 + \xi_2$ is symmetric. Then the following statements are valid.*

   1(i) *If $p > 2$ and $c_0 \neq p - 1$, then $\mu_j = m_K * E_{x_j}$, where K is a compact subgroup of Δ_p and $x_j \in \Delta_p$. Moreover, if $c_0 = 1$, then μ_1 and μ_2 are degenerate distributions.*

   1(ii) *If $p = 2$, $c_0 = 1$ and $c_1 = 0$, then μ_1 and μ_2 are degenerate distributions.*

2. *If one of the following conditions holds:*

   2(i) *$p > 2$, $c_0 = p - 1$;*

   2(ii) *$p = 2$, $c_0 = c_1 = 1$,*

   then there exist independent random variables ξ_1 and ξ_2 with values in the group Δ_p and distributions μ_1 and μ_2 such that the conditional distribution of the linear form $L_2 = \xi_1 + \alpha\xi_2$ given $L_1 = \xi_1 + \xi_2$ is symmetric, whereas $\mu_1, \mu_2 \notin I(\Delta_p)$.

PROOF. We recall that each automorphism $\alpha \in \mathrm{Aut}(\Delta_p)$ is the multiplication by an element $c_\alpha \in \Delta_p^\times$. Let ξ_1 and ξ_2 be independent random variables with values in the group Δ_p. Then we can consider ξ_j as independent random variables with values in the group Ω_p. Moreover, the multiplication by an element $c \in \Delta_p^\times$ is a topological isomorphism of the group Ω_p. This implies that statements 1(i) and 1(ii) in Theorem 13.2 are also valid for the group Δ_p. To complete the proof of the theorem, we note that in the proof of statements 2(i) and 2(ii) in Theorem 13.2 the corresponding independent random variables take values in the subgroup $\Delta_p \subset \Omega_p$. □

Observe that Lemma 10.11 follows from Theorem 13.3.

REMARK 13.4. Assume that in Theorem 13.2 $p > 2$, $k = 0$, $c_0 \neq 1$, and $c_0 \neq p-1$. This implies in particular that $I - \alpha \in \mathrm{Aut}(\Delta_p)$ and hence

$$I - \widetilde{\alpha} \in \mathrm{Aut}(\mathbb{Z}(p^\infty)). \tag{13.29}$$

Let ξ_1 and ξ_2 be independent identically distributed random variables with values in the group Δ_p and distribution m_{Δ_p}. We check that the conditional distribution of the linear form $L_2 = \xi_1 + \alpha\xi_2$ given $L_1 = \xi_1 + \xi_2$ is symmetric. By Corollary 9.2, it suffices to verify that the characteristic functions $f(y) = g(y) = \widehat{m}_{\Delta_p}(y)$ satisfy equation (13.20). It follows from (2.3) that

$$\widehat{m}_{\Delta_p}(y) = \begin{cases} 1, & \text{if } y = 0, \\ 0, & \text{if } y \neq 0. \end{cases} \tag{13.30}$$

It suffices to check that equation (13.20) holds when $u \neq 0$, $v \neq 0$. Thus, assume that $u \neq 0$, $v \neq 0$. If $u + v = 0$ and $u + \widetilde{\alpha}v = 0$, then $(I - \widetilde{\alpha})v = 0$. Taking into account (13.29), this implies that $v = 0$, contrary to the assumption. Assume that either $u + v \neq 0$ or $u + \widetilde{\alpha}v \neq 0$. Then the left-hand side of equation (13.20) is equal to zero. Reasoning analogously we see that the right-hand side of equation (13.20) is also equal to zero. So, equation (13.20) holds. This example shows that statement 1(i) in Theorem 13.2 cannot be strengthened to the statement that μ_j are degenerate distributions.

Assume that conditions 1(iii) of Theorem 13.2 hold, i.e., $p > 2$ and $|k| = 1$. We can assume without loss of generality that $k = 1$. Let ξ_1 and ξ_2 be independent random variables with values in the group Δ_p and distributions $\mu_1 = am_{\Delta_p} + (1-a)m_{p\Delta_p}$, where $0 < a < 1$, and $\mu_2 = m_{\Delta_p}$. Then the characteristic function $f(y) = \hat{\mu}_1(y)$ is defined by formula (13.21) and the characteristic function $g(y) = \hat{\mu}_2(y)$ is defined by formula (13.30). Arguing as in the proof of statement 2(iv) it is easy to check that the characteristic functions $f(y)$ and $g(y)$ satisfy equation (13.20). Thus, statement 1(iii) cannot be strengthened to the statement that μ_j are shifts of idempotent distributions.

We complement Theorem 13.2 with the following assertion.

PROPOSITION 13.5. *Consider the group of p-adic numbers Ω_p. Let α be a topological automorphism of Ω_p. Assume that $\alpha \neq -I$. Let ξ_1 and ξ_2 be independent random variables with values in the group Ω_p and distributions μ_1 and μ_2. If the conditional distribution of the linear form $L_2 = \xi_1 + \alpha\xi_2$ given $L_1 = \xi_1 + \xi_2$ is symmetric, then either μ_1 and μ_2 are degenerate distributions or there exists a closed subgroup $M \subset \Omega_p$ such that $\hat{\mu}_j(y) = 0$ for all $y \notin M$, $j = 1, 2$.*

PROOF. Let $\alpha = p^k c$, where $k \in \mathbb{Z}$ and $c = (c_0, c_1, \dots, c_n, \dots) \in \Delta_p^\times$. We can assume without loss of generality that $k \geq 0$. Taking into account that in cases 1(i) and 1(ii) of Theorem 13.2 the proposition holds, it remains to prove the proposition in the following cases:

1. $p \geq 2$, $k \geq 1$;
2. $p > 2$, $k = 0$, $c_0 = p - 1$;
3. $p = 2$, $k = 0$, $c_0 = c_1 = 1$.

We argue in the same way as in the proof of Theorem 13.2 and keep the same notation. Note that in proving (13.17) and (13.18) we used only the fact that $k \geq 1$.

1. $p \geq 2$, $k \geq 1$. Assume that there exists a sequence of elements $y_n \in \mathbb{Z}(p^\infty)$ satisfying the conditions:

(a) the order of the element y_n is equal to p^{i_n}, $i_n \to \infty$;
(b) $\hat{g}(y_n) \neq 0$.

Then it follows from (13.14) and (13.17) that,

$$\hat{f}(2\hat{\gamma}^{-1}y_n) = 1. \tag{13.31}$$

First suppose $p > 2$. Then (13.31) implies that $\hat{f}(y) = 1$ for all $y \in \mathbb{Z}(p^{i_n})$, because the order of the element $2\hat{\gamma}^{-1}y_n$ is equal to p^{i_n}, and hence the element $2\hat{\gamma}^{-1}y_n$ generates the subgroup $\mathbb{Z}(p^{i_n})$.

If $p = 2$, then (13.31) implies that $\hat{f}(y) = 1$ for all $y \in \mathbb{Z}(2^{i_n-1})$, because the order of the element $2\hat{\gamma}^{-1}y_n$ is equal to 2^{i_n-1}, and the element $2\hat{\gamma}^{-1}y_n$ generates the subgroup $\mathbb{Z}(2^{i_n-1})$. Thus, for $p \geq 2$ we have $\hat{f}(y) = 1$ for all $y \in \mathbb{Z}(p^\infty)$ and equation (13.8) implies that

$$\hat{g}(u + \hat{\alpha}v) = \hat{g}(u - \hat{\alpha}v), \quad u, v \in \mathbb{Z}(p^\infty).$$

It follows from this that $\hat{g}(2\hat{\alpha}y) = 1$ for all $y \in \mathbb{Z}(p^\infty)$ and hence $\hat{g}(y) = 1$ for all $y \in \mathbb{Z}(p^\infty)$, because $2\hat{\alpha}$ is an epimorphism. We proved that μ_1 and μ_2 are degenerate distributions.

Similar reasoning shows that if there exists a sequence of elements $z_n \in \mathbb{Z}(p^\infty)$ satisfying the conditions:

(a) the order of element z_n is equal to p^{j_n}, $j_n \to \infty$;
(b) $\hat{f}(z_n) \neq 0$,

then μ_1 and μ_2 are also degenerate distributions.

From what has been said it follows that if μ_1 and μ_2 are nondegenerate distributions, then there exists n such that $\hat{f}(y) = \hat{g}(y) = 0$ for all $y \notin \mathbb{Z}(p^n)$. The proposition in case 1 is proved.

2. $p > 2$, $k = 0$, $c_0 = p - 1$. Inasmuch as $p > 2$, we have $\hat{\beta} \in \mathrm{Aut}(\mathbb{Z}(p^\infty))$. We find from equation (13.12) that

$$\hat{g}(y) = \hat{f}(2\hat{\beta}^{-1}y)\hat{g}(\hat{\gamma}\hat{\beta}^{-1}y), \quad y \in \mathbb{Z}(p^\infty). \tag{13.32}$$

Since $0 \leq \hat{g}(y) \leq 1$, equation (13.32) implies the inequality

$$\hat{g}(y) \leq \hat{f}(2\hat{\beta}^{-1}y), \quad y \in \mathbb{Z}(p^\infty). \tag{13.33}$$

Reasoning similarly we get from (13.13) that

$$\hat{f}(y) = \hat{f}(\hat{\gamma}\hat{\beta}^{-1}y)g(2\hat{\alpha}\hat{\beta}^{-1}y), \quad y \in \mathbb{Z}(p^\infty). \tag{13.34}$$

Taking into account that $0 \leq \hat{f}(y) \leq 1$, it follows from this that

$$\hat{f}(y) \leq \hat{g}(2\hat{\alpha}\hat{\beta}^{-1}y), \quad y \in \mathbb{Z}(p^\infty). \tag{13.35}$$

Inequalities (13.33) and (13.35) imply the inequalities

$$\begin{gathered} \hat{g}(y) \leq \hat{f}(2\hat{\beta}^{-1}y) \leq \hat{g}(4\hat{\alpha}\hat{\beta}^{-2}y), \quad y \in \mathbb{Z}(p^\infty), \\ \hat{f}(y) \leq \hat{g}(2\hat{\alpha}\hat{\beta}^{-1}y) \leq \hat{f}(4\hat{\alpha}\hat{\beta}^{-2}y), \quad y \in \mathbb{Z}(p^\infty). \end{gathered} \tag{13.36}$$

Reasoning as in the proof of case 1, we obtain from (13.36) that

$$\hat{g}(y) = \hat{g}(4\hat{\alpha}\hat{\beta}^{-2}y), \quad \hat{f}(y) = \hat{f}(4\hat{\alpha}\hat{\beta}^{-2}y), \quad y \in \mathbb{Z}(p^\infty). \tag{13.37}$$

We find from (13.36) and (13.37) that

$$\hat{g}(y) = \hat{f}(2\hat{\beta}^{-1}y), \quad \hat{f}(y) = \hat{g}(2\hat{\alpha}\hat{\beta}^{-1}y), \quad y \in \mathbb{Z}(p^\infty). \tag{13.38}$$

It follows from (13.32), (13.34), and (13.38) that if $\hat{g}(y_0) \neq 0$ for some $y_0 \in \mathbb{Z}(p^\infty)$, then $\hat{g}(\hat{\gamma}\hat{\beta}^{-1}y_0) = 1$ and if $\hat{f}(y_0) \neq 0$, then $\hat{f}(\hat{\gamma}\hat{\beta}^{-1}y_0) = 1$. We complete the proof as in case 1. The proposition in case 2 is proved.

3. $p = 2$, $k = 0$, $c_0 = c_1 = 1$. Put $\beta = 2\beta_1$, $\gamma = 2\gamma_1$. Then $\hat{\beta}_1 \in \mathrm{Aut}(\mathbb{Z}(2^\infty))$, and $\hat{\gamma}_1$ is an epimorphism. It follows from (13.12) that

$$\hat{f}(y)\hat{g}(\hat{\gamma}_1 y) = \hat{g}(\hat{\beta}_1 y), \quad y \in \mathbb{Z}(2^\infty). \tag{13.39}$$

Similarly, we find from (13.13) that

$$\hat{f}(\hat{\gamma}_1 y)\hat{g}(\hat{\alpha} y) = \hat{f}(\hat{\beta}_1 y), \quad y \in \mathbb{Z}(2^\infty). \tag{13.40}$$

It follows from (13.39) and (13.40) that

$$\hat{g}(y) = \hat{f}(\hat{\beta}_1^{-1}y)\hat{g}(\hat{\gamma}_1\hat{\beta}_1^{-1}y), \quad \hat{f}(y) = \hat{f}(\hat{\gamma}_1\hat{\beta}_1^{-1}y)\hat{g}(\hat{\alpha}\hat{\beta}_1^{-1}y), \quad y \in \mathbb{Z}(2^\infty).$$

Hence

$$\hat{g}(y) \leq \hat{f}(\hat{\beta}_1^{-1}y), \quad \hat{f}(y) \leq \hat{g}(\hat{\alpha}\hat{\beta}_1^{-1}y).$$

This implies that

$$\hat{g}(y) \leq \hat{f}(\hat{\beta}_1^{-1}y) \leq \hat{g}(\hat{\alpha}\beta_1{}^{-2}y), \quad \hat{f}(y) \leq \hat{g}(\hat{\alpha}\hat{\beta}_1^{-1}y) \leq \hat{f}(\hat{\alpha}\hat{\beta}_1^{-2}y).$$

Taking into account these inequalities, we complete the proof as in case 2. □

14. Characterization of shifts of idempotent distributions through the independence of sum and difference squared

The following characterization of the Gaussian distribution on the real line is well known: Let ξ and η be independent identically distributed random variables with distribution μ. If $S = \xi + \eta$ and $D = (\xi - \eta)^2$ are independent, then μ is a Gaussian distribution. This statement is a particular case of the well-known theorem, where the Gaussian distribution is characterized by the independence of the sample mean and the sample variance of n independent identically distributed random variables.

In this section, first for countable discrete fields, next for the field of p-adic numbers Ω_p, we study distributions μ which are characterized by the independence of S and D. In the case of a discrete field, we prove that μ is a shift of an idempotent distribution. A similar result is proved for the field of p-adic numbers Ω_p, where $p > 2$, assuming that μ has a density with respect to a Haar measure on Ω_p and this density is continuous.

Let X be a locally compact field. Denote by e the unit of the field X. The additive group of the field X is a locally compact Abelian group. We also denote this group by X. Denote by (x, y), where $x, y \in X$, elements of the group X^2. Denote by T the mapping $T : X^2 \to X^2$ defined by the formula

$$T(x, y) = (x + y, (x - y)^2).$$

Let A be a subset of X. Put

$$A^{[2]} = \{x \in X : x = t^2,\ t \in A\}.$$

We recall that if ξ and η are random variables with values in X, then we denote by μ_ξ the distribution of the random variable ξ, and by $\mu_{(\xi,\eta)}$ the distribution of the random vector (ξ, η).

Note that if the characteristic of a field X differs from 2, then X is an Abelian group with unique division by 2, i.e., for every element $x \in X$ the element $\frac{x}{2} \in X$ is uniquely determined.

LEMMA 14.1. *Let X be a countable discrete field of characteristic p, where $p \neq 2$. Let ξ and η be independent identically distributed random variables with values in X and distribution μ such that $\mu(0) > 0$. The random variables $S = \xi + \eta$ and $D = (\xi - \eta)^2$ are independent if and only if the function $\mu(x)$ satisfies the equation*

$$\mu^2(u)\mu(v)\mu(-v) = \mu^2(0)\mu(u+v)\mu(u-v), \quad u, v \in X. \tag{14.1}$$

PROOF. Note that $\mu_{(S,D)} = T(\mu_{(\xi,\eta)})$. It is obvious that

$$\mu_{(S,D)}(u, 0) = T(\mu_{(\xi,\eta)})(u, 0) = \mu_{(\xi,\eta)}(T^{-1}(u, 0)) = \mu_{(\xi,\eta)}\left(\frac{u}{2}, \frac{u}{2}\right). \tag{14.2}$$

Let $v = t^2$, $t \in X$, $t \neq 0$. We have

$$\begin{aligned} \mu_{(S,D)}(u, v) &= \mu_{(S,D)}(u, t^2) = \mu_{(\xi,\eta)}(T^{-1}(u, t^2)) \\ &= \mu_{(\xi,\eta)}\left(\frac{u+t}{2}, \frac{u-t}{2}\right) + \mu_{(\xi,\eta)}\left(\frac{u-t}{2}, \frac{u+t}{2}\right), \quad u, t \in X, \ t \neq 0. \end{aligned} \tag{14.3}$$

Taking into account the independence of the random variables ξ and η, it follows from (14.2) and (14.3) that

$$\mu_{(S,D)}(u, 0) = \mu^2\left(\frac{u}{2}\right), \quad u \in X, \tag{14.4}$$

$$\mu_{(S,D)}(u, v) = 2\mu\left(\frac{u+t}{2}\right)\mu\left(\frac{u-t}{2}\right), \quad u, t \in X, \ v = t^2, \ t \neq 0. \tag{14.5}$$

Necessity. Taking into account the independence of the random variables S and D, it follows from (14.4) and (14.5) that

$$\mu_S(u)\mu_D(0) = \mu^2\left(\frac{u}{2}\right), \quad u \in X, \tag{14.6}$$

$$\mu_S(u)\mu_D(t^2) = 2\mu\left(\frac{u+t}{2}\right)\mu\left(\frac{u-t}{2}\right), \quad u, t \in X, \ t \neq 0. \tag{14.7}$$

Moreover, (14.6) implies that $\mu_D(0) \neq 0$, because otherwise $\mu(x) = 0$ for all $x \in X$, but this contradicts the condition of the theorem. We find $\mu_S(u)$ from equation (14.6) and substitute it in equation (14.7). We get

$$\mu^2\left(\frac{u}{2}\right)\mu_D(t^2) = 2\mu_D(0)\mu\left(\frac{u+t}{2}\right)\mu\left(\frac{u-t}{2}\right), \quad u, t \in X, \ t \neq 0. \tag{14.8}$$

Rewrite equation (14.8) in the form

$$\mu^2(u)\mu_D(4t^2) = 2\mu_D(0)\mu\left(u+t\right)\mu\left(u-t\right), \quad u, t \in X, \ t \neq 0. \tag{14.9}$$

Substituting $u = 0$ in equation (14.9), we obtain

$$\mu_D(4t^2) = \frac{2\mu_D(0)\mu(t)\mu(-t)}{\mu^2(0)}, \quad u, t \in X, \ t \neq 0. \tag{14.10}$$

It follows from (14.9) and (14.10) that

$$\mu^2(u)\mu(t)\mu(-t) = \mu^2(0)\mu(u+t)\mu(u-t), \quad u, t \in X, \ t \neq 0. \tag{14.11}$$

Note that equation (14.11) also holds true when $t = 0$. Therefore, the function $\mu(x)$ satisfies equation (14.1). So, the necessity is proved.

Sufficiency. First we note that random variables ξ_1 and ξ_2 with values in X are independent if and only if there exist functions $a(x)$ and $b(x)$ on X such that

$$\mu_{(\xi_1,\xi_2)}(x, y) = a(x)b(y), \quad x, y \in X. \tag{14.12}$$

Let ξ and η be independent identically distributed random variables with values in the field X and distribution μ such that $\mu(0) > 0$. It is obvious that if $v \notin X^{[2]}$, then $\mu_{(S,D)}(u, v) = 0$. It follows from (14.4) and (14.5) that we have the representation

$$\mu_{(S,D)}(u, v) = \begin{cases} \mu^2\left(\frac{u}{2}\right), & \text{if} \quad u \in X,\ v = 0, \\ 2\mu\left(\frac{u+t}{2}\right)\mu\left(\frac{u-t}{2}\right), & \text{if} \quad u, t \in X,\ v = t^2,\ t \neq 0, \\ 0, & \text{if} \quad u, v \in X,\ v \notin X^{[2]}. \end{cases}$$

This representation implies that if a function $\mu(x)$ satisfies equation (14.1), then the random variables S and D are independent, because in this case we can put in (14.12)

$$a(x) = \mu^2\left(\frac{x}{2}\right), \quad b(x) = \begin{cases} 1, & \text{if} \quad x = 0, \\ \dfrac{2\mu\left(\frac{t}{2}\right)\mu\left(-\frac{t}{2}\right)}{\mu^2(0)}, & \text{if} \quad x = t^2,\ t \neq 0, \\ 0, & \text{if} \quad x \notin X^{[2]}. \end{cases}$$

□

LEMMA 14.2. *Let X be an Abelian group with unique division by 2. Let $\mu(x)$ be a function on X, satisfying equation* (14.1) *and the condition $\mu(0) \neq 0$. Then the set $K = \{x \in X : \mu(x) \neq 0\}$ is a subgroup of X.*

PROOF. Assume that $\mu(x) \neq 0$ at a point $x \in X$. Substituting $u = v = \frac{x}{2}$ in equation (14.1), we get

$$\mu^3\left(\frac{x}{2}\right)\mu\left(-\frac{x}{2}\right) = \mu^3(0)\mu(x). \tag{14.13}$$

Inasmuch as $\mu(0) \neq 0$ and $\mu(x) \neq 0$, it follows from (14.13) that

$$\mu\left(\frac{x}{2}\right)\mu\left(-\frac{x}{2}\right) \neq 0. \tag{14.14}$$

Substituting $u = v = -\frac{x}{2}$ in equation (14.1), we obtain

$$\mu^3\left(-\frac{x}{2}\right)\mu\left(\frac{x}{2}\right) = \mu^3(0)\mu(-x). \tag{14.15}$$

Taking into account (14.14), we deduce from (14.15) that $\mu(-x) \neq 0$. So, we proved that if $\mu(x) \neq 0$, then also $\mu(-x) \neq 0$. This fact together with (14.1) implies that the set $K = \{x \in X : \mu(x) \neq 0\}$ is a subgroup of X. □

First we prove a theorem on characterization of shifts of idempotent distributions for a discrete field of nonzero characteristic.

THEOREM 14.3. *Let X be a countable discrete field of characteristic $p \neq 0$. Then the following statements hold.*

1. *Assume $p > 2$. Let ξ and η be independent identically distributed random variables with values in X and distribution μ. The random variables $S = \xi + \eta$ and $D = (\xi - \eta)^2$ are independent if and only if $\mu \in I(X)$.*
2. *Assume $p = 2$. Let ξ and η be independent random variables with values in X and distributions μ and ν. The random variables $S = \xi + \eta$ and $D = (\xi - \eta)^2$ are independent if and only if μ and ν are degenerate distributions.*

PROOF. Statement 1. Necessity. Replacing if necessary the random variables ξ and η by the new independent random variables $\xi + x$ and $\eta + x$, $x \in X$, we can assume that $\mu(0) > 0$. Then by Lemma 14.1, the function $\mu(x)$ satisfies equation (14.1). Inasmuch as $p > 2$, X is an Abelian group with unique division by 2. Hence by Lemma 14.2, the set $K = \{x \in X : \mu(x) \neq 0\}$ is a subgroup of X. Let $x_0 \in K$, $x_0 \neq 0$. Denote by L the subgroup generated by x_0. Then $L \cong \mathbb{Z}(p)$ and $L \subset K$. Consider the restriction of equation (14.1) to L. Put $\varphi(x) = \ln \mu(x)$, $x \in L$. It follows from (14.1) that

$$2\varphi(u) + \varphi(v) + \varphi(-v) = 2\varphi(0) + \varphi(u + v) + \varphi(u - v), \quad u, v \in L.$$

Integrate both sides of this equality over the subgroup L with respect to the Haar distribution m_L in the variable v. Using that the Haar distribution m_L is L-invariant we conclude that $\varphi(u) = \varphi(0)$ for all $u \in L$. It follows that $\mu(x) = \mu(0)$ for all $x \in L$. Hence $\mu(x) = \mu(0)$ for all $x \in K$. This implies that K is a finite subgroup and $\mu = m_K$. The necessity is proved.

Sufficiency. Let K be a finite subgroup of X. Let ξ and η be independent identically distributed random variables with values in X and distribution $\mu = m_K$. Since all nonzero elements of the field X have order p and $p > 2$, the following statement holds: if $2x \in K$, then $x \in K$. This easily implies that the function

$$\mu(x) = m_K(x) = \begin{cases} |K|^{-1}, & \text{if } x \in K, \\ 0, & \text{if } x \notin K, \end{cases}$$

satisfies equation (14.1). Hence by Lemma 14.1, the random variables S and D are independent.

Statement 2. Necessity. Since X is a field of characteristic 2, the equality $(\xi - \eta)^2 = (\xi + \eta)^2$ holds, i.e., $D = S^2$. As it easily seen, in a field of characteristic 2 if μ_S is a nondegenerate distribution, then μ_{S^2} is also a nondegenerate distribution. Hence in a field of characteristic 2 the independence of S and S^2 implies that S has a degenerate distribution (compare below with Example 14.5). Then μ and ν are also degenerate distributions.

Sufficiency is obvious. □

We prove now a theorem on the characterization of degenerate distributions for a discrete field of characteristic 0.

THEOREM 14.4. *Let X be a countable discrete field of characteristic 0. Let ξ and η be independent identically distributed random variables with values in X and distribution μ. The random variables $S = \xi + \eta$ and $D = (\xi - \eta)^2$ are independent if and only if μ is a degenerate distribution.*

PROOF. Necessity. Replacing if necessary the random variables ξ and η by the new independent random variables $\xi + x$ and $\eta + x$, $x \in X$, we can assume that $\mu(0) > 0$. Then by Lemma 14.1, the function $\mu(x)$ satisfies equation (14.1). Since

$p = 0$, X is an Abelian group with unique division by 2. Then by Lemma 14.2, the set $K = \{x \in X : \mu(x) \neq 0\}$ is a subgroup of X.

Assume that $\mu(x_0) \neq 0$ at a point $x_0 \in X$, $x_0 \neq 0$. Consider a subgroup G of X of the form

$$G = \left\{ \frac{mx_0}{2^n} : n = 0, 1, \ldots; \ m \in \mathbb{Z} \right\}.$$

Taking into account that K is a subgroup, equation (14.1) implies that $\mu(x) \neq 0$ for all $x \in G$. Consider the restriction of equation (14.1) to the subgroup G. Taking into account that $\mu(x) > 0$ for all $x \in G$, put $\varphi(x) = \ln \mu(x)$, $x \in G$. It follows from (14.1) that the function $\varphi(x)$ satisfies the equation

$$2\varphi(0) + \varphi(u+v) + \varphi(u-v) = 2\varphi(u) + \varphi(v) + \varphi(-v), \quad u, v \in G. \tag{14.16}$$

We use the finite difference method to solve equation (14.16). Let h be an arbitrary element of the group G. Replacing u by $u + h$ and v by $v + h$ in equation (14.16), we obtain

$$\begin{aligned} 2\varphi(0) + \varphi(u+v+2h) &+ \varphi(u-v) \\ &= 2\varphi(u+h) + \varphi(v+h) + \varphi(-v-h), \quad u, v, h \in G. \end{aligned} \tag{14.17}$$

Subtracting (14.16) from (14.17) we find

$$\Delta_{2h}\varphi(u+v) = 2\Delta_h\varphi(u) + \Delta_h\varphi(v) + \Delta_{-h}\varphi(-v), \quad u, v, h \in G. \tag{14.18}$$

Substituting $u = 0$ in equation (14.18), we get

$$\Delta_{2h}\varphi(v) = 2\Delta_h\varphi(0) + \Delta_h\varphi(v) + \Delta_{-h}\varphi(-v), \quad v, h \in G. \tag{14.19}$$

Subtracting (14.19) from (14.18), we obtain

$$\Delta_{2h}\Delta_u\varphi(v) = 2\Delta_h\Delta_u\varphi(0), \quad u, v, h \in G. \tag{14.20}$$

Let k be an arbitrary element of the group G. Replacing v by $v + k$ in equation (14.20) and subtracting equation (14.20) from the obtained equation, we find

$$\Delta_{2h}\Delta_u\Delta_k\varphi(v) = 0, \quad u, v, h, k \in G. \tag{14.21}$$

Since u, h and k are arbitrary elements of G, it follows from (14.21) that the function $\varphi(x)$ satisfies the equation

$$\Delta_u^3\varphi(v) = 0, \quad u, v \in G. \tag{14.22}$$

Let x be an arbitrary element of the group G. Then $x = rx_0$, where $r = \frac{m}{2^n}$. As follows from (14.22), $\varphi(x) = ar^2 + br + c$, where a, b, c are some real numbers. Hence

$$\mu(x) = e^{ar^2+br+c}.$$

But this contradicts the fact that

$$\sum_{x \in G} \mu(x) \leq 1.$$

Thus, there does not exist a point $x_0 \in X$, $x_0 \neq 0$, such that $\mu(x_0) \neq 0$. Hence $\mu = E_0$.

Sufficiency is obvious. □

EXAMPLE 14.5. Let X be a countable discrete field of characteristic p, where $p \neq 2$. Let ξ and η be independent random variables with values in X and distributions μ and ν. Assume that the random variables $S = \xi + \eta$ and $D = (\xi - \eta)^2$ are independent. Generally speaking, this does not imply that both distributions μ and ν are idempotent. Here is a corresponding example.

Inasmuch as $p \neq 2$, we have $e \neq -e$. Let ξ and η be independent random variables with values in X and distributions μ and ν such that $\mu = \frac{1}{2}(E_{-e} + E_e)$, $\nu = E_0$. Then $S = \xi$ and $D = \xi^2$. Since μ_D is a degenerate distribution, the random variables S and D are independent.

Consider the field Ω_p. We recall that the norm $|x|_p$ on Ω_p is defined as follows. If $x \in \Omega_p$ and $x = p^k c$, where $k \in \mathbb{Z}$ and $c \in \Delta_p^\times$, we put $|x|_p = p^{-k}$, $|0|_p = 0$. The norm $|x|_p$ satisfies the conditions

$$|x+y|_p \le \max(|x|_p, |y|_p), \quad |xy|_p = |x|_p |y|_p.$$

Choose a Haar measure m_{Ω_p} on Ω_p in such a way that $m_{\Omega_p}(\Delta_p) = 1$. Then $m_{\Omega_p}(p^k \Delta_p) = p^{-k}$. We will also assume that $m_{\Omega_p^2} = m_{\Omega_p} \times m_{\Omega_p}$.

We prove now a theorem on characterization of shifts of idempotent distributions on the field of p-adic numbers Ω_p. We need some lemmas.

LEMMA 14.6. *Consider the field Ω_p. Then on the set $\Omega_p^{[2]}$ there is a continuous function $\mathfrak{s}(x)$ satisfying the equation*

$$\mathfrak{s}^2(x) = x, \quad x \in \Omega_p^{[2]}.$$

The set $\Omega_p^{[2]} \backslash \{0\}$ is a disjoint union of balls in each of which the function $\mathfrak{s}(x)$ is expressed by a convergent power series.

PROOF. The representation

$$\Omega_p = \{0\} \cup \bigcup_{l=-\infty}^{\infty} p^l \Delta_p^\times$$

implies that

$$\Omega_p^{[2]} = \{0\} \cup \bigcup_{l=-\infty}^{\infty} p^{2l} (\Delta_p^\times)^{[2]}. \tag{14.23}$$

First assume that $p > 2$. It is well known that in the multiplicative group $\Delta_p^\times$ there is an element ε of order $p-1$ such that the elements $0, \varepsilon, \varepsilon^2, \dots, \varepsilon^{p-1} = e$ form a complete set of coset representatives of the subgroup $p\Delta_p$ in the group Δ_p. Put

$$A_k = \varepsilon^k + p\Delta_p = \left\{x : |x - \varepsilon^k|_p \le \frac{1}{p}\right\}, \quad k = 1, 2, \dots, p-1.$$

Then

$$\Delta_p^\times = \bigcup_{k=1}^{p-1} A_k.$$

First define the function $\mathfrak{s}(x)$ on the set $(\Delta_p^\times)^{[2]}$. Since $(e + p\Delta_p)^{[2]} = e + p\Delta_p$, we have $A_k^{[2]} = A_{2k}$, if $k = 1, 2, \dots, \frac{p-1}{2}$ and $A_k^{[2]} = A_{2k-p+1}$, if $k = \frac{p+1}{2}, \frac{p+3}{2}, \dots, p-1$. Note that

$$(\Delta_p^\times)^{[2]} = \bigcup_{k=1}^{\frac{p-1}{2}} A_{2k}$$

and define the function $\mathfrak{s}(x)$ on each coset A_{2k}. Take $x \in A_{2k}$. Then the equation $x = t^2$ has two solutions $t_1 \in A_k$ and $-t_1 \in A_{k+\frac{p-1}{2}}$. These solutions belong to different cosets. The coset A_k is a compact set, the function $g(x) = x^2$ is continuous on A_k, and $g(x)$ is a one-to-one mapping of the set A_k on A_{2k}. This implies that the mapping $\mathfrak{s}_k : A_{2k} \to A_k$, which is inverse to $g(x)$, is also continuous. Hence $\mathfrak{s}_k$ is a homeomorphism between A_{2k} and A_k. Put $\mathfrak{s}(x) = \mathfrak{s}_k(x)$, if $x \in A_{2k}$, $k = 1, 2, \ldots, \frac{p-1}{2}$. Inasmuch as A_{2k} is an open set in Ω_p, the function $\mathfrak{s}(x)$ is continuous and satisfies equation $\mathfrak{s}^2(x) = x$ on $(\Delta_p^\times)^{[2]}$. Taking into account (14.23), put

$$\mathfrak{s}(x) = \begin{cases} p^l \mathfrak{s}(c), & \text{if } x = p^{2l}c, \ c \in (\Delta_p^\times)^{[2]}, \\ 0, & \text{if } x = 0. \end{cases}$$

It is not difficult to verify that the constructed function $\mathfrak{s}(x)$ in each ball A_{2k} for $k = 1, 2, \ldots, \frac{p-1}{2}$ is expressed by a convergent power series

$$\mathfrak{s}(x) = \varepsilon^k \left(e + \frac{\varepsilon^{-2k}}{2}(x - \varepsilon^{2k}) + \sum_{n=2}^{\infty} (-1)^{n-1} \frac{(2n-3)!!}{2n!!} \varepsilon^{-2kn}(x - \varepsilon^{2k})^n \right).$$

This implies that in each ball $p^{2l}A_{2k}$, $l = \pm 1, \pm 2, \ldots$, the function $\mathfrak{s}(x)$ is also expressed by a convergent power series. It is obvious that the set $\Omega_p^{[2]} \backslash \{0\}$ is a disjoint union of balls $p^{2l}A_{2k}$, $k = 1, 2, \ldots, \frac{p-1}{2}$, $l \in \mathbb{Z}$, and $\mathfrak{s}(x)$ is the required function.

If $p = 2$, we change the argument. Put $B_k = ke + 4\Delta_2$, $k = 0, 1, 2, 3$. Then the elements $0, e, 2e, 3e$ form a complete set of coset representatives of the subgroup $4\Delta_2$ in the group Δ_2. We have $(\Delta_2^\times)^{[2]} = B_1^{[2]} = B_3^{[2]} = e + 8\Delta_2$. Take $x \in e + 8\Delta_2$. Then the equation $x = t^2$ has two solutions $t_1 \in B_1$ and $-t_1 \in B_3$. These solutions belong to different cosets. The remaining part of the proof is similar to the case when $p > 2$. □

LEMMA 14.7. *Consider the field Ω_p, where $p > 2$. Take $(x_0, y_0) \in \Omega_p^2$ such that $|x_0 - y_0|_p = p^{-l}$. Then for $k \geq l + 1$ the equality*

$$T\{(x_0, y_0) + (p^k\Delta_p)^2\} = (x_0 + y_0, (x_0 - y_0)^2) + (p^k\Delta_p) \times (p^{k+l}\Delta_p) \tag{14.24}$$

holds.

PROOF. Since $|x_0 - y_0|_p = p^{-l}$, we have $x_0 - y_0 = p^l c$, where $c \in \Delta_p^\times$. Note that on the one hand, the equality

$$\begin{aligned} T\{(x_0, y_0) + (p^k\Delta_p)^2\} &= T\{(x_0 + p^k x, y_0 + p^k y) : x, y \in \Delta_p\} \\ = \{(x_0 + y_0 + p^k(x+y), &(x_0 - y_0)^2 + 2p^k(x_0 - y_0)(x - y) + p^{2k}(x - y)^2) : x, y \in \Delta_p\} \\ &= \{(x_0 + y_0 + p^k s, (x_0 - y_0)^2 + 2p^{k+l}ct + p^{2k}t^2) : s, t \in \Delta_p\} \end{aligned} \tag{14.25}$$

holds true for all k. On the other hand, the equality

$$\{2ct + p^{k-l}t^2 : t \in \Delta_p\} = \Delta_p \tag{14.26}$$

holds true for $k \geq l + 1$. Indeed, note that the equality $(e + p^m\Delta_p)^{[2]} = e + p^m\Delta_p$ is fulfilled for all $m \geq 1$. This implies that $(c + p^m\Delta_p)^{[2]} = c^2 + p^m\Delta_p$ for all $c \in \Delta_p^\times$, i.e., $\{c^2 + 2cp^m t + p^{2m}t^2 : t \in \Delta_p\} = c^2 + p^m\Delta_p$ and hence $\{2ct + p^m t^2 : t \in \Delta_p\} = \Delta_p$. For $k \geq l + 1$ this equality implies (14.26). Taking into account (14.26), equality (14.24) follows from (14.25). □

LEMMA 14.8. *Consider the field Ω_p, where $p > 2$. Let $\mathfrak{s}(x)$ be the function constructed in the proof of Lemma 14.6. Consider the mappings $S_j(u,v)$ from $\Omega_p \times \Omega_p^{[2]}$ to Ω_p^2 of the form*

$$S_1(u,v) = \left(\frac{u+\mathfrak{s}(v)}{2}, \frac{u-\mathfrak{s}(v)}{2}\right),$$

$$S_2(u,v) = \left(\frac{u-\mathfrak{s}(v)}{2}, \frac{u+\mathfrak{s}(v)}{2}\right), \quad u \in \Omega_p,\ v \in \Omega_p^{[2]}.$$

Let $(u_0, v_0) \in \Omega_p \times \Omega_p^{[2]}$ and $|\mathfrak{s}(v_0)|_p = p^{-l}$. Put

$$E_k = \{(u_0, v_0) + (p^k\Delta_p) \times (p^{k+l}\Delta_p)\}.$$

Then for $k \geq l+1$ the following statements are valid:

(i) $E_k \subset \Omega_p \times \Omega_p^{[2]}$;

(ii) $S_1(E_k) \cap S_2(E_k) = \emptyset$;

(iii) $\int\limits_{S_j(E_k)} \Phi_j(x,y)dm_{\Omega_p^2}(x,y) = \int\limits_{E_k} \Phi_j(S_j(u,v))|\mathfrak{s}(v)|_p^{-1}dm_{\Omega_p^2}(u,v), \quad j=1,2,$

for any continuous function $\Phi_j(x,y)$ on $S_j(E_k)$.

PROOF. Statement (i). Note that $|v_0|_p = p^{-2l}$. It follows from the proof of Lemma 14.6 that if $w_0 \in \Omega_p^{[2]}$ and $|w_0|_p = p^{-2l}$, then $w_0 + w \in \Omega_p^{[2]}$ for $w \in p^{2l+1}\Delta_p$. In view of $k \geq l+1$, it follows from what has been said that (i) is fulfilled.

Statement (ii). Assume that $S_1(E_k) \cap S_2(E_k) \neq \emptyset$. Then as easily seen, there exist elements $v_1, v_2 \in \Delta_p$ such that

$$\mathfrak{s}(v_0 + p^{k+l}v_1) + \mathfrak{s}(v_0 + p^{k+l}v_2) \in p^k\Delta_p. \tag{14.27}$$

Since $v_0 \in \Omega_p^{[2]}$ and $|v_0|_p = p^{-2l}$, we have $v_0 = p^{2l}c$, where $c \in (\Delta_p^\times)^{[2]}$ and $v_0 + p^{k+l}v_i = p^{2l}(c + p^{k-l}v_i)$, $i = 1,2$. This implies that $\mathfrak{s}(v_0 + p^{k+l}v_1) + \mathfrak{s}(v_0 + p^{k+l}v_2) = p^l(\mathfrak{s}(c + p^{k-l}v_1) + \mathfrak{s}(c + p^{k-l}v_2))$. It follows from the definition of the function $\mathfrak{s}$ that the elements $\mathfrak{s}(c + p^{k-l}v_i)$, $i = 1,2$, are in the same coset of the subgroup $p\Delta_p$ in the group Δ_p. Hence $\mathfrak{s}(c+p^{k-l}v_1)+\mathfrak{s}(c+p^{k-l}v_2) \notin p\Delta_p$. Therefore $\mathfrak{s}(v_0 + p^{k+l}v_1) + \mathfrak{s}(v_0 + p^{k+l}v_2) \notin p^{l+1}\Delta_p$, but this contradicts (14.27) for $k \geq l+1$. So, we proved (ii).

Statement (iii). We will prove that equality (iii) holds true for $S_1(u,v)$. For $S_2(u,v)$ the reasoning is similar. Put

$$(x_0, y_0) = S_1(u_0, v_0) = \left(\frac{u_0 + \mathfrak{s}(v_0)}{2}, \frac{u_0 - \mathfrak{s}(v_0)}{2}\right)$$

and check that

$$S_1(E_k) = (x_0, y_0) + (p^k\Delta_p)^2. \tag{14.28}$$

We have

$$S_1(u_0 + u, v_0 + v) = \left(\frac{u_0 + u + \mathfrak{s}(v_0 + v)}{2}, \frac{u_0 + u - \mathfrak{s}(v_0 + v)}{2}\right).$$

It is easily seen that

$$\left|\frac{u_0 + u + \mathfrak{s}(v_0 + v)}{2} - \frac{u_0 + \mathfrak{s}(v_0)}{2}\right|_p \leq \max\{|u|_p, |\mathfrak{s}(v_0 + v) - \mathfrak{s}(v_0)|_p\}. \tag{14.29}$$

Under the condition of the theorem

$$|u|_p \le p^{-k}. \tag{14.30}$$

Inasmuch as $\mathfrak{s}^2(x) = x$, we have

$$|\mathfrak{s}(v_0 + v) - \mathfrak{s}(v_0)|_p = \frac{|v|_p}{|\mathfrak{s}(v_0 + v) + \mathfrak{s}(v_0)|_p}. \tag{14.31}$$

Note that $v_0+v = p^{2l}c+p^{k+l}t$ for some $t \in \Delta_p$. This implies that $\mathfrak{s}(v_0+v) = p^l\mathfrak{s}(c+p^{k-l}t)$ and moreover, $\mathfrak{s}(v_0) = p^l\mathfrak{s}(c)$. Inasmuch as the points $\mathfrak{s}(c+p^{k-l}t)$ and $\mathfrak{s}(c)$ are in the same coset of the subgroup $p\Delta_p$ in the group Δ_p, so is $|\mathfrak{s}(c+p^{k-l}t)+\mathfrak{s}(c)|_p = 1$ and hence $|\mathfrak{s}(v_0 + v) + \mathfrak{s}(v_0)|_p = p^{-l}$. We recall that $|v|_p \le p^{-k-l}$, and we get from (14.31) that

$$|\mathfrak{s}(v_0 + v) - \mathfrak{s}(v_0)|_p \le p^{-k}. \tag{14.32}$$

Taking into account (14.30) and (14.32), it follows from (14.29) that inequality

$$\left|\frac{u_0 + u + \mathfrak{s}(v_0 + v)}{2} - \frac{u_0 + \mathfrak{s}(v_0)}{2}\right|_p \le p^{-k}$$

holds and hence

$$S_1(E_k) \subset (x_0, y_0) + (p^k\Delta_p)^2. \tag{14.33}$$

We note that if $T(a, b) = T(a', b')$, then either $(a, b) = (a', b')$ or $(a, b) = (b', a')$. Since $|x_0 - y_0|_p = |\mathfrak{s}(v_0)|_p = p^{-l}$ and $k \ge l + 1$, the restriction of the mapping T to the set $(x_0, y_0) + (p^k\Delta_p)^2$ is injective. Taking this into account, (14.28) follows from Lemma 14.7 and (14.33). Moreover, it follows that the mappings T and S_1 are inverse homeomorphisms of the sets $S_1(E_k)$ and E_k.

Observe that the set E_k is a product of the balls

$$E_k = \left\{u : |u - u_0|_p \le \frac{1}{p^k}\right\} \times \left\{v : |v - v_0|_p \le \frac{1}{p^{k+l}}\right\}.$$

Put

$$x(u, v) = \frac{u + \mathfrak{s}(v)}{2}, \quad y(u, v) = \frac{u - \mathfrak{s}(v)}{2}, \ (u, v) \in E_k.$$

The mapping S_1 is a homeomorphism of the open compacts E_k and $S_1(E_k)$. By Lemma 14.6, the functions $x(u, v)$ and $y(u, v)$ on E_k are expressed by convergent power series in u and v. We have

$$\frac{\mathfrak{d}x}{\mathfrak{d}u} = \frac{\mathfrak{d}y}{\mathfrak{d}u} = \frac{1}{2}, \quad \frac{\mathfrak{d}x}{\mathfrak{d}v} = \frac{1}{4\mathfrak{s}(v)}, \quad \frac{\mathfrak{d}y}{\mathfrak{d}v} = -\frac{1}{4\mathfrak{s}(v)}.$$

It follows from this that

$$\left|\det\begin{pmatrix} \frac{\mathfrak{d}x}{\mathfrak{d}u} & \frac{\mathfrak{d}x}{\mathfrak{d}v} \\ \frac{\mathfrak{d}y}{\mathfrak{d}u} & \frac{\mathfrak{d}y}{\mathfrak{d}v} \end{pmatrix}\right|_p = \left|\det\begin{pmatrix} \frac{1}{2} & \frac{1}{4\mathfrak{s}(v)} \\ \frac{1}{2} & -\frac{1}{4\mathfrak{s}(v)} \end{pmatrix}\right|_p \tag{14.34}$$

$$= |\mathfrak{s}(v)|_p^{-1} \ne 0, \quad (u, v) \in E_k.$$

It is obvious that (iii) follows from the change of variables formula in integrals [**97**, §4] and (14.34). □

LEMMA 14.9. *Consider the field Ω_p, where $p > 2$. Let ξ and η be independent identically distributed random variables with values in Ω_p and distribution μ. Suppose μ has a density $\rho(x)$ with respect to m_{Ω_p} such that $\rho(x)$ is continuous and $\rho(0) > 0$. The random variables $S = \xi + \eta$ and $D = (\xi - \eta)^2$ are independent if and only if the density $\rho(x)$ satisfies the equation*

$$\rho^2(u)\rho(v)\rho(-v) = \rho^2(0)\rho(u+v)\rho(u-v), \quad u, v \in \Omega_p. \tag{14.35}$$

PROOF. Necessity. Inasmuch as $\mu_{(S,D)} = T(\mu_{(\xi,\eta)})$ and the distribution $\mu_{(\xi,\eta)}$ is absolutely continuous with respect to $m_{\Omega_p^2}$, so is $\mu_{(\xi,\eta)}\{(t,t) : t \in \Omega_p\} = 0$. Therefore, the distribution $\mu_{(S,D)}$ is concentrated at the set $\Omega_p \times (\Omega_p^{[2]} \backslash \{0\})$. Let the mappings $S_j(u,v)$ and the sets E_k be the same as in Lemma 14.8. Take $(u_0, v_0) \in \Omega_p \times (\Omega_p^{[2]} \backslash \{0\})$ and represent the element $\mathfrak{s}(v_0)$ in the form $\mathfrak{s}(v_0) = p^l c$, where $l \in \mathbb{Z}$ and $c \in \Delta_p^\times$. By Lemma 14.8, statements (i) and (ii) of Lemma 14.8 are fulfilled for $k \geq l+1$. We have

$$\begin{aligned} \mu_{(S,D)}\{E_k\} &= T(\mu_{(\xi,\eta)})\{E_k\} = \mu_{(\xi,\eta)}\{T^{-1}(E_k)\} \\ &= \int\limits_{T^{-1}(E_k)} \rho(x)\rho(y)dm_{\Omega_p^2}(x,y) = \int\limits_{S_1(E_k)} \rho(x)\rho(y)dm_{\Omega_p^2}(x,y) \\ &\quad + \int\limits_{S_2(E_k)} \rho(x)\rho(y)dm_{\Omega_p^2}(x,y). \end{aligned} \tag{14.36}$$

Taking into account equalities (iii) of Lemma 14.8, we transform integrals in the right-hand side of equality (14.36) and obtain

$$\int\limits_{S_1(E_k)} \rho(x)\rho(y)dm_{\Omega_p^2}(x,y) = \int\limits_{E_k} \rho\left(\frac{u+\mathfrak{s}(v)}{2}\right)\rho\left(\frac{u-\mathfrak{s}(v)}{2}\right)|\mathfrak{s}(v)|_p^{-1}dm_{\Omega_p^2}(u,v),$$

$$\int\limits_{S_2(E_k)} \rho(x)\rho(y)dm_{\Omega_p^2}(x,y) = \int\limits_{E_k} \rho\left(\frac{u-\mathfrak{s}(v)}{2}\right)\rho\left(\frac{u+\mathfrak{s}(v)}{2}\right)|\mathfrak{s}(v)|_p^{-1}dm_{\Omega_p^2}(u,v).$$

Then (14.36) implies that

$$\mu_{(S,D)}\{E_k\} = 2\int\limits_{E_k} \rho\left(\frac{u+\mathfrak{s}(v)}{2}\right)\rho\left(\frac{u-\mathfrak{s}(v)}{2}\right)|\mathfrak{s}(v)|_p^{-1}dm_{\Omega_p^2}(u,v).$$

This equality means that the distribution $\mu_{(S,D)}$ has a density with respect to $m_{\Omega_p^2}$, and this density is equal to

$$2\rho\left(\frac{u+\mathfrak{s}(v)}{2}\right)\rho\left(\frac{u-\mathfrak{s}(v)}{2}\right)|\mathfrak{s}(v)|_p^{-1}, \quad u \in \Omega_p,\ v \in \Omega_p^{[2]} \backslash \{0\}. \tag{14.37}$$

Note that when we got the representation (14.37) for the density of the distribution $\mu_{(S,D)}$, we did not use the independence of the random variables S and D.

By the assumption of the lemma the random variables S and D are independent. Therefore, there exist integrable functions $r_j(x)$ on Ω_p with respect to m_{Ω_p} such that the equality

$$r_1(u)r_2(v) = 2\rho\left(\frac{u+\mathfrak{s}(v)}{2}\right)\rho\left(\frac{u-\mathfrak{s}(v)}{2}\right)|\mathfrak{s}(v)|_p^{-1} \tag{14.38}$$

holds true a.e. on $\Omega_p \times (\Omega_p^{[2]}\backslash\{0\})$ with respect to $m_{\Omega_p^2}$. Since the function in the right-hand side of equality (14.38) is continuous, we can assume without loss of generality that the functions r_j are also continuous and equality (14.38) holds true everywhere on $\Omega_p \times (\Omega_p^{[2]}\backslash\{0\})$. Inasmuch as $\rho(0) > 0$, it is easily seen that $r_1(0) > 0$. Put $v = t^2$, $t \neq 0$. It follows from (14.38) that

$$(14.39)\qquad r_2(t^2) = 2r_1^{-1}(0)\rho\left(\frac{\mathfrak{s}(t^2)}{2}\right)\rho\left(-\frac{\mathfrak{s}(t^2)}{2}\right)|\mathfrak{s}(t^2)|_p^{-1}, \quad t \in \Omega_p,\ t \neq 0.$$

Note that (14.38) and (14.39) imply the equality

$$(14.40)\quad r_1(u)\rho\left(\frac{\mathfrak{s}(t^2)}{2}\right)\rho\left(-\frac{\mathfrak{s}(t^2)}{2}\right) = r_1(0)\rho\left(\frac{u+\mathfrak{s}(t^2)}{2}\right)\rho\left(\frac{u-\mathfrak{s}(t^2)}{2}\right), \quad (u,t) \in \Omega_p^2,\ t \neq 0.$$

It follows from the continuity of $\rho(x)$ and $r_1(x)$ that equality (14.40) holds true for all $u, t \in \Omega_p$. Put $t = 0$ in (14.40). We deduce from the resulting equality that

$$(14.41)\qquad r_1(u) = \frac{r_1(0)}{\rho^2(0)}\rho^2\left(\frac{u}{2}\right), \quad u \in \Omega_p.$$

Substituting (14.41) into (14.40), we find that the equality

$$(14.42)\quad \rho^2\left(\frac{u}{2}\right)\rho\left(\frac{\mathfrak{s}(t^2)}{2}\right)\rho\left(-\frac{\mathfrak{s}(t^2)}{2}\right) = \rho^2(0)\rho\left(\frac{u+\mathfrak{s}(t^2)}{2}\right)\rho\left(\frac{u-\mathfrak{s}(t^2)}{2}\right)$$

is fulfilled for all $u, t \in \Omega_p$. Since either $\mathfrak{s}(t^2) = t$ or $\mathfrak{s}(t^2) = -t$, the equalities

$$(14.43)\qquad \rho\left(\frac{\mathfrak{s}(t^2)}{2}\right)\rho\left(-\frac{\mathfrak{s}(t^2)}{2}\right) = \rho\left(\frac{t}{2}\right)\rho\left(-\frac{t}{2}\right), \quad t \in \Omega_p,$$

and

$$(14.44)\qquad \rho\left(\frac{u+\mathfrak{s}(t^2)}{2}\right)\rho\left(\frac{u-\mathfrak{s}(t^2)}{2}\right) = \rho\left(\frac{u+t}{2}\right)\rho\left(\frac{u-t}{2}\right), \quad u, t \in \Omega_p$$

hold true. Substituting (14.43) and (14.44) into (14.42), we get that the density $\rho(x)$ satisfies equation (14.35). The necessity is proved.

Sufficiency. Taking into account (14.37) and (14.44) we have the following representation for the density $\varrho(u,v)$ of the distribution $\mu_{(S,D)}$:

$$\varrho(u,v) = \begin{cases} 2\rho\left(\frac{u+t}{2}\right)\rho\left(\frac{u-t}{2}\right)|\mathfrak{s}(t^2)|_p^{-1}, & \text{if } \ u \in \Omega_p,\ v = t^2,\ t \neq 0, \\ 0, & \text{if } \ u \in \Omega_p,\ v \notin (\Omega_p^{[2]}\backslash\{0\}). \end{cases}$$

If a density $\rho(x)$ satisfies equation (14.35), it is easily seen that the density $\varrho(u,v)$ is represented as a product of a function in u and a function in v. This implies the independence of S and D. □

Now we can prove the main theorem of this section.

THEOREM 14.10. *Consider the field Ω_p, where $p > 2$. Let ξ and η be independent identically distributed random variables with values in Ω_p and distribution μ. Suppose μ has a density $\rho(x)$ with respect to m_{Ω_p} such that $\rho(x)$ is continuous. The random variables $S = \xi + \eta$ and $D = (\xi - \eta)^2$ are independent if and only if $\mu \in I(\Omega_p)$.*

PROOF. Necessity. We argue as in the proof of Theorem 14.3. Replacing if necessary the random variables ξ and η by the new independent random variables $\xi + x$ and $\eta + x$, $x \in X$, we can assume that $\rho(0) > 0$. Applying Lemma 14.9, we find that the density $\rho(x)$ satisfies equation (14.35). Since the field Ω_p is an Abelian group with unique division by 2, by Lemma 14.2, the set $K = \{x \in \Omega_p : \rho(x) \neq 0\}$ is a subgroup of Ω_p. Obviously, K is an open subgroup. Hence K is a closed subgroup.

First suppose $K \neq \Omega_p$. Inasmuch as $K \neq \{0\}$, the subgroup K is of the form $K = p^m\Delta_p$ for some integer m. Hence K is a compact group. Consider the restriction of equation (14.35) to K. Put $\varphi(x) = \ln \rho(x)$, $x \in K$. It follows from (14.35) that the function $\varphi(x)$ satisfies the equation

$$2\varphi(u) + \varphi(v) + \varphi(-v) = 2\varphi(0) + \varphi(u+v) + \varphi(u-v), \quad u, v \in K.$$

Integrate both sides of this equation over the group K with respect to the Haar distribution m_K in the variable v. Using that the Haar distribution m_K is K-invariant, we get that $\varphi(u) = \varphi(0)$ for all $u \in K$. Hence $\rho(x) = \rho(0)$ for all $x \in K$. This means that $\mu = m_K$.

If $K = \Omega_p$, then we consider the restriction of equation (14.35) to an arbitrary subgroup $G = p^m\Delta_p$ of Ω_p. The reasoning above shows that $\rho(x) = \rho(0)$ for all $x \in G$ and hence $\rho(x) = \rho(0)$ for all $x \in \Omega_p$. But this contradicts the integrability of the density $\rho(x)$ with respect to m_{Ω_p}. So, the case when $K = \Omega_p$ is impossible. The necessity is proved.

Sufficiency. Let K be a nonzero compact subgroup of Ω_p. As has been stated above, K is of the form $K = p^m\Delta_p$ for some integer m. Let ξ and η be independent identically distributed random variables with values in Ω_p and distribution $\mu = m_K$. Since $p > 2$, K is a Corwin group. As proven in Remark 4.4, this implies that the random variables $\xi + \eta$ and $\xi - \eta$ are independent. Hence the random variables S and D are also independent. □

REMARK 14.11. Let us discuss the case of the field Ω_2. A lemma similar to Lemma 14.7 holds true also for the field Ω_2. Unlike equality (14.24), the equality

$$T\{(x_0, y_0) + (2^k\mathbb{Z}_2)^2\} = (x_0 + y_0, (x_0 - y_0)^2) + (2^k\mathbb{Z}_2) \times (2^{k+l+1}\mathbb{Z}_2) \tag{14.45}$$

holds true for $k \geq l + 2$. Respectively, taking (14.45) into account, we can also reformulate Lemma 14.8. It allows us to prove Lemma 14.9 for the field Ω_2. Next, reasoning as in the proof of the necessity in Theorem 14.10, we get that if ξ and η are independent identically distributed random variables with values in Ω_2 and distribution μ such that μ has a density $\rho(x)$ with respect to m_{Ω_2}, where the density $\rho(x)$ is continuous, $\rho(0) > 0$, and the random variables $S = \xi + \eta$ and $D = (\xi - \eta)^2$ are independent, then $\mu = m_K$, where $K = 2^m\Delta_2$ for some integer m. But in this case the corresponding density

$$\rho(x) = \begin{cases} 2^m, & \text{if } x \in 2^m\Delta_2, \\ 0, & \text{if } x \notin 2^m\Delta_2, \end{cases}$$

does not satisfy equation (14.35). To see this, substitute $u = v \in 2^{m-1}\Delta_2 \backslash 2^m\Delta_2$ in equation (14.35). Then the left-hand side of equation (14.35) is equal to zero, whereas the right-hand side is not. Thus, there do not exist independent identically distributed random variables ξ and η with values in the field Ω_2 and distribution μ such that μ has a density $\rho(x)$ with respect to a Haar measure m_{Ω_2}, where the density $\rho(x)$ is continuous and the random variables $S = \xi + \eta$ and $D = (\xi - \eta)^2$ are independent.

Notes

The results of Sections 12 and 13 were obtained by G.M. Feldman in [**41**] and [**42**] respectively. We note that in [**79**] M.V. Myronyuk studied the Skitovich–Darmois theorem on the group Ω_p in the case of three independent random variables and three linear forms. The results of Sections 14 were obtained by G.M. Feldman and M.V. Myronyuk in [**55**].

CHAPTER VI

Miscellaneous characterization theorems

15. Rao theorems

C.R. Rao proved the following theorem: Let ξ_j, $j = 1, 2, 3$, be independent random variables with nonvanishing characteristic functions. Let a_j, b_j be real numbers such that $a_i/b_i \neq a_j/b_j$ for all $i \neq j$. Put $L_1 = a_1\xi_1 + a_2\xi_2 + a_3\xi_3$ and $L_2 = b_1\xi_1 + b_2\xi_2 + b_3\xi_3$. Then the distribution of the random vector (L_1, L_2) determines the distributions of the random variables ξ_j up to a shift. C.R. Rao also proved that if four independent random variables ξ_j are considered, then the distribution of the random vector (L_1, L_2) determines the distributions of the random variables ξ_j up to a convolution with a Gaussian distribution.

Let X be a second countable locally compact Abelian group. In this section we prove an analogue of the first Rao theorem for independent random variables with values in X. We also prove an analogue of the second Rao theorem for independent random variables with values in an $\boldsymbol{a}$-adic solenoid. In both cases coefficients of linear forms are continuous endomorphisms of the corresponding group.

THEOREM 15.1. *Let X be a locally compact Abelian group. Let β_j, $j = 1, 2, 3$, be continuous endomorphisms of X. Let ξ_j be independent random variables with values in the group X with nonvanishing characteristic functions. Put $L_2 = \beta_1\xi_1 + \beta_2\xi_2 + \beta_3\xi_3$. Consider two cases:*

(I) *$L_1 = \xi_1 + \xi_2 + \xi_3$ and β_j satisfy the conditions*

$$\text{(i)} \qquad \operatorname{Ker}(\beta_1 - \beta_2) = \operatorname{Ker}(\beta_1 - \beta_3) = \operatorname{Ker}(\beta_2 - \beta_3) = \{0\},$$

(II) *$L_1 = \xi_1 + \xi_2$ and β_j satisfy the conditions*

$$\text{(ii)} \qquad \operatorname{Ker}(\beta_1 - \beta_2) = \operatorname{Ker} \beta_3 = \{0\}.$$

In each of these cases the distribution of the random vector (L_1, L_2) determines the distributions of the random variables ξ_j up to a shift.

PROOF. Denote by Y the character group of the group X. Let η_j, $j = 1, 2, 3$, be independent random variables with values in X with nonvanishing characteristic functions. Denote by μ_j the distributions of the random variables ξ_j and by ν_j the distributions of the random variables η_j. Put

$$M_2 = \beta_1\eta_1 + \beta_2\eta_2 + \beta_3\eta_3, \quad f_j(y) = \hat{\nu}_j(y)/\hat{\mu}_j(y), \ b_j = \widetilde{\beta}_j, \ j = 1, 2, 3.$$

Assume that (I) holds. Put $M_1 = \eta_1 + \eta_2 + \eta_3$. Taking into account that the random variables ξ_j are independent, the characteristic function of the random vector (L_1, L_2) is of the form

$$\begin{aligned}(15.1) \qquad \mathbf{E}[(L_1, u)(L_2, v)] &= \mathbf{E}[(\xi_1 + \xi_2 + \xi_3, u)(\beta_1\xi_1 + \beta_2\xi_2 + \beta_3\xi_3, v)] \\ &= \mathbf{E}[(\xi_1, u + b_1v)(\xi_2, u + b_2v)(\xi_3, u + b_3v)]\end{aligned}$$

$$= \mathbf{E}[(\xi_1, u + b_1 v)]\mathbf{E}[(\xi_2, u + b_2 v)]\mathbf{E}[(\xi_3, u + b_3 v)]$$
$$= \hat{\mu}_1(u + b_1 v)\hat{\mu}_2(u + b_2 v)\hat{\mu}_3(u + b_3 v), \quad u, v \in Y.$$

Analogically, the characteristic function of the random vector (M_1, M_2) is of the form

$$\mathbf{E}[(M_1, u)(M_2, v)] = \hat{\nu}_1(u + b_1 v)\hat{\nu}_2(u + b_2 v)\hat{\nu}_3(u + b_3 v), \quad u, v \in Y. \tag{15.2}$$

It follows from (15.1) and (15.2) that the random vectors (L_1, L_2) and (M_1, M_2) have the same characteristic functions and hence they are identically distributed if and only if the characteristic functions $\hat{\mu}_j(y)$ and $\hat{\nu}_j(y)$ satisfy the equation

$$\hat{\mu}_1(u + b_1 v)\hat{\mu}_2(u + b_2 v)\hat{\mu}_3(u + b_3 v) = \hat{\nu}_1(u + b_1 v)\hat{\nu}_2(u + b_2 v)\hat{\nu}_3(u + b_3 v), \quad u, v \in Y. \tag{15.3}$$

Obviously, (15.3) is equivalent to the fact that the functions $f_j(y)$ satisfy the equation

$$f_1(u + b_1 v)f_2(u + b_2 v)f_3(u + b_3 v) = 1, \quad u, v \in Y. \tag{15.4}$$

We will prove that there exist elements $x_j \in X$ such that $f_j(y) = (x_j, y)$, $y \in Y$, $j = 1, 2, 3$. For the proof we use the finite difference method.

Let k_1 be an arbitrary element of the group Y. Put $h_1 = -b_1 k_1$. Then $h_1 + b_1 k_1 = 0$. Substitute $u + h_1$ for u and $v + k_1$ for v in equation (15.4). Dividing the resulting equation by equation (15.4), we obtain

$$\frac{f_2(u + b_2 v + (b_2 - b_1)k_1)f_3(u + b_3 v + (b_3 - b_1)k_1)}{f_2(u + b_2 v)f_3(u + b_3 v)} = 1, \quad u, v \in Y. \tag{15.5}$$

Let k_2 be an arbitrary element of the group Y. Put $h_2 = -b_2 k_2$. Then $h_2 + b_2 k_2 = 0$. Substitute $u + h_2$ for u and $v + k_2$ for v in equation (15.5). Dividing the resulting equation by equation (15.5), we find

$$\frac{f_3(u + b_3 v + (b_3 - b_1)k_1 + (b_3 - b_2)k_2)f_3(u + b_3 v)}{f_3(u + b_3 v + (b_3 - b_1)k_1)f_3(u + b_3 v + (b_3 - b_2)k_2)} = 1, \quad u, v \in Y. \tag{15.6}$$

Substitute $u = v = 0$, $(b_3 - b_1)k_1 = k$, and $(b_3 - b_2)k_2 = l$ in equation (15.6). Note that since

$$\operatorname{Ker}(\beta_1 - \beta_3) = \operatorname{Ker}(\beta_2 - \beta_3) = \{0\}, \tag{15.7}$$

it follows from statement (b) of Theorem 1.20 that the subgroups $(b_3 - b_1)(Y)$ and $(b_3 - b_2)(Y)$ are dense in Y. Inasmuch as k_1 and k_2 are arbitrary elements of the group Y, we get that the function $f_3(y)$ satisfies the equation

$$\frac{f_3(k + l)}{f_3(k)f_3(l)} = 1, \quad k, l \in Y.$$

Hence

$$f_3(k + l) = f_3(k)f_3(l), \quad k, l \in Y. \tag{15.8}$$

Taking into account that $f_3(-y) = \overline{f_3(y)}$, $y \in Y$, and $f_3(0) = 1$, we find from equation (15.8) that $|f_3(y)| = 1$, $y \in Y$. Thus, we proved that the function $f_3(y)$ is a character of the group Y. By the Pontryagin duality theorem there is an element $x_3 \in X$ such that $f_3(y) = (x_3, y)$, $y \in Y$. Note that in the proof we used only conditions (15.7), but not the condition $\operatorname{Ker}(b_1 - b_2) = \{0\}$ (compare below with Remark 15.5, where we discuss the necessity of conditions (i) and (ii) for the theorem to be valid).

Arguing in a similar way we get that there exist elements $x_j \in X$ such that $f_j(y) = (x_j, y)$, $y \in Y$, $j = 1, 2$. Thus, $\hat{\nu}_j(y) = \hat{\mu}_j(y)(x_j, y)$, $y \in Y$. This implies that $\nu_j = \mu_j * E_{x_j}$, $j = 1, 2, 3$. Thus, in case (I) the theorem is proved.

Assume that (II) holds. Put $M_1 = \eta_1 + \eta_2$. Arguing as in the proof of the theorem in case (I), we get that the characteristic function of the random vectors (L_1, L_2) and (M_1, M_2) are of the form

$$\mathbf{E}[(L_1, u)(L_2, v)] = \hat{\mu}_1(u + b_1 v)\hat{\mu}_2(u + b_2 v)\hat{\mu}_3(b_3 v), \quad u, v \in Y, \tag{15.9}$$

$$\mathbf{E}[(M_1, u)(M_2, v)] = \hat{\nu}_1(u + b_1 v)\hat{\nu}_2(u + b_2 v)\hat{\nu}_3(b_3 v), \quad u, v \in Y, \tag{15.10}$$

respectively. It follows from (15.9) and (15.10) that the random vectors (L_1, L_2) and (M_1, M_2) have the same characteristic functions and hence they are identically distributed if and only if the functions $f_j(y)$ satisfy the equation

$$f_1(u + b_1 v)f_2(u + b_2 v)f_3(b_3 v) = 1, \quad u, v \in Y. \tag{15.11}$$

As noted in the proof of the theorem in case (I), the theorem will be proved if we prove that the functions $f_j(y)$, $j = 1, 2, 3$, are characters of the group Y. First verify that the function $f_3(y)$ is a character of the group Y. We follow the scheme of the proof of the theorem in case (I). Eliminating from equation (15.11) the function $f_1(y)$, we obtain

$$\frac{f_2(u + b_2 v + (b_2 - b_1)k_1)f_3(b_3 v + b_3 k_1)}{f_2(u + b_2 v)f_3(b_3 v)} = 1, \quad u, v \in Y. \tag{15.12}$$

Eliminating from equation (15.12) the function $f_2(y)$, we get

$$\frac{f_3(b_3 v + b_3 k_1 + b_3 k_2)f_3(b_3 v)}{f_3(b_3 v + b_3 k_1)f_3(b_3 v + b_3 k_2)} = 1, \quad u, v \in Y. \tag{15.13}$$

Substituting $v = 0$ in equation (15.13), we obtain that the function $f_3(y)$ satisfies the equation for the character on the subgroup $b_3(Y)$. In view of $\operatorname{Ker} \beta_3 = \{0\}$, it follows from statement (b) of Theorem 1.20 that the subgroup $b_3(Y)$ is dense in Y. Hence the function $f_3(y)$ satisfies the equation for the character on the group Y. Taking into account that $f_3(-y) = \overline{f_3(y)}$, $y \in Y$ and $f_3(0) = 1$, this implies that $|f_3(y)| = 1$, $y \in Y$. Thus, the function $f_3(y)$ is a character of the group Y.

Verify now that the function $f_2(y)$ is also a character of the group Y. Let h be an arbitrary element of the group Y. Substitute $u + h$ for u in equation (15.12). Dividing the resulting equation by equation (15.12), we obtain

$$\frac{f_2(u + b_2 v + (b_2 - b_1)k_1 + h)f_2(u + b_2 v)}{f_2(u + b_2 v + h)f_2(u + b_2 v + (b_2 - b_1)k_1)} = 1, \quad u, v \in Y.$$

Substituting here $u = v = 0$ and arguing in the same way as for the function $f_3(y)$, we make sure that the function $f_2(y)$ is a character of the group Y. Eliminating from equation (15.11) the function $f_2(y)$ and arguing in a similar way, we prove that the function $f_1(y)$ is a character of the group Y. □

Suppose that in Theorem 15.1 $L_1 = \xi_1 + \xi_2$ and $L_2 = \xi_2 + \xi_3$. It is obvious that in this case conditions (ii) of Theorem 15.1 are satisfied. It is easy to see that Theorem 15.1 implies the following statement.

COROLLARY 15.2. *Let X be a locally compact Abelian group. Let ξ_j, $j = 1, 2, 3$, be independent random variables with values in X with nonvanishing characteristic* *functions. Put* $L_1 = \xi_1 - \xi_3$, $L_2 = \xi_2 - \xi_3$. *Then the distribution of the random vector* (L_1, L_2) *determines the distributions of the random variables* ξ_j *up to a shift.*

It is easy to see that the following statement also follows from Theorem 15.1.

COROLLARY 15.3. *Let* ξ_j, $j = 1, 2, 3$, *be independent random vectors with values in the space* $\mathbb{R}^n$ *with nonvanishing characteristic functions. Let* A_j, B_j *be* $n \times n$ *matrices satisfying the conditions:*

(i) A_j *is either the zero matrix or a nonsingular matrix and only one of the matrices* A_j *may be the zero matrix;*
(ii) B_j *is either the zero matrix or a nonsingular matrix and only one of the matrices* B_j *may be the zero matrix;*
(iii) *the matrices* A_j *and* B_j *cannot be both the zero matrices;*
(iv) *if the matrix* $B_iA_i^{-1} - B_jA_j^{-1}$, $i \neq j$, *is defined, then it is nonsingular.*

Let $L_1 = A_1\xi_1 + A_2\xi_2 + A_3\xi_3$ *and* $L_2 = B_1\xi_1 + B_2\xi_2 + B_3\xi_3$. *Then the distribution of* (L_1, L_2) *determines the distributions of the random vectors* ξ_j *up to a shift.*

We complement Theorem 15.1 with the following assertion.

PROPOSITION 15.4. *Let* X *be a locally compact Abelian group. Let* β_1 *and* β_2 *be continuous endomorphisms of* X *satisfying the condition*

$$\text{(i)} \qquad \operatorname{Ker}(\beta_1 - \beta_2) = \{0\}.$$

Let ξ_1 *and* ξ_2 *be independent random variables with values in the group* X *with nonvanishing characteristic functions. Let* $L_1 = \xi_1 + \xi_2$ *and* $L_2 = \beta_1\xi_1 + \beta_2\xi_2$. *Then the distribution of the random vector* (L_1, L_2) *uniquely determines the distributions of the random variables* ξ_j.

PROOF. Retaining the notation used in the proof of Theorem 15.1, we get that the functions $f_j(y)$ satisfy the equation

$$f_1(u + b_1v)f_2(u + b_2v) = 1, \quad u, v \in Y.$$

Let y be an arbitrary element of the group Y. Substituting here $u = -b_2y$, $v = y$, we obtain

$$f_1((b_1 - b_2)y) = 1, \quad y \in Y. \tag{15.14}$$

In view of (i), it follows from statement (b) of Theorem 1.20 that the subgroup $(b_1 - b_2)(Y)$ is dense in Y. Then equation (15.14) implies that $f_1(y) = 1$ for all $y \in Y$. Thus, $\hat{\nu}_1(y) = \hat{\mu}_1(y)$ for all $y \in Y$. Hence $\nu_1 = \mu_1$. Arguing in a similar way we get that $\nu_2 = \mu_2$. □

REMARK 15.5. Let X be a locally compact Abelian group with character group Y. Theorem 15.1 and Proposition 15.4 fail if we omit either conditions (i) or (ii) of Theorem 15.1, or condition (i) of Proposition 15.4 respectively. Construct corresponding examples.

Recall the definition of the *Poisson distribution* on the group X. Take $x_0 \in X$ and $\lambda > 0$ and define the Poisson distribution $e(\lambda E_{x_0})$ as follows:

$$e(\lambda E_{x_0}) = e^{-\lambda}\left(E_0 + \lambda E_{x_0} + \frac{\lambda^2}{2!}E_{2x_0} + \dots + \frac{\lambda^n}{n!}E_{nx_0} + \dots\right).$$

The characteristic function of the Poisson distribution $e(\lambda E_{x_0})$ is of the form

$$\widehat{e(\lambda E_{x_0})}(y) = \exp\{\lambda((x_0, y) - 1)\}, \quad y \in Y. \tag{15.15}$$

Let β_j, $j = 1, 2, 3$, be continuous endomorphisms of the group X. Assume that conditions (i) of Theorem 15.1 are not satisfied. For definiteness, let $\mathrm{Ker}(\beta_1 - \beta_2) \neq \{0\}$. Take $x_0 \in \mathrm{Ker}(\beta_1 - \beta_2)$, $x_0 \neq 0$ and $\lambda > 0$. Consider the distributions $\mu_1 = \mu_2 = e(2\lambda E_{x_0})$, $\nu_1 = e(\lambda E_{x_0})$, $\nu_2 = e(3\lambda E_{x_0})$, $\mu_3 = \nu_3 = \mu$, where μ is an arbitrary distribution on X. Let ξ_j and η_j, $j = 1, 2, 3$, be independent random variables with values in the group X and distributions μ_j and ν_j respectively. It is obvious that ν_1 and ν_2 are not shifts of μ_1 and μ_2. Put $L_1 = \xi_1 + \xi_2 + \xi_3$, $L_2 = \beta_1\xi_1 + \beta_2\xi_2 + \beta_3\xi_3$, $M_1 = \eta_1 + \eta_2 + \eta_3$ and $M_2 = \beta_1\eta_1 + \beta_2\eta_2 + \beta_3\eta_3$. Since $x_0 \in \mathrm{Ker}(\beta_1 - \beta_2)$, we have $\beta_1 x_0 = \beta_2 x_0 = \tilde{x}$. Put $b_j = \widetilde{\beta}_j$, $j = 1, 2, 3$. It follows from (15.1), (15.2), and (15.15) that

$$\begin{aligned}\mathbf{E}[(L_1, u)(L_2, v)] &= \exp\{2\lambda((x_0, u + b_1 v) - 1)\}\exp\{2\lambda((x_0, u + b_2 v) - 1)\} \\ &\quad\times \hat{\mu}(u + b_3 v) = e^{-4\lambda}\exp\{4\lambda(x_0, u)(\tilde{x}, v)\}\hat{\mu}(u + b_3 v), \quad u, v \in Y,\end{aligned}$$

$$\begin{aligned}\mathbf{E}[(M_1, u)(M_2, v)] &= \exp\{\lambda((x_0, u + b_1 v) - 1)\}\exp\{3\lambda((x_0, u + b_2 v) - 1)\} \\ &\quad\times \hat{\mu}(u + b_3 v) = e^{-4\lambda}\exp\{4\lambda(x_0, u)(\tilde{x}, v)\}\hat{\mu}(u + b_3 v), \quad u, v \in Y.\end{aligned}$$

Thus, the characteristic functions of the random vectors (L_1, L_2) and (M_1, M_2) coincide. Hence the distributions of the random vectors (L_1, L_2) and (M_1, M_2) also coincide. Thus, if $L_1 = \xi_1 + \xi_2 + \xi_3$ and conditions (i) of Theorem 15.1 are not satisfied, then Theorem 15.1 is false. Note also that if $\mathrm{Ker}(\beta_1 - \beta_2) \neq \{0\}$ and $\mathrm{Ker}(\beta_1 - \beta_3) \neq \{0\}$, then using the above arguments, it is not difficult to construct independent random variables ξ_j and η_j, $j = 1, 2, 3$, with values in the group X and distributions μ_j and ν_j respectively such that the distributions of the random vectors (L_1, L_2) and (M_1, M_2) coincide, but none of the distributions ν_j is a shift of the distribution μ_j.

Arguing in a similar way we convince ourselves that Theorem 15.1 is false in the case when $L_1 = \xi_1 + \xi_2$ if we omit the condition $\mathrm{Ker}(\beta_1 - \beta_2) = \{0\}$. For Proposition 15.4 we argue similarly.

Assume now that $G = \mathrm{Ker}\,\beta_3 \neq \{0\}$. Let μ_3 and ν_3 be arbitrary distributions on G. Consider μ_3 and ν_3 as distributions on the group X. Then $\hat{\mu}_3(y) = \hat{\nu}_3(y) = 1$ for all $y \in A(Y, G)$. It follows from Theorems 1.8, 1.19, and statement (a) of Theorem 1.20 that $\overline{b_3(Y)} = A(Y, G)$. Thus,

$$\hat{\mu}_3(b_3 y) = \hat{\nu}_3(b_3 y) = 1, \quad y \in Y. \tag{15.16}$$

Let $\mu_1 = \nu_1$ and $\mu_2 = \nu_2$ be arbitrary distributions on the group X. Assume that ν_3 is not a shift of μ_3. Let ξ_j and η_j, $j = 1, 2, 3$, be independent random variables with values in the group X and distributions μ_j and ν_j respectively. Put $L_1 = \xi_1 + \xi_2$, $L_2 = \beta_1\xi_1 + \beta_2\xi_2 + \beta_3\xi_3$, $M_1 = \eta_1 + \eta_2$, and $M_2 = \beta_1\eta_1 + \beta_2\eta_2 + \beta_3\eta_3$. It follows from (15.9) and (15.16) that the characteristic functions of the random vectors (L_1, L_2) and (M_1, M_2) coincide. Hence the distributions of the random vectors (L_1, L_2) and (M_1, M_2) also coincide. Thus, Theorem 15.1 fails in the case when $L_1 = \xi_1 + \xi_2$ if we omit the condition $\mathrm{Ker}\,\beta_3 = \{0\}$.

Consider now the case when four independent random variables take values in an $\boldsymbol{a}$-adic solenoid. We need the following lemmas.

LEMMA 15.6. *Let Y be a locally compact Abelian group. Let b_j, $j = 1, 2, \ldots, n$, $n \geq 2$, be continuous endomorphisms of Y such that the subgroups $(b_i - b_j)(Y)$ are* *dense in Y for all $i \neq j$. Consider on the group Y the equation*

$$\sum_{j=1}^{n} \psi_j(u + b_j v) = B(v), \quad u, v \in Y, \tag{15.17}$$

where $\psi_j(y)$ and $B(y)$ are continuous functions on Y. Then $\psi_j(y)$ are polynomial on Y of degree at most $n-1$. If $B(y) = 0$ for all $y \in Y$, then $\psi_j(y)$ are of degree at most $n-2$.

PROOF. For the proof we use the finite difference method. Let k_1 be an arbitrary element of the group Y. Put $h_1 = -b_n k_1$. Then $h_1 + b_n k_1 = 0$. Substitute $u + h_1$ for u and $v + k_1$ for v in equation (15.17). Subtracting equation (15.17) from the resulting equation we get

$$\sum_{j=1}^{n-1} \Delta_{l_{1j}} \psi_j(u + b_j v) = \Delta_{k_1} B(v), \quad u, v \in Y, \tag{15.18}$$

where $l_{1j} = h_1 + b_j k_1 = (b_j - b_n)k_1$, $j = 1, 2, \ldots, n-1$. Let k_2 be an arbitrary element of the group Y. Put $h_2 = -b_{n-1}k_2$. Then $h_2 + b_{n-1}k_2 = 0$. Substitute $u + h_2$ for u and $v + k_2$ for v in equation (15.18). Subtracting equation (15.18) from the resulting equation we obtain

$$\sum_{j=1}^{n-2} \Delta_{l_{2j}} \Delta_{l_{1j}} \psi_j(u + b_j v) = \Delta_{k_2} \Delta_{k_1} B(v), \quad u, v \in Y,$$

where

$$l_{2j} = h_2 + b_j k_2 = (b_j - b_{n-1})k_2, \quad j = 1, 2, \ldots, n-2.$$

Arguing in a similar way we arrive at the equation

$$\Delta_{l_{n-1,1}} \Delta_{l_{n-2,1}} \cdots \Delta_{l_{11}} \psi_1(u + b_1 v) = \Delta_{k_{n-1}} \Delta_{k_{n-2}} \cdots \Delta_{k_1} B(v), \quad u, v \in Y, \tag{15.19}$$

where k_m are arbitrary elements of the group Y,

$$h_m = -b_{n-m+1}k_m, \quad m = 1, 2, \ldots, n-1,$$

$$l_{mj} = h_m + b_j k_m = (b_j - b_{n-m+1})k_m, \quad j = 1, 2, \ldots, n-m.$$

Let h be an arbitrary element of the group Y. Substitute $u + h$ for u in equation (15.19). Subtracting equation (15.19) from the resulting equation we obtain

$$\Delta_h \Delta_{l_{n-1,1}} \Delta_{l_{n-2,1}} \cdots \Delta_{l_{11}} \psi_1(u + b_1 v) = 0, \quad u, v \in Y. \tag{15.20}$$

Substituting $v = 0$ in equation (15.20) and taking into account that all subgroups $(b_i - b_j)(Y)$ are dense in Y for $i \neq j$, we obtain that the function $\psi_1(y)$ satisfies the equation

$$\Delta_h^n \psi_1(y) = 0, \quad y, h \in Y,$$

i.e., $\psi_1(y)$ is a polynomial on Y of degree at most $n-1$. If $B(y) = 0$ for all $y \in Y$, then in (15.19) the right-hand side is equal to zero and the polynomial $\psi_1(y)$ is of degree at most $n-2$. For the functions $\psi_j(y)$, $j = 2, 3, \ldots, n$, we argue in a similar way. □

LEMMA 15.7. *Let X be a locally compact Abelian group with character group Y. Assume that the group X contains no more than one element of order 2. Let $l(y)$ be a continuous function on the group Y satisfying equation* (4.11) *and conditions* (4.12). *Then $l(y)$ is a character of the group Y.*

The proof of the lemma is carried out in the same way as in the proof of Lemma 6.18.

THEOREM 15.8. *Consider an $\boldsymbol{a}$-adic solenoid $\Sigma_{\boldsymbol{a}}$. Let β_j, $j = 1,2,3,4$, be continuous endomorphisms of $\Sigma_{\boldsymbol{a}}$. Let ξ_j be independent random variables with values in the group $\Sigma_{\boldsymbol{a}}$ with nonvanishing characteristic functions. Put $L_2 = \beta_1\xi_1 + \beta_2\xi_2 + \beta_3\xi_3 + \beta_4\xi_4$. Consider two cases:*

(I) *$L_1 = \xi_1 + \xi_2 + \xi_3 + \xi_4$ and β_j satisfy the conditions*

$$(\beta_i - \beta_j) \in \mathrm{Aut}(\Sigma_{\boldsymbol{a}}), \quad i \neq j, \ i,j \in \{1,2,3,4\}, \tag{i}$$

(II) *$L_1 = \xi_1 + \xi_2 + \xi_3$ and β_j satisfy the conditions*

$$(\beta_i - \beta_j) \in \mathrm{Aut}(\Sigma_{\boldsymbol{a}}), \quad i \neq j, \ i,j \in \{1,2,3\}, \quad \beta_4 \in \mathrm{Aut}(\Sigma_{\boldsymbol{a}}). \tag{ii}$$

In each of these cases the distribution of the random vector (L_1, L_2) determines the distributions of the random variables ξ_j up to a convolution with a Gaussian distribution.

PROOF. Let $H_{\boldsymbol{a}}$ be the character group of the group $\Sigma_{\boldsymbol{a}}$. Let η_j, $j = 1,2,3,4$, be independent random variables with values in the group $\Sigma_{\boldsymbol{a}}$ with nonvanishing characteristic functions. Denote by μ_j the distributions of the random variables ξ_j, and by ν_j the distributions of the random variables η_j. Put $M_2 = \beta_1\eta_1 + \beta_2\eta_2 + \beta_3\eta_3 + \beta_4\eta_4$ and

$$f_j(y) = \hat{\nu}_j(y)/\hat{\mu}_j(y), \quad \psi_j(y) = \ln|f_j(y)|, \tag{15.21}$$

$$g_j(y) = f_j(y)/|f_j(y)|, \quad b_j = \widetilde{\beta}_j, \ j = 1,2,3,4.$$

Assume that (I) holds. Put $M_1 = \eta_1 + \eta_2 + \eta_3 + \eta_4$. Suppose that the distributions of the random vectors (L_1, L_2) and (M_1, M_2) coincide. We prove that either $f_j(y)$ or $(f_j(y))^{-1}$ is the characteristic function of a Gaussian distribution. Thus, in case (I) the theorem will be proved.

Arguing as in the proof of Theorem 15.1, we obtain that the functions $f_j(y)$ satisfy the equation

$$f_1(u + b_1v)f_2(u + b_2v)f_3(u + b_3v)f_4(u + b_4v) = 1, \quad u,v \in H_{\boldsymbol{a}}. \tag{15.22}$$

It follows from (15.22) that the functions $\psi_j(y)$ satisfy the equation

$$\psi_1(u + b_1v) + \psi_2(u + b_2v) + \psi_3(u + b_3v) + \psi_4(u + b_4v) = 0, \quad u,v \in H_{\boldsymbol{a}}.$$

In view of (i), by statement (c) of Theorem 1.20, we have $(b_i - b_j) \in \mathrm{Aut}(H_{\boldsymbol{a}})$ for all $i \neq j$, $i,j \in \{1,2,3,4\}$. By Lemma 15.6, the functions $\psi_j(y)$ are polynomials of degree at most 2. Since $\psi_j(-y) = \psi_j(y)$ for all $y \in H_{\boldsymbol{a}}$ and $\psi_j(0) = 0$, it is easy to see that each of the functions $\psi_j(y)$ satisfies equation (3.2). Hence

$$\psi_j(y) = \sigma_j y^2, \quad y \in H_{\boldsymbol{a}},$$

for some real σ_j. It follows from (3.14) that either $|f_j(y)|$ or $|f_j(y)|^{-1}$ is the characteristic function of a Gaussian distribution on the group $\Sigma_{\boldsymbol{a}}$.

We will prove that each of the functions $g_j(y)$ satisfies equation (4.11). Taking into account that each of the functions $g_j(y)$ satisfies conditions (4.12) and every $\boldsymbol{a}$-adic solenoid $\Sigma_{\boldsymbol{a}}$ contains no more than one element of order 2, by Lemma 15.7, all functions $g_j(y)$ are characters of the group $H_{\boldsymbol{a}}$. Taking into account (15.21), this implies the assertion of the theorem in case (I).

The functions $g_j(y)$ satisfy the equation

$$g_1(u+b_1v)g_2(u+b_2v)g_3(u+b_3v)g_4(u+b_4v)=1, \quad u,v\in H_{\boldsymbol{a}}. \tag{15.23}$$

Arguing as in the proof of Theorem 15.1 and retaining the notation used in the proof of Theorem 15.1, we eliminate from equation (15.23) successively the functions $g_1(y)$, $g_2(y)$, and $g_3(y)$. We obtain that the function $g_4(y)$ satisfies the equation

$$\begin{aligned}\frac{g_4(u+b_4v+(b_4-b_1)k_1+(b_4-b_2)k_2+(b_4-b_3)k_3)}{g_4(u+b_4v+(b_4-b_2)k_2+(b_4-b_3)k_3)}&\\ \times\frac{g_4(u+b_4v+(b_4-b_3)k_3)}{g_4(u+b_4v+(b_4-b_1)k_1+(b_4-b_3)k_3)}&\\ \times\frac{g_4(u+b_4v+(b_4-b_2)k_2)g_4(u+b_4v+(b_4-b_1)k_1)}{g_4(u+b_4v+(b_4-b_1)k_1+(b_4-b_2)k_2)g_4(u+b_4v)}&=1, \quad u,v\in H_{\boldsymbol{a}}.\end{aligned} \tag{15.24}$$

Put $v=0$, $(b_4-b_1)k_1=k$, $(b_4-b_2)k_2=l$, $(b_4-b_3)k_3=m$ in equation (15.24). As noted above, $(b_i-b_j)\in \mathrm{Aut}(H_{\boldsymbol{a}})$ for all $i\neq j$, $i,j\in\{1,2,3,4\}$. In particular, $(b_4-b_j)\in \mathrm{Aut}(H_{\boldsymbol{a}})$, $j=1,2,3$. Since k_1, k_2 and k_3 are arbitrary elements of the group $H_{\boldsymbol{a}}$, it follows from this that the function $g_4(y)$ satisfies the equation

$$\frac{g_4(u+k+l+m)g_4(u+m)g_4(u+l)g_4(u+k)}{g_4(u+l+m)g_4(u+k+m)g_4(u+k+l)g_4(u)}=1, \quad u,k,l,m\in H_{\boldsymbol{a}}. \tag{15.25}$$

Put $u=-k$, $m=k$ in equation (15.25). We get

$$\frac{g_4(l+k)g_4(l-k)}{g_4(l)g_4(l)}=1, \quad k,l\in H_{\boldsymbol{a}}.$$

Hence the function $g_4(y)$ satisfies equation (4.11). For the functions $g_j(y)$, $j=1,2,3$, we argue in a similar way.

Assume that (II) holds. Put $M_1=\eta_1+\eta_2+\eta_3$. Arguing as in the proof of the theorem in case (I), we get that the identical distributiveness of the random vectors (L_1,L_2) and (M_1,M_2) implies that the functions $f_j(y)$ satisfy the equation

$$f_1(u+b_1v)f_2(u+b_2v)f_3(u+b_3v)f_4(b_4v)=1, \quad u,v\in H_{\boldsymbol{a}}. \tag{15.26}$$

It follows from (15.26) that the functions $\psi_j(y)$ satisfy the equation

$$\psi_1(u+b_1v)+\psi_2(u+b_2v)+\psi_3(u+b_3v)=-\psi_4(b_4v), \quad u,v\in H_{\boldsymbol{a}}. \tag{15.27}$$

In view of $(b_i-b_j)\in\mathrm{Aut}(H_{\boldsymbol{a}})$ for all $i\neq j$, $i,j\in\{1,2,3\}$, by Lemma 15.6, it follows from (15.27) that the functions $\psi_j(y)$, $j=1,2,3$, are polynomials of degree at most 2. We note that since $\beta_4\in\mathrm{Aut}(\Sigma_{\boldsymbol{a}})$, by statement (c) of Theorem 1.20, we have $b_4\in\mathrm{Aut}(H_{\boldsymbol{a}})$. Taking into account and substituting $u=0$ in equation (15.27), we obtain that the function $\psi_4(y)$ is also a polynomial of degree at most 2. As in case (I), it follows from this that either $|f_j(y)|$ or $|f_j(y)|^{-1}$ is the characteristic function of a Gaussian distribution.

The functions $g_j(y)$ satisfy the equation

$$g_1(u+b_1v)g_2(u+b_2v)g_3(u+b_3v)g_4(b_4v)=1, \quad u,v\in H_{\boldsymbol{a}}. \tag{15.28}$$

We will prove that each of the functions $g_j(y)$ satisfies equation (4.11). Taking into account that each of the functions $g_j(y)$ satisfies conditions (4.12) and every $\boldsymbol{a}$-adic solenoid $\Sigma_{\boldsymbol{a}}$ contains no more than one element of order 2, by Lemma 15.7, all functions $g_j(y)$ are characters of the group $H_{\boldsymbol{a}}$. Taking into account (15.21), this implies the assertion of the theorem in case (II).

First verify that the function $g_4(y)$ satisfies equation (4.11). Arguing as in the proof of the theorem in case (I), we eliminate from equation (15.28) the functions $g_1(y)$ and $g_2(y)$. We get

$$\frac{g_3(u+b_3v+(b_3-b_1)k_1+(b_3-b_2)k_2)}{g_3(u+b_3v+(b_3-b_2)k_2)g_4(b_4v+b_4k_2)} \times \frac{g_4(b_4v+b_4k_1+b_4k_2)g_3(u+b_3v)g_4(b_4v)}{g_3(u+b_3v+(b_3-b_1)k_1)g_4(b_4v+b_4k_1)} = 1, \quad u,v\in H_{\boldsymbol{a}}. \tag{15.29}$$

Eliminating from equation (15.29) the function $g_3(y)$, we obtain

$$\frac{g_4(b_4v+b_4k_1+b_4k_2+b_4k_3)g_4(b_4v+b_4k_3)}{g_4(b_4v+b_4k_2+b_4k_3)g_4(b_4v+b_4k_1+b_4k_3)} \times \frac{g_4(b_4v+b_4k_2)g_4(b_4v+b_4k_1)}{g_4(b_4v+b_4k_1+b_4k_2)g_4(b_4v)} = 1, \quad u,v\in H_{\boldsymbol{a}}. \tag{15.30}$$

Taking into account that $b_4\in \mathrm{Aut}(H_{\boldsymbol{a}})$, substitute $b_4v=u$, $b_4k_1=k$, $b_4k_2=l$, $b_4k_3=m$ in equation (15.30). Since v,k_1,k_2,k_3 are arbitrary elements of the group $H_{\boldsymbol{a}}$, we get that the function $g_4(y)$ satisfies equation (15.25). The desired assertion follows from this.

Make sure now that the functions $g_j(y)$, $j=1,2,3$, also satisfy equation (4.11). Let k be an arbitrary element of the group $H_{\boldsymbol{a}}$. Substitute $u+k$ for u in equation (15.29). Dividing the resulting equation by equation (15.29), we find

$$\frac{g_3(u+k+b_3v+(b_3-b_1)k_1+(b_3-b_2)k_2)g_3(u+k+b_3v)}{g_3(u+k+b_3v+(b_3-b_2)k_2)g_3(u+k+b_3v+(b_3-b_1)k_1)} \times \frac{g_3(u+b_3v+(b_3-b_2)k_2)g_3(u+b_3v+(b_3-b_1)k_1)}{g_3(u+b_3v+(b_3-b_1)k_1+(b_3-b_2)k_2)g_3(u+b_3v)} = 1, \quad u,v\in H_{\boldsymbol{a}}. \tag{15.31}$$

Put $v=0$, $(b_3-b_1)k_1=l$, $(b_3-b_2)k_2=m$ in (15.31). Taking into account that $(b_3-b_j)\in \mathrm{Aut}(H_{\boldsymbol{a}})$, $j=1,2$, the resulting equation implies that the function $g_3(y)$ satisfies equation (15.25). The desired assertion follows from this. For the functions $g_j(y)$, $j=1,2$, we argue in a similar way. □

REMARK 15.9. The reasoning similar to that in Remark 15.5 shows that Theorem 15.8 fails if we omit either conditions (i) or (ii) in Theorem 15.8.

EXAMPLE 15.10. Generally speaking, Theorem 15.8 is false for an arbitrary locally compact Abelian group. Here is an example.

Consider the group $\mathbb{R}^2$. Denote by $y=(y_1,y_2)$, where $y_j\in\mathbb{R}$, elements of the character group of the group $\mathbb{R}^2$. Put

$$\beta_j=\begin{pmatrix} j & 0\\ 0 & -j\end{pmatrix}, \quad j=1,2,3,4.$$

Then β_j are continuous endomorphisms of the group $\mathbb{R}^2$ satisfying conditions (i) of Theorem 15.8. Put $\mu_1=\mu_2=\mu_3=\mu_4=\mu$, where μ is the distribution on the group $\mathbb{R}^2$ with the characteristic function

$$\hat\mu(y_1,y_2)=\exp\{-4(y_1^2+y_2^2)\}, \quad y_1,y_2\in\mathbb{R}.$$

Let ν_j be the distributions on the group $\mathbb{R}^2$ with the characteristic functions

$$\hat\nu_1(y_1,y_2)=\exp\{-3y_1^2-5y_2^2\}, \quad \hat\nu_2(y_1,y_2)=\exp\{-7y_1^2-y_2^2\}, \quad y_1,y_2\in\mathbb{R},$$
$$\hat\nu_3(y_1,y_2)=\exp\{-y_1^2-7y_2^2\}, \quad \hat\nu_4(y_1,y_2)=\exp\{-5y_1^2-3y_2^2\}, \quad y_1,y_2\in\mathbb{R}.$$

Let ξ_j and η_j, $j = 1, 2, 3, 4$, be independent random variables with values in the group $\mathbb{R}^2$ and distributions μ_j and ν_j respectively. Put $L_1 = \xi_1+\xi_2+\xi_3+\xi_4$, $L_2 = \beta_1\xi_1+\beta_2\xi_2+\beta_3\xi_3+b_4\xi_4$, $M_1 = \eta_1+\eta_2+\eta_3+\eta_4$, and $M_2 = \beta_1\eta_1+\beta_2\eta_2+\beta_3\eta_3+b_4\eta_4$. On the one hand, it is easy to verify that (15.22) holds. Hence the random vectors (L_1, L_2) and (M_1, M_2) are identically distributed. On the other hand, there are no Gaussian distributions γ_j on $\mathbb{R}^2$ such that either $\mu_j = \nu_j * \gamma_j$ or $\nu_j = \mu_j * \gamma_j$, $j = 1, 2, 3, 4$.

16. Generalized Pólya theorem

The following statement is known as the generalized Pólya theorem: Let ξ_j, $j = 1, 2, \dots, n$, $n \geq 2$, be independent identically distributed random variables. Let α_j be nonzero real numbers such that

$$\alpha_1^2 + \cdots + \alpha_n^2 = 1. \tag{16.1}$$

If ξ_j and $\alpha_1\xi_1 + \cdots + \alpha_n\xi_n$ are identically distributed, then the random variables ξ_j are Gaussian.

In this section, we will study an analogue of the generalized Pólya theorem on $\boldsymbol{a}$-adic solenoids. We need the following lemma, which is true for an arbitrary second countable locally compact Abelian group.

LEMMA 16.1. *Let X be a locally compact Abelian group with character group Y. Let α_j, $j = 1, 2, \dots, n$, $n \geq 2$, be continuous endomorphisms of X. Let ξ_j be independent identically distributed random variables with values in the group X and distribution μ. Then ξ_j and $\alpha_1\xi_1+\cdots+\alpha_n\xi_n$ are identically distributed if and only if the characteristic function $\hat{\mu}(y)$ satisfies the equation*

$$\hat{\mu}(y) = \hat{\mu}(\widetilde{\alpha}_1 y)\cdots\hat{\mu}(\widetilde{\alpha}_n y), \quad y \in Y. \tag{16.2}$$

PROOF. Taking into account that $\hat{\mu}(y) = \mathbf{E}[(\xi_j, y)]$ and the independence of the random variables ξ_j, we have

$$\begin{aligned}
(16.2) \iff& \mathbf{E}[(\xi_j, y)] = \mathbf{E}[(\xi_1, \widetilde{\alpha}_1 y)]\cdots\mathbf{E}[(\xi_n, \widetilde{\alpha}_n y)], \quad y \in Y.\\
\iff& \mathbf{E}[(\xi_j, y)] = \mathbf{E}[(\xi_1, \widetilde{\alpha}_1 y)\cdots(\xi_n, \widetilde{\alpha}_n y)], \quad y \in Y.\\
\iff& \mathbf{E}[(\xi_j, y)] = \mathbf{E}[(\alpha_1\xi_1, y)\cdots(\alpha_n\xi_n, y)], \quad y \in Y.\\
\iff& \mathbf{E}[(\xi_j, y)] = \mathbf{E}[\alpha_1\xi_1 + \cdots + \alpha_n\xi_n, y)], \quad y \in Y.
\end{aligned}$$

We see that (16.2) is equivalent to the fact that the characteristic functions of ξ_j and $\alpha_1\xi_1 + \cdots + \alpha_n\xi_n$ coincide, i.e., ξ_j and $\alpha_1\xi_1 + \cdots + \alpha_n\xi_n$ are identically distributed. □

THEOREM 16.2. *Let $\Sigma_{\boldsymbol{a}}$ be an $\boldsymbol{a}$-adic solenoid. Let α_j, $j = 1, 2, \dots, n$, $n \geq 2$, be topological automorphisms of the group $\Sigma_{\boldsymbol{a}}$. Assume that the following conditions are satisfied:*

(i) *there is a unique prime number p such that the group $\Sigma_{\boldsymbol{a}}$ contains no elements of order p;*
(ii) $\alpha_1^2 + \cdots + \alpha_n^2 = I$.

Let ξ_j be independent identically distributed random variables with values in $\Sigma_{\boldsymbol{a}}$ and distribution μ. If ξ_j and $\alpha_1\xi_1 + \cdots + \alpha_n\xi_n$ are identically distributed, then $\mu = \gamma * m_K$, where $\gamma \in \Gamma(\Sigma_{\boldsymbol{a}})$ and K is a compact subgroup of the group $\Sigma_{\boldsymbol{a}}$. Moreover, $f_p(K) = K$.

PROOF. Let $H_{\boldsymbol{a}}$ be the character group of the group $\Sigma_{\boldsymbol{a}}$. We will consider $H_{\boldsymbol{a}}$ as a subset of $\mathbb{R}$. By Lemma 16.1, the characteristic function $\hat{\mu}(y)$ satisfies equation (16.2). First, we will prove the theorem assuming that $\hat{\mu}(y) \geq 0$ for all $y \in H_{\boldsymbol{a}}$. This implies that $\hat{\mu}(-y) = \hat{\mu}(y)$ for all $y \in H_{\boldsymbol{a}}$.

Inasmuch as $\Sigma_{\boldsymbol{a}}$ is a compact connected Abelian group, by Theorem 1.13, the mapping f_n is an epimorphism for each nonzero integer n. Since $\Sigma_{\boldsymbol{a}}$ has no elements of order p, we conclude that f_p is a monomorphism. In view of the fact that f_p is an epimorphism, $f_p \in \mathrm{Aut}(\Sigma_{\boldsymbol{a}})$. By statement (c) of Theorem 1.20, this implies that $f_p \in \mathrm{Aut}(H_{\boldsymbol{a}})$. It means that both $\Sigma_{\boldsymbol{a}}$ and $H_{\boldsymbol{a}}$ are groups with unique division by p. Note that in view of condition (i), if q is a prime number and $q \neq p$, then $f_q \notin \mathrm{Aut}(\Sigma_{\boldsymbol{a}})$. It follows from this that each topological automorphism of the group $\Sigma_{\boldsymbol{a}}$ is of the form $\pm f_{p^m}^{\pm 1}$, where m is a nonnegative integer. Taking into account condition (ii), we conclude that each topological automorphism α_j in the theorem is of the form $\alpha_j = \pm f_{p^{m_j}}^{-1}$, where m_j is a natural number, i.e., α_j is a multiplication by a number $\pm p^{-m_j}$. Put $f(y) = \hat{\mu}(y)$. If necessary, changing the numbering of the random variables ξ_j and taking into account that $f(-y) = f(y)$, we can rewrite equation (16.2) in the form

$$f(y) = (f(p^{-1}y))^{k_1} \cdots (f(p^{-l}y))^{k_l}, \quad y \in H_{\boldsymbol{a}}, \tag{16.3}$$

where $k_j \geq 0$, $j = 1, 2, \ldots, l$, $k_1 + \cdots + k_l = n$. It follows from condition (ii) that

$$\sum_{j=1}^{l} \frac{k_j}{p^{2j}} = 1. \tag{16.4}$$

Assume that for each j the inequality $k_j \leq p^j$ is valid. Then we have

$$\sum_{j=1}^{l} \frac{k_j}{p^{2j}} \leq \sum_{j=1}^{l} \frac{1}{p^j} < 1,$$

but this contradicts (16.4). Hence there is j_0 such that

$$k_{j_0} > p^{j_0}. \tag{16.5}$$

Taking this into account and the fact that $0 \leq f(y) \leq 1$, (16.3) implies the inequalities

$$f(y) \leq (f(p^{-j_0}y))^{k_{j_0}} \leq (f(p^{-j_0}y))^{p^{j_0}}, \quad y \in H_{\boldsymbol{a}}. \tag{16.6}$$

Set $m = p^{j_0}$ and rewrite (16.6) in the form

$$f(my) \leq (f(y))^m, \quad y \in H_{\boldsymbol{a}}. \tag{16.7}$$

Observe that we can assume that $m > 2$ in (16.7). Indeed, if $m = 2$, then $p = 2$ and $j_0 = 1$. In this case (16.5) implies that $k_1 > 2$. Hence in view of (16.4), either $k_1 = 4$ or $k_1 = 3$. Let $k_1 = 4$. Then (16.3) takes the form $f(y) = (f(2^{-1}y))^4$. This implies that $f(4y) = (f(y))^{16}$ and we may suppose that $m = 4$ in (16.7). Let $k_1 = 3$. Then (16.3) can be written as $f(y) = (f(2^{-1}y))^3\phi(y)$, where $\phi(y)$ is a characteristic function. It follows from this that $f(4y) = (f(y))^9(\phi(2y))^3\phi(4y)$, and we may also suppose $m = 4$ in (16.7).

It may be noted that if $\mu = m_{\Sigma_{\boldsymbol{a}}}$, then the theorem is proved. Therefore, suppose that $\mu \neq m_{\Sigma_{\boldsymbol{a}}}$. Then there is an element $y_0 \in H_{\boldsymbol{a}}$, $y_0 \neq 0$, such that

$f(y_0) > 0$. Consider the discrete additive group F of the rational numbers of the form

$$F = \left\{ \frac{m}{p^n} : n = 0, 1, \ldots; \ m \in \mathbb{Z} \right\}.$$

Since $H_{\boldsymbol{a}}$ is a group with unique division by p, we can consider on the subgroup F the function $g(r) = f(ry_0)$, $r \in F$. We verify that (16.7) implies the uniform continuity of the function $g(r)$ in the topology induced on F by the topology of $\mathbb{R}$. In view of statement (vii) of Theorem 2.5 and taking into account that $g(r)$ is a characteristic function on F, it suffices to show that the function $g(r)$ is continuous at the zero.

It follows from (16.7) that $f(m^k y_0) \le (f(y_0))^{m^k}$ for all natural k. Hence

$$f(m^{-k} y_0) \ge (f(y_0))^{1/m^k}.$$

This implies that

$$g\left(\frac{1}{m^k}\right) \ge (g(1))^{1/m^k}. \tag{16.8}$$

Let $a > 0$. Note that for all $x > 0$ the inequality

$$1 - e^{-ax} \le ax$$

holds. Taking this into account we get from (16.8) that

$$1 - g\left(\frac{1}{m^k}\right) \le \frac{C}{m^k}, \tag{16.9}$$

where $C = -\ln g(1)$.

In view of the fact that $f(y)$ is a real-valued characteristic function, we have from (2.2) that the function $f(y)$ satisfies the inequality

$$1 - f(y_1 + y_2) \le 2[(1 - f(y_1)) + (1 - f(y_2))], \quad y_1, y_2 \in H_{\boldsymbol{a}}. \tag{16.10}$$

By induction, this implies the inequality

$$1 - f(y_1 + \cdots + y_q) \le \sum_{j=1}^{q} 2^j (1 - f(y_j)), \quad y_j \in H_{\boldsymbol{a}}. \tag{16.11}$$

Let i be a natural number such that $0 < i < m^k$. We have

$$\frac{i}{m^k} = \frac{i_1}{m} + \cdots + \frac{i_k}{m^k}, \tag{16.12}$$

where i_l are integers such that $0 \le i_l < m$, $l = 1, 2, \ldots, k$. Substituting $q = i_l$, $y_1 = \cdots = y_{i_l} = \frac{y_0}{m^l}$ in (16.11), we get

$$1 - g\left(\frac{i_l}{m^l}\right) \le \sum_{j=1}^{i_l} 2^j \left(1 - g\left(\frac{1}{m^l}\right)\right) \le 2^m \left(1 - g\left(\frac{1}{m^l}\right)\right). \tag{16.13}$$

Suppose

$$\frac{i}{m^k} < \frac{1}{m^s}. \tag{16.14}$$

It follows from this that $i_l = 0$ for all $l \leq s$ in (16.12). In this case, taking into account (16.9), (16.12), and (16.13), we find from (16.11)

$$(16.15)\quad 1 - g\left(\frac{i}{m^k}\right) = 1 - g\left(\frac{i_{s+1}}{m^{s+1}} + \cdots + \frac{i_k}{m^k}\right) \leq \sum_{j=1}^{k-s} 2^j \left(1 - g\left(\frac{i_{s+j}}{m^{s+j}}\right)\right)$$
$$\leq \sum_{j=1}^{k-s} 2^j 2^m \left(1 - g\left(\frac{1}{m^{s+j}}\right)\right) \leq \sum_{j=1}^{k-s} \frac{C2^{m+j}}{m^{s+j}} = \frac{C2^m}{m^s} \sum_{j=1}^{k-s} \left(\frac{2}{m}\right)^j.$$

In view of $m > 2$, we find from (16.15) that the inequality

$$(16.16)\qquad 1 - g\left(\frac{i}{m^k}\right) \leq \frac{C2^{m+1}}{(m-2)m^s}$$

is valid. Obviously, (16.14) and (16.16) imply the continuity at the zero of the function $g(r)$ in the topology induced on F by the topology of $\mathbb{R}$, and hence the uniform continuity of the function $g(r)$ on F in this topology.

It follows from (16.3) that the function $g(r)$ satisfies the equation

$$(16.17)\qquad g(r) = (g(p^{-1}r))^{k_1} \cdots (g(p^{-l}r))^{k_l}, \quad r \in F.$$

We extend the function $g(r)$ by continuity from F to a continuous positive definite function on $\mathbb{R}$. We keep the notation $g(s)$ for the extended function. It follows from (16.17) that the function $g(s)$ satisfies the equation

$$(16.18)\qquad g(s) = (g(p^{-1}s))^{k_1} \cdots (g(p^{-l}s))^{k_l}, \quad s \in \mathbb{R}.$$

By the generalized Pólya theorem, it follows from (16.4) and (16.18) that $g(s)$ is the characteristic function of a symmetric Gaussian distribution on the real line. Hence we get the representation

$$(16.19)\qquad f(y) = \exp\{-\sigma(y_0)y^2\}, \quad y = ry_0,\ r \in F,$$

where $\sigma(y_0) \geq 0$.

When we obtained (16.19) we fixed an element $y_0 \in H_{\boldsymbol{a}}$, $y_0 \neq 0$, such that $f(y_0) > 0$. If we fix another element $z_0 \in H_{\boldsymbol{a}}$, $z_0 \neq 0$, such that $f(z_0) > 0$, we obtain the representation

$$f(y) = \exp\{-\sigma(z_0)y^2\}, \quad y = rz_0,\ r \in F,$$

where $\sigma(z_0) \geq 0$. Since the subgroups $\{y = ry_0 : r \in F\}$ and $\{y = rz_0 : r \in F\}$ have a nonzero intersection, this implies that $\sigma(y_0) = \sigma(z_0) = \sigma$. Put

$$E = \{y \in H_{\boldsymbol{a}} : f(y) \neq 0\}.$$

Thus, we have the representation

$$(16.20)\qquad f(y) = \exp\{-\sigma y^2\}, \quad y \in E.$$

We will verify that E is a subgroup of $H_{\boldsymbol{a}}$. First, observe that when we got (16.19) we proved that if $y \in E$, then $ry \in E$ for all $r \in F$. Take $y_1, y_2 \in E$. It follows from (16.20) that $f(p^{-k}y_j) \to 1$, $j = 1, 2$, as $k \to \infty$. Then (16.10) implies that $p^{-k}(y_1 + y_2) \in E$ for sufficiently large natural numbers k. Hence $y_1 + y_2 \in E$.

Thus, we obtain the representation

$$(16.21)\qquad f(y) = \begin{cases} \exp\{-\sigma y^2\}, & \text{if } y \in E, \\ 0, & \text{if } y \notin E. \end{cases}$$

Put $K = A(\Sigma_{\boldsymbol{a}}, E)$. Taking into account (3.14), let γ be the Gaussian distribution on the group $\Sigma_{\boldsymbol{a}}$ with the characteristic function

$$\hat{\gamma}(y) = \exp\{-\sigma y^2\}, \quad y \in H_{\boldsymbol{a}}.$$

Since $E = A(H_{\boldsymbol{a}}, K)$, it follows from (2.3) and (16.21) that $f(y) = \hat{\mu}(y) = \hat{\gamma}(y)\widehat{m}_K(y)$. Hence $\mu = \gamma * m_K$. Note that the subgroup E has the property: if $py \in E$, then $y \in E$. This implies that $f_p(K) = K$.

Get rid of the restriction that $\hat{\mu}(y) \geq 0$. Put $\nu = \mu * \bar{\mu}$. This implies that $\hat{\nu}(y) = |\hat{\mu}(y)|^2 \geq 0$ for all $y \in H_{\boldsymbol{a}}$. The characteristic function $\hat{\nu}(y)$ also satisfies equation (16.2). Then, as has been proved above, the function $\hat{\nu}(y)$ is represented in the form (16.21). On the one hand, the subgroup E possesses the property: if $y \in E$, then $\frac{1}{p}y \in E$. Therefore, E is not isomorphic to the group of integers $\mathbb{Z}$. Note that if a subgroup of the group of rational numbers is not isomorphic to $\mathbb{Z}$, then it is isomorphic to a group of the form

$$H_{\boldsymbol{b}} = \left\{ \frac{m}{b_0 b_1 \cdots b_n} : n = 0, 1, \ldots; \ m \in \mathbb{Z} \right\}$$

for some $\boldsymbol{b} = (b_0, b_1, \ldots, b_n, \ldots)$, where all $b_j \in \mathbb{Z}$, $b_j > 1$. Hence the group E is isomorphic to a discrete additive group $H_{\boldsymbol{b}}$. On the other hand, since $E = A(H_{\boldsymbol{a}}, K)$, by Theorem 1.9, the group E is topologically isomorphic to the character group of the factor-group $\Sigma_{\boldsymbol{a}}/K$. By the Pontryagin duality theorem, the factor-group $\Sigma_{\boldsymbol{a}}/K$ is topologically isomorphic to the corresponding $\boldsymbol{a}$-adic solenoid $\Sigma_{\boldsymbol{a}}$. Hence the group $\Sigma_{\boldsymbol{a}}/K$ contains no subgroups topologically isomorphic to the circle group $\mathbb{T}$. Applying Theorem 3.14 to the group $\Sigma_{\boldsymbol{a}}/K$ and taking into account that by Theorem 1.9, each character of the subgroup E can be written as (x, y), $x \in \Sigma_{\boldsymbol{a}}$, $y \in E$, we get the statement of the theorem in the general case from representation (16.21) for the function $\hat{\nu}(y)$. □

Condition (ii) of Theorem 16.2 for $\boldsymbol{a}$-adic solenoids is an analogue of condition (16.1) for the real line. Thus, Theorem 16.2 can be considered as an analogue of the generalized Pólya theorem for $\boldsymbol{a}$-adic solenoids satisfying condition (i) of Theorem 16.2 .

COROLLARY 16.3. *Let $\Sigma_{\boldsymbol{a}}$ be an $\boldsymbol{a}$-adic solenoid satisfying condition* (i) *of Theorem* 16.2. *Let α_j, $j = 1, 2, \ldots, n$, $n \geq 2$, be topological automorphisms of the group $\Sigma_{\boldsymbol{a}}$ satisfying condition* (ii) *of Theorem* 16.2. *Let ξ_j be independent identically distributed random variables with values in $\Sigma_{\boldsymbol{a}}$ and distribution μ with a nonvanishing characteristic function. If ξ_j and $\alpha_1\xi_1 + \cdots + \alpha_n\xi_n$ are identically distributed, then $\mu \in \Gamma(\Sigma_{\boldsymbol{a}})$.*

REMARK 16.4. Let $\boldsymbol{a}$=$(a_0, a_1, \ldots, a_n, \ldots)$, where all $a_j \in \mathbb{Z}$, $a_j > 1$, and let $\Sigma_{\boldsymbol{a}}$ be the corresponding $\boldsymbol{a}$-adic solenoid. It is easy to verify that the following statements are equivalent:

(1) condition (i) of Theorem 16.2 is satisfied;
(2) there is a unique prime number p such that infinite number of a_j are divided by p.

It is easy to see that the following statements are equivalent:

(a) there are topological automorphisms α_j, $j = 1, 2, \ldots, n$, $n \geq 2$, of the group $\Sigma_{\boldsymbol{a}}$ satisfying condition (ii) of Theorem 16.2;

(b) there is a prime number p such that the group $\Sigma_{\boldsymbol{a}}$ contains no elements of order p;

(c) there is a prime number p such that infinite number of a_j are divided by p.

For the proof we note that there is an alternative: either $f_p \in \mathrm{Aut}(\Sigma_{\boldsymbol{a}})$ for a prime number p or $\mathrm{Aut}(\Sigma_{\boldsymbol{a}}) = \{\pm I\}$.

EXAMPLE 16.5. An example of an $\boldsymbol{a}$-adic solenoid $\Sigma_{\boldsymbol{a}}$ that satisfies condition (i) of Theorem 16.2 is an $\boldsymbol{a}$-adic solenoid $\Sigma_{\boldsymbol{a}}$, where $\boldsymbol{a} = (p, p, p, \dots)$ with character group

$$H_{\boldsymbol{a}} = \left\{ \frac{m}{p^n} : \ n = 0, 1, \dots; \ m \in \mathbb{Z} \right\},$$

i.e., the p-adic solenoid Σ_p.

REMARK 16.6. Let $\Sigma_{\boldsymbol{a}}$ be an $\boldsymbol{a}$-adic solenoid with character group $H_{\boldsymbol{a}}$. In view of (3.14), let γ be the Gaussian distribution on $\Sigma_{\boldsymbol{a}}$ with the characteristic function of the form

$$\hat{\gamma}(y) = \exp\{-\sigma y^2\}, \quad y \in H_{\boldsymbol{a}}, \tag{16.22}$$

where $\sigma \geq 0$. Let α_j, $j = 1, 2, \dots, n$, $n \geq 2$, be topological automorphisms of the group $\Sigma_{\boldsymbol{a}}$ satisfying condition (ii) of Theorem 16.2. This implies that the characteristic function $\hat{\gamma}(y)$ satisfies equation (16.2).

Let $\Sigma_{\boldsymbol{a}}$ be an $\boldsymbol{a}$-adic solenoid with the property that there is a prime number p such that $\Sigma_{\boldsymbol{a}}$ has no elements of order p. This implies that $\Sigma_{\boldsymbol{a}}$ is a group with unique division by p. Let α_j, $j = 1, 2, \dots, n$, $n \geq 2$, be topological automorphisms of the group $\Sigma_{\boldsymbol{a}}$ of the form $\alpha_j = \pm f_{p^{k_j}}^{-1}$, where k_j are natural numbers. Let K be a compact subgroup of $\Sigma_{\boldsymbol{a}}$ such that $f_p(K) = K$. Note that the condition $f_p(K) = K$ implies that if $py \in A(H_{\boldsymbol{a}}, K)$, then $y \in A(H_{\boldsymbol{a}}, K)$. Taking this into account and using (2.3) it is not difficult to verify that the characteristic function $\widehat{m}_K(y)$ satisfies equation (16.2). Put $\mu = \gamma * m_K$. Then the characteristic function $\hat{\mu}(y)$ also satisfies equation (16.2).

It follows from Lemma 16.1 that if ξ_j, $j = 1, 2, \dots, n$, $n \geq 2$, are independent identically distributed random variables with values in the group $\Sigma_{\boldsymbol{a}}$ and distribution μ, then ξ_j and $\alpha_1\xi_1 + \cdots + \alpha_n\xi_n$ are identically distributed. Hence we cannot narrow the class of distributions in Theorem 16.2 which is characterized by the property of equidistribution of ξ_j and $\alpha_1\xi_1 + \cdots + \alpha_n\xi_n$.

It turns out that Theorem 16.2 is false if condition (i) of Theorem 16.2 is not satisfied. Namely, the following statement holds.

PROPOSITION 16.7. *Let $\Sigma_{\boldsymbol{a}}$ be an $\boldsymbol{a}$-adic solenoid satisfying the condition:*

(i) *there are prime numbers p and q such that the group $\Sigma_{\boldsymbol{a}}$ contains neither elements of order p nor q.*

Then there are topological automorphisms α_j, $j = 1, 2, \dots, n$, $n \geq 2$, of the group $\Sigma_{\boldsymbol{a}}$ satisfying condition (ii) *of Theorem* 16.2, *and independent identically distributed random variables ξ_j with values in $\Sigma_{\boldsymbol{a}}$ and distribution μ such that ξ_j and $\alpha_1\xi_1 + \cdots + \alpha_n\xi_n$ are identically distributed, whereas $\mu \notin \Gamma(\Sigma_{\boldsymbol{a}}) * I(\Sigma_{\boldsymbol{a}})$.*

PROOF. Let $H_{\boldsymbol{a}}$ be the character group of the group $\Sigma_{\boldsymbol{a}}$. Suppose for definiteness that $p < q$. Condition (i) implies that $f_p, f_q \in \mathrm{Aut}(\Sigma_{\boldsymbol{a}})$ and hence

$f_p, f_q \in \mathrm{Aut}(H_{\boldsymbol{a}})$. Therefore, both $\Sigma_{\boldsymbol{a}}$ and $H_{\boldsymbol{a}}$ are groups with unique division by p and by q. Denote by H a subgroup of $H_{\boldsymbol{a}}$ of the form

$$H = \left\{\frac{m}{n} \in H_{\boldsymbol{a}} : n \text{ is not divided by } p;\ m \in \mathbb{Z}\right\}.$$

Let k be a natural number. Set

$$H_k = \left\{\frac{m}{p^k n} \in H_{\boldsymbol{a}} : m, n \text{ are not divided by } p\right\}$$

and

$$L = H \cup H_1.$$

It is obvious that

$$H_{\boldsymbol{a}} = H \cup \bigcup_{k=1}^{\infty} H_k$$

and L is a subgroup of $H_{\boldsymbol{a}}$. Denote by G the character group of the group L and put $F = A(G, H)$. It is easy to see that $L/H \cong \mathbb{Z}(p)$. By Theorem 1.9, the character group of the factor-group L/H is topologically isomorphic to the annihilator $A(G, H)$. We have $F \cong \mathbb{Z}(p)$. Take $0 < c < 1$. Let ω be the distribution on the group F of the form $\omega = cE_0 + (1-c)m_F$. Consider ω as a distribution on the group G. In view of (2.3), the characteristic function $\hat{\omega}(l)$, $l \in L$, is of the form

$$\hat{\omega}(l) = \begin{cases} 1, & \text{if } \ l \in H, \\ c, & \text{if } \ l \in H_1. \end{cases}$$

Obviously, $\hat{\omega}(l)$ is a positive definite function. Consider on the group $H_{\boldsymbol{a}}$ the function

$$f(y) = \begin{cases} \hat{\omega}(y), & \text{if } \ y \in L, \\ 0, & \text{if } \ y \notin L. \end{cases}$$

By Proposition 2.9, $f(y)$ is a positive definite function. By Bochner's theorem, there is a probability distribution μ on the group $\Sigma_{\boldsymbol{a}}$ such that $f(y) = \hat{\mu}(y)$. It is obvious that $\mu \notin \Gamma(\Sigma_{\boldsymbol{a}}) * I(\Sigma_{\boldsymbol{a}})$.

Take natural numbers l and m, where $l < m$, such that the remainders of the division of the numbers q^{2l} and q^{2m} by p^2 are equal. Then $q^{2m} - q^{2l}$ is divided by p^2. Hence $q^{2m-2l} - 1$ is also divided by p^2 and the remainder of the division q^{2m-2l} by p^2 is equal to 1. Put $a = m - l$. We have

$$q^{2a} = p^2 b + 1. \tag{16.23}$$

Put $\alpha_1 = \cdots = \alpha_b = f_p f_{q^a}^{-1}$, $\alpha_{b+1} = f_{q^a}^{-1}$, $n = b+1$. In view of (16.23), the topological automorphisms α_j, $j = 1, 2, \dots, n$, of the group $\Sigma_{\boldsymbol{a}}$ satisfy condition (i) of Theorem 16.2.

Let ξ_j, $j = 1, \dots n$, be independent identically distributed random variables with values in the group $\Sigma_{\boldsymbol{a}}$ and distribution μ. Verify that the function $f(y)$ satisfies the equation

$$f(y) = \left(f\left(\frac{p}{q^a}y\right)\right)^b f\left(\frac{1}{q^a}y\right), \quad y \in H_{\boldsymbol{a}}. \tag{16.24}$$

Indeed, if $y \in H$, then $\frac{p}{q^a}y, \frac{1}{q^a}y \in H$, and both parts of equation (16.24) are equal to 1. If $y \in H_1$, then $\frac{p}{q^a}y \in H$, $\frac{1}{q^a}y \in H_1$, and both parts of equation (16.24) are equal to c. If $y \in H_k$, $k \geq 2$, then $\frac{1}{q^a}y \in H_k$ and both parts of equation (16.24) are equal to zero. Thus, the function $f(y)$ satisfies equation (16.24) and by Lemma 16.1, ξ_j and $\alpha_1\xi_1 + \cdots + \alpha_n\xi_n$ are identically distributed. □

Theorem 16.2 and Proposition 16.7 imply the following statement.

COROLLARY 16.8. *Consider an* $\boldsymbol{a}$*-adic solenoid* $\Sigma_{\boldsymbol{a}}$. *Let* α_j, $j = 1, 2, \dots, n$, $n \geq 2$, *be topological automorphisms of the group* $\Sigma_{\boldsymbol{a}}$ *satisfying condition* (ii) *of Theorem* 16.2. *Let* ξ_j *be independent identically distributed random variables with values in* $\Sigma_{\boldsymbol{a}}$ *and distribution* μ. *The equidistribution of* ξ_j *and* $\alpha_1\xi_1 + \cdots + \alpha_n\xi_n$ *implies that* $\mu \in \Gamma(\Sigma_{\boldsymbol{a}}) * I(\Sigma_{\boldsymbol{a}})$ *if and only if condition* (i) *of Theorem* 16.2 *is satisfied.*

REMARK 16.9. We keep the notation used in the proof of Proposition 16.7. In view of Proposition 2.10, the distribution μ constructed in the proof of Proposition 16.7 is supported in the annihilator $A(\Sigma_{\boldsymbol{a}}, H)$, i.e., in a proper closed subgroup of the group $\Sigma_{\boldsymbol{a}}$. We can strengthen Proposition 16.7 as follows.

Let $\Sigma_{\boldsymbol{a}}$ be an $\boldsymbol{a}$-adic solenoid satisfying condition (i) of Theorem 16.7. Let γ be a nondegenerate Gaussian distribution on the group $\Sigma_{\boldsymbol{a}}$ with the characteristic function of the form (16.22). Set $\lambda = \gamma * \mu$, where the distribution μ is constructed in the proof of Proposition 16.7. Then the characteristic function of the distribution λ is of the form

$$\hat{\lambda}(y) = \begin{cases} \exp\{-\sigma y^2\}, & \text{if} \quad y \in H, \\ c\exp\{-\sigma y^2\}, & \text{if} \quad y \in H_1, \\ 0, & \text{if} \quad y \notin L. \end{cases}$$

It is obvious that $\lambda \notin \Gamma(\Sigma_{\boldsymbol{a}}) * I(\Sigma_{\boldsymbol{a}})$. Let ξ_j be independent identically distributed random variables with values in the group $\Sigma_{\boldsymbol{a}}$ and distribution λ. Arguing as in the proof of Proposition 16.7 and using Lemma 16.1, we are convinced that ξ_j and $\alpha_1\xi_1 + \cdots + \alpha_n\xi_n$ are identically distributed. As is easily seen, the distribution λ is not supported in a coset of any proper closed subgroup of the group $\Sigma_{\boldsymbol{a}}$.

Notes

The Rao theorem [**88**], where the distribution of the random vector (L_1, L_2), where $L_1 = a_1\xi_1 + a_2\xi_2 + a_3\xi_3$, $L_2 = b_1\xi_1 + b_2\xi_2 + b_3\xi_3$, determines the distributions of the random variables ξ_j up to a shift, is a generalisation of the following theorem by I. Kotlarski [**70**]: Let ξ_j, $j = 1, 2, 3$, be independent random variables with nonvanishing characteristic functions. Let $L_1 = \xi_1 - \xi_3$, $L_2 = \xi_2 - \xi_3$. Then the distribution of the random vector (L_1, L_2) determines the distributions of the random variables ξ_j up to a shift. In the article [**87**] B.L.S. Prakasa Rao noted that the proof by I. Kotlarski of the above-mentioned theorem can be extended without any changes to locally compact Abelian groups. The proof of Theorem 15.1 is new even for the group of real numbers $\mathbb{R}$ and differs from the proof of the Rao theorem given in [**88**]. Theorem 15.8 is an analogue for $\boldsymbol{a}$-adic solenoids of another Rao theorem [**88**]: Let ξ_j, $j = 1, 2, 3, 4$, be independent random variables with nonvanishing characteristic functions and let a_j, b_j be real numbers such that $a_i/b_i \neq a_j/b_j$ for all $i \neq j$. Let $L_1 = a_1\xi_1 + a_2\xi_2 + a_3\xi_3 + a_4\xi_4$ and $L_2 = b_1\xi_1 + b_2\xi_2 + b_3\xi_3 + b_4\xi_4$. Then the distribution of the random vector (L_1, L_2) determines the distributions of the random variables ξ_j up to a convolution with a Gaussian distribution.

The results of Section 15, with the exception of Corollaries 15.2 and 15.3, were obtained by G.M. Feldman in [**45**]. Corollary 15.2 was proved by B.L.S. Prakasa Rao in [**87**]. Corollary 15.3 was proved by C.R. Rao in [**88**].

In 1923 G. Pólya proved the following theorem: If ξ_1 and ξ_2 are independent identically distributed random variables such that ξ_j and $\frac{\xi_1+\xi_2}{\sqrt{2}}$ are identically distributed, then ξ_j are Gaussian. Subsequently, the characterization of the Gaussian distribution on the real line by the property of equidistribution of two linear forms of independent identically distributed random variables were studied by J. Marcinkiewicz, Yu. V. Linnik, A. A. Zinger and others (see [**69**, Chapter 2]). The results of Section 16 were obtained by G.M. Feldman in [**49**].

Bibliography

[1] David L. Armacost, *The structure of locally compact abelian groups*, Monographs and Textbooks in Pure and Applied Mathematics, vol. 68, Marcel Dekker, Inc., New York, 1981. MR637201

[2] Wojciech Banaszczyk, *Additive subgroups of topological vector spaces*, Lecture Notes in Mathematics, vol. 1466, Springer-Verlag, Berlin, 1991, DOI 10.1007/BFb0089147. MR1119302

[3] Jean-René Barra, *Notions fondamentales de statistique mathématique*, Dunod, Paris, 1971. Préface de Yu. V. Linnik; Maîtrise de mathématiques et applications fondamentales. MR0402992

[4] Y. Baryshnikov, B. Eisenberg, and W. Stadje, *Independent variables with independent sum and difference: S^1-case*, J. Multivariate Anal. **45** (1993), no. 2, 161–170, DOI 10.1006/jmva.1993.1031. MR1221915

[5] A. D. Bendikov, *Spatially homogeneous continuous Markov processes on abelian groups, and harmonic structures* (Russian), Uspehi Mat. Nauk **29** (1974), no. 5(179), 215–216. MR0410956

[6] A. D. Bendikov, *On Functions that are harmonic for a certain class of diffusion processes on a group* (Russian, with English summary), Teor. Veroyatnost. i Primenen. **20** (1975), no. 4, 773–784. MR0397891; English transl., Theory Probab. Appl. **20** (1976), no. 4, 759–769

[7] A. D. Bendikov, *Symmetric stable semigroups on the infinite-dimensional torus*, Exposition. Math. **13** (1995), no. 1, 39–79. MR1339813

[8] Alexander Bendikov, *Potential theory on infinite-dimensional abelian groups*, De Gruyter Studies in Mathematics, vol. 21, Walter de Gruyter & Co., Berlin, 1995. Translated from the Russian manuscript by Carol Regher, DOI 10.1515/9783110876840. MR1445559

[9] A. Bendikov and L. Saloff-Coste, *Elliptic diffusions on infinite products*, J. Reine Angew. Math. **493** (1997), 171–220, DOI 10.1515/crll.1997.493.171. MR1491812

[10] A. Bendikov and L. Saloff-Coste, *Central Gaussian convolution semigroups on compact groups: a survey*, Infin. Dimens. Anal. Quantum Probab. Relat. Top. **6** (2003), no. 4, 629–659, DOI 10.1142/S0219025703001456. MR2030215

[11] A. Bendikov and L. Saloff-Coste, *On the sample paths of diagonal Brownian motions on the infinite dimensional torus* (English, with English and French summaries), Ann. Inst. H. Poincaré Probab. Statist. **40** (2004), no. 2, 227–254, DOI 10.1016/S0246-0203(03)00048-7. MR2044817

[12] Christian Berg, *Fonctions harmoniques sur T^∞* (French, with English summary), C. R. Acad. Sci. Paris Sér. A-B **280** (1975), no. 22, Aii, A1519–A1521. MR385139

[13] Christian Berg, *Potential theory on the infinite dimensional torus*, Invent. Math. **32** (1976), no. 1, 49–100, DOI 10.1007/BF01389771. MR402093

[14] S. N. Bernstein, *On a property characteristic of the normal low* (Russian), Trudy Leningrad. Polytech. Inst., **217**, no. 3, (1941) 21–22.

[15] C. Bessaga and A. Pełczyński, *On a class of B_0-spaces* (English, with Russian summary), Bull. Acad. Polon. Sci. Cl. III. **5** (1957), 375–377, XXX. MR0088680

[16] Nicolas Bourbaki, *General topology. Chapters 5–10*, Elements of Mathematics (Berlin), Springer-Verlag, Berlin, 1989. Translated from the French; Reprint of the 1966 edition. MR979295

[17] Halina Byczkowska and Tomasz Byczkowski, *A generalization of the zero-one law for Gaussian measures on metric abelian groups* (English, with Russian summary), Bull. Acad. Polon. Sci. Sér. Sci. Math. **29** (1981), no. 3-4, 187–191. MR638763

[18] T. Byczkowski, *Gaussian measures on metric abelian groups*, Probability measures on groups (Proc. Fifth Conf., Oberwolfach, 1978), Lecture Notes in Math., vol. 706, Springer, Berlin, 1979, pp. 41–53. MR536974

[19] T. Byczkowski, *Zero-one laws for Gaussian measures on metric abelian groups*, Studia Math. **69** (1980/81), no. 2, 159–189, DOI 10.4064/sm-69-2-159-189. MR604349

[20] G. Darmois, *Analyse générale des liaisons stochastiques. Etude particulière de l'analyse factorielle linéaire* (French), Rev. Inst. Internat. Statist. **21** (1953), 2–8. MR61322

[21] D. Ž. Djoković, *A representation theorem for* $(X_1-1)(X_2-1)\cdots(X_n-1)$ *and its applications*, Ann. Polon. Math. **22** (1969/70), 189–198, DOI 10.4064/ap-22-2-189-198. MR265798

[22] G. M. Fel'dman, *On the decomposition of Gaussian distributions on groups* (Russian, with English summary), Teor. Veroyatnost. i Primenen. **22** (1977), no. 1, 136–143. MR0436250; English transl., Theory Probab. Appl., **22** (1977), no. 1, 133–140

[23] G. M. Fel'dman, *Gaussian distributions on locally compact Abelian groups* (Russian, with English summary), Teor. Veroyatnost. i Primenen. **23** (1978), no. 3, 548–563. MR0509729; English transl., Theory Probab. Appl., **23** (1979), no. 3, 529–542

[24] G. M. Fel'dman, *Bernstein Gaussian distributions on groups* (Russian, with English summary), Teor. Veroyatnost. i Primenen. **31** (1986), no. 1, 47–58. MR0836951; English transl., Theory Probab. Appl., **31** (1987), no. 1, 40–49

[25] G. M. Fel'dman, *Marcinkiewicz and Lukacs theorems on Abelian groups* (Russian), Teor. Veroyatnost. i Primenen. **34** (1989), no. 2, 330–339; English transl., Theory Probab. Appl., **34** (1990), no. 2, 290–297, DOI 10.1137/1134024. MR1005737

[26] G. M. Fel'dman, *Characterization of the Gaussian distribution on groups by the independence of linear statistics* (Russian), Sibirsk. Mat. Zh., **31** (1990), no. 2, 180–190; English transl., Siberian Math. J., **31** (1990), no. 2, 336–345. MR1065592

[27] G. M. Fel'dman, *On the Skitovich–Darmois theorem on Abelian groups* (Russian), Teor. Veroyatnost. i Primenen. **37** (1992), no. 4, 695–708; English transl., Theory Probab. Appl., **37** (1993), no. 4, 621–631, DOI 10.1137/1137118. MR1210054

[28] G.M. Fel′dman, *Arithmetic of probability distributions, and characterization problems on Abelian groups*, Translation of Mathematical Monographs, vol. 116, Providence, RI, 1993.

[29] G. M. Fel'dman, *On the Skitovich-Darmois theorem on compact groups* (Russian), Teor. Veroyatnost. i Primenen. **41** (1996), no. 4, 901–906; English transl., Theory Probab. Appl., **41** (1997), no. 4, 768–773. MR1687152

[30] G. M. Fel'dman, *The Skitovich–Darmois theorem for discrete periodic Abelian groups* (Russian), Teor. Veroyatnost. i Primenen. **42** (1997), no. 4, 747–756; English transl., Theory Probab. Appl., **42** (1998), no. 4, 611–617. MR1618734

[31] G. M. Feldman, *More on the Skitovich-Darmous theorem for finite Abelian groups* (Russian), Teor. Veroyatnost. i Primenen. **45** (2000), no. 3, 747–756; English transl., Theory Probab. Appl., **45** (2001), no. 3, 507–511, DOI 10.1137/S0040585X97978452. MR1967795

[32] G. M. Feldman, *A characterization of the Gaussian distribution on abelian groups*, Probab. Theory Related Fields **126** (2003), no. 1, 91–102, DOI 10.1007/s00440-003-0256-4. MR1981634

[33] G. M. Feldman, *On the Heyde theorem for finite abelian groups*, J. Theoret. Probab. **17** (2004), no. 4, 929–941, DOI 10.1007/s10959-004-0583-0. MR2105741

[34] G. M. Feldman, *On a characterization theorem for locally compact abelian groups*, Probab. Theory Related Fields **133** (2005), no. 3, 345–357, DOI 10.1007/s00440-005-0429-4. MR2198016

[35] G. M. Feldman, *On the Heyde theorem for discrete abelian groups*, Studia Math. **177** (2006), no. 1, 67–79, DOI 10.4064/sm177-1-5. MR2283708

[36] Gennadiy Feldman, *Functional equations and characterization problems on locally compact abelian groups*, EMS Tracts in Mathematics, vol. 5, European Mathematical Society (EMS), Zürich, 2008, DOI 10.4171/045. MR2412878

[37] G. M. Feldman, *On a theorem of K. Schmidt*, Bull. Lond. Math. Soc. **41** (2009), no. 1, 103–108, DOI 10.1112/blms/bdn111. MR2481995

[38] G. M. Feldman, *The Heyde theorem for locally compact abelian groups*, J. Funct. Anal. **258** (2010), no. 12, 3977–3987, DOI 10.1016/j.jfa.2010.03.005. MR2609534

[39] G. M. Feldman, *Gaussian measures in the sense of Bernstein: factorization, supports, zero-one law* (Russian, with Russian summary), Teor. Veroyatnost. i Primenen. **56** (2011),

no. 3, 494–513, English transl., Theory Probab. Appl. **56** (2012), no. 3, 359–375, DOI 10.1137/S0040585X97985510. MR3136462

[40] Gennadiy Feldman, *On a characterization of convolutions of Gaussian and Haar distributions*, Math. Nachr. **286** (2013), no. 4, 340–348, DOI 10.1002/mana.201100320. MR3028778

[41] Gennadiy Feldman, *On the Skitovich-Darmois theorem for the group of p-adic numbers*, J. Theoret. Probab. **28** (2015), no. 2, 539–549, DOI 10.1007/s10959-013-0525-9. MR3370665

[42] Gennadiy Feldman, *On a characterization theorem for the group of p-adic numbers*, Publ. Math. Debrecen **87** (2015), no. 1-2, 147–166, DOI 10.5486/PMD.2015.7100. MR3367917

[43] G. M. Feldman, *Independent random variables in Abelian groups with independent sum and difference* (Russian, with Russian summary), Teor. Veroyatnost. i Primenen. **61** (2016), no. 2, 404–414, English transl., Theory Probab. Appl. **61** (2017), no. 2, 335–345, DOI 10.1137/S0040585X97T988198. MR3626791

[44] G. M. Feldman, *Heyde's characterization theorem for some locally compact Abelian groups* (Russian, with Russian summary), Teor. Veroyatn. Primen. **62** (2017), no. 3, 499–517; English transl., Theory Probab. Appl. **62** (2018), no. 3, 399–412. MR3684646

[45] G. M. Feldman, *On analogues of C. R. Rao's theorems for locally compact Abelian groups*, J. Difference Equ. Appl. **24** (2018), no. 12, 1967–1980, DOI 10.1080/10236198.2018.1551381. MR3890842

[46] G. M. Feldman, *On a characterization theorem for probability distributions on discrete abelian groups* (Russian, with Russian summary), Teor. Veroyatn. Primen. **63** (2018), no. 4, 730–754; English transl., Theory Probab. Appl. **63** (2019), no. 4, 594–612. MR3869629

[47] Gennadiy Feldman, *On a characterization theorem for connected locally compact Abelian groups*, J. Fourier Anal. Appl. **26** (2020), no. 1, Paper No. 14, 1–22, DOI 10.1007/s00041-019-09721-w. MR4056842

[48] Gennadiy Feldman, *A characterization theorem on compact Abelian groups*, J. Fourier Anal. Appl. **27** (2021), no. 5, Paper No. 86, 1–23, DOI 10.1007/s00041-021-09889-0. MR4323329

[49] Gennadiy Feldman, *Generalized Pólya's theorem on a-adic solenoids*, Results Math. **76** (2021), no. 3, Paper No. 124, 14, DOI 10.1007/s00025-021-01443-0. MR4266687

[50] G. M. Feldman, *On a characterization theorem for locally compact Abelian groups containing an element of order 2*, Potential Anal. **56** (2022), no. 2, 297–315, DOI 10.1007/s11118-020-09885-x. MR4367942

[51] G. M. Feldman, *Generalization of the Heyde theorem to some locally compact Abelian groups*, Results Math. **77** (2022), no. 4, Paper No. 177, 1–20, DOI 10.1007/s00025-022-01719-z. MR4453363

[52] G. M. Feldman and P. Graczyk, *On the Skitovich-Darmois theorem for compact abelian groups*, J. Theoret. Probab. **13** (2000), no. 3, 859–869, DOI 10.1023/A:1007870814570. MR1785533

[53] G. M. Feldman and P. Graczyk, *On the Skitovich-Darmois theorem for discrete abelian groups* (Russian, with Russian summary), Teor. Veroyatn. Primen. **49** (2004), no. 3, 596–601; English transl., Theory Probab. Appl. **49** (2005), no. 3, 527–531. MR2144871

[54] Gennadiy Feldman and Margaryta Myronyuk, *Independent linear forms on connected abelian groups*, Math. Nachr. **284** (2011), no. 2-3, 255–265, DOI 10.1002/mana.200810101. MR2790886

[55] Gennadiy Feldman and Margaryta Myronyuk, *On a characterization of idempotent distributions on discrete fields and on the field of p-adic numbers*, J. Theoret. Probab. **30** (2017), no. 2, 608–623, DOI 10.1007/s10959-015-0657-1. MR3647072

[56] Gennadiy Feldman and Margaryta Myronyuk, *On the Skitovich-Darmois theorem for some locally compact abelian groups*, Aequationes Math. **92** (2018), no. 6, 1129–1147, DOI 10.1007/s00010-018-0580-5. MR3873044

[57] U. Franz, D. Neuenschwander, and R. Schott, *Gauss laws in the sense of Bernstein and uniqueness of embedding into convolution semigroups on quantum groups and braided groups*, Probab. Theory Related Fields **109** (1997), no. 1, 101–127, DOI 10.1007/s004400050127. MR1469922

[58] László Fuchs, *Infinite abelian groups. Vol. I*, Pure and Applied Mathematics, Vol. 36, Academic Press, New York-London, 1970. MR0255673

[59] László Fuchs, *Infinite abelian groups. Vol. II*, Pure and Applied Mathematics. Vol. 36-II, Academic Press, New York-London, 1973. MR0349869

[60] S. G. Ghurye and Ingram Olkin, *A characterization of the multivariate normal distribution*, Ann. Math. Statist. **33** (1962), 533–541, DOI 10.1214/aoms/1177704579. MR137201

[61] Piotr Graczyk and Jean-Jacques Loeb, *A Bernstein property of measures on groups and symmetric spaces*, Probab. Math. Statist. **20** (2000), no. 1, Acta Univ. Wratislav. No. 2246, 141–149. MR1785243

[62] Wilfried Hazod and Eberhard Siebert, *Stable probability measures on Euclidean spaces and on locally compact groups*, Mathematics and its Applications, vol. 531, Kluwer Academic Publishers, Dordrecht, 2001. Structural properties and limit theorems, DOI 10.1007/978-94-017-3061-7. MR1860249

[63] Edwin Hewitt and Kenneth A. Ross, *Abstract harmonic analysis. Vol. I*, 2nd ed., Grundlehren der Mathematischen Wissenschaften [Fundamental Principles of Mathematical Sciences], vol. 115, Springer-Verlag, Berlin-New York, 1979. Structure of topological groups, integration theory, group representations. MR551496

[64] Edwin Hewitt and Kenneth A. Ross, *Abstract harmonic analysis. Vol. II: Structure and analysis for compact groups. Analysis on locally compact Abelian groups*, Die Grundlehren der mathematischen Wissenschaften, Band 152, Springer-Verlag, New York-Berlin, 1970. MR0262773

[65] C. C. Heyde, *Characterization of the normal law by the symmetry of a certain conditional distribution*, Sankhyā Ser. A **32** (1970), 115–118. MR293753

[66] Herbert Heyer, *Probability measures on locally compact groups*, Ergebnisse der Mathematik und ihrer Grenzgebiete, Band 94, Springer-Verlag, Berlin-New York, 1977. MR0501241

[67] Herbert Heyer and Christian Rall, *Gausssche Wahrscheinlichkeitsmasse auf Corwinschen Gruppen*, Math. Z. **128** (1972), 343–361, DOI 10.1007/BF01111573. MR428375

[68] M. Kac, *On a characterization of the normal distribution*, Amer. J. Math. **61** (1939), 726–728, DOI 10.2307/2371328. MR371

[69] A. M. Kagan, Yu. V. Linnik, and C. Radhakrishna Rao, *Characterization problems in mathematical statistics*, Wiley Series in Probability and Mathematical Statistics, John Wiley & Sons, New York-London-Sydney, 1973. Translated from the Russian by B. Ramachandran. MR0346969

[70] Ignacy Kotlarski, *On characterizing the gamma and the normal distribution*, Pacific J. Math. **20** (1967), 69–76. MR203769

[71] K. Kuratowski, *Topology. Vol. I*, Academic Press, New York-London; Państwowe Wydawnictwo Naukowe [Polish Scientific Publishers], Warsaw, 1966. New edition, revised and augmented; Translated from the French by J. Jaworowski. MR0217751

[72] K. Lajkó, *On the functional equation* $f(x)g(y) = h(ax+by)k(cx+dy)$, Period. Math. Hungar. **11** (1980), no. 3, 187–195, DOI 10.1007/BF02026615. MR590383

[73] Ju. V. Linnik and Ĭ. V. Ostrovs′kiĭ, *Decomposition of random variables and vectors*, Translations of Mathematical Monographs, Vol. 48, American Mathematical Society, Providence, R.I., 1977. Translated from the Russian. MR0428382

[74] I. P. Mazur, *Skitovich–Darmois theorem for finite Abelian groups* (Russian, with English and Ukrainian summary), Ukraïn. Mat. Zh. **63** (2011), no. 11, 1512–1523; English transl., Ukrainian Math. J. **63** (2012), no. 11, 1719–1732, DOI 10.1007/s11253-012-0608-2. MR3109683

[75] I. P. Mazur, *Skitovich–Darmois theorem for discrete and compact totally disconnected Abelian groups* (Russian, with English and Ukrainian summary), Ukraïn. Mat. Zh. **65** (2013), no. 7, 946–960; English transl., Ukrainian Math. J. **65** (2013), no. 7, 1054–1070, DOI 10.1007/s11253-013-0841-3. MR3145890

[76] Margaryta Myronyuk, *An analogue of the Bernstein theorem for the cylinder*, Aequationes Math. **71** (2006), no. 1-2, 54–69, DOI 10.1007/s00010-005-2773-y. MR2214813

[77] M. V. Myronyuk, *On the Skitovich–Darmous theorem and Heyde theorem in a Banach space* (Ukrainian, with English and Russian summary), Ukraïn. Mat. Zh. **60** (2008), no. 9, 1234–1242, English transl., Ukrainian Math. J. **60** (2008), no. 9, 1437–1447, DOI 10.1007/s11253-009-0142-z. MR2489747

[78] Margaryta Myronyuk, *The Heyde theorem on a-adic solenoids*, Colloq. Math. **132** (2013), no. 2, 195–210, DOI 10.4064/cm132-2-3. MR3125137

[79] Margaryta Myronyuk, *Independent linear forms on the group* Ω_p, J. Theoret. Probab. **33** (2020), no. 1, 1–21, DOI 10.1007/s10959-019-00888-y. MR4064291

[80] M. Myronyuk. *The Heyde theorem on a group* $\mathbb{R} \times D$*, where* D *is a discrete Abelian group*, J. Theoretical Probab, (2022). https://doi.org/10.1007/s10959-022-01168-y

[81] M.V. Myronyuk and G.M. Feldman, *Independent linear statistics on the two-dimensional torus* (Russian, with Russian summary), Teor. Veroyatnost. i Primenen. **59** (2014), no. 2, 252–275; English transl., Theory Probab. Appl. **59** (2015), no. 2, 335–345, DOI 10.1137/S0040585X97T987065. MR3416050

[82] M.V. Myronyuk and G.M. Feldman, *Independent linear statistics on the cylinders* (Russian, with Russian summary), Teor. Veroyatnost. i Primenen. **52** (2007), no. 1, 3–20; English transl., Theory Probab. Appl. **52** (2008), no. 1, 78–92, DOI 10.1137/S0040585X97982785. MR2354566

[83] Daniel Neuenschwander, *Gauss measures in the sense of Bernstein on the Heisenberg group*, Probab. Math. Statist. **14** (1993), no. 2, 253–256 (1994). MR1321764

[84] Daniel Neuenschwander, Bernard Roynette, and René Schott, *Characterizations of Gaussian distributions on simply connected nilpotent Lie groups and symmetric spaces* (English, with English and French summaries), C. R. Acad. Sci. Paris Sér. I Math. **324** (1997), no. 1, 87–92, DOI 10.1016/S0764-4442(97)80109-7. MR1435593

[85] K. R. Parthasarathy, *Probability measures on metric spaces*, Probability and Mathematical Statistics, No. 3, Academic Press, Inc., New York-London, 1967. MR0226684

[86] K. R. Parthasarathy, R. Ranga Rao, and S. R. S. Varadhan, *Probability distributions on locally compact abelian groups*, Illinois J. Math. **7** (1963), 337–369. MR190968

[87] B. L. S. Prakasa Rao, *On a characterization of probability distributions on locally compact abelian groups*, Z. Wahrscheinlichkeitstheorie und Verw. Gebiete **9** (1968), 98–100, DOI 10.1007/BF01851000. MR228029

[88] C. Radhakrishna Rao, *Characterization of probability laws by linear functions*, Sankhyā Ser. A **33** (1971), 265–270. MR303638

[89] A.L. Rukhin, *Some statistical and probabilistic problems on groups* (Russian), Trudy Mat. Inst. Steklov. **111** (1970), 52–109. MR0292221; English transl., Proc. Steklov Inst. Math. **111** (1970), 52–109. (errata insert)

[90] A.L. Rukhin, *On a theorem of S. N. Bernshtein* (Russian), Mat. Zametki, **6** (1969), 301–307. MR0253455; English transl., Math. Notes **6** (1969), 638–641

[91] V. V. Sazonov and V. N. Tutubalin, *Probability distributions on topological groups* (Russian, with English summary), Teor. Veroyatnost. i Primenen. **11** (1966), no. 1, 3–55. MR0199872; English transl., Theory Probab. Appl. **11** (1966), no. 1, 1–45

[92] K. Schmidt, *On a characterisation of certain infinitely divisible positive definite functions and measures*, J. London Math. Soc. (2) **4** (1971/72), 401–407, DOI 10.1112/jlms/s2-4.3.401. MR294997

[93] Eberhard Siebert, *Einige Bemerkungen zu den Gauss-Verteilungen auf lokalkompakten abelschen Gruppen* (German, with English summary), Manuscripta Math. **14** (1974), 41–55, DOI 10.1007/BF01637621. MR391215

[94] V. P. Skitovič, *On a property of the normal distribution* (Russian), Doklady Akad. Nauk SSSR (N.S.) **89** (1953), 217–219. MR0055597

[95] S. M. Srivastava, *A course on Borel sets*, Graduate Texts in Mathematics, vol. 180, Springer-Verlag, New York, 1998, DOI 10.1007/978-3-642-85473-6. MR1619545

[96] James H. Stapleton, *A characterization of the uniform distribution on a compact topological group*, Ann. Math. Statist. **34** (1963), 319–326, DOI 10.1214/aoms/1177704269. MR144373

[97] V. S. Vladimirov, I. V. Volovich, and E. I. Zelenov, *p-adic analysis and mathematical physics*, Series on Soviet and East European Mathematics, vol. 1, World Scientific Publishing Co., Inc., River Edge, NJ, 1994, DOI 10.1142/1581. MR1288093

Index of terms

Index of symbols

Selected Published Titles in This Series

273 **Gennadiy Feldman,** Characterization of Probability Distributions on Locally Compact Abelian Groups, 2023

271 **Hideto Asashiba,** Categories and Representation Theory, 2022

270 **Leslie Hogben, Jephian C.-H. Lin, and Bryan L. Shader,** Inverse Problems and Zero Forcing for Graphs, 2022

269 **Ralph S. Freese, Ralph N. McKenzie, George F. McNulty, and Walter F. Taylor,** Algebras, Lattices, Varieties, 2022

268 **Ralph S. Freese, Ralph N. McKenzie, George F. McNulty, and Walter F. Taylor,** Algebras, Lattices, Varieties, 2022

267 **Dragana S. Cvetković Ilić,** Completion Problems on Operator Matrices, 2022

266 **Kate Juschenko,** Amenability of Discrete Groups by Examples, 2022

265 **Nabil H. Mustafa,** Sampling in Combinatorial and Geometric Set Systems, 2022

264 **J. Scott Carter and Seiichi Kamada,** Diagrammatic Algebra, 2021

263 **Vugar E. Ismailov,** Ridge Functions and Applications in Neural Networks, 2021

262 **Ragnar-Olaf Buchweitz,** Maximal Cohen–Macaulay Modules and Tate Cohomology, 2021

261 **Shiri Artstein-Avidan, Apostolos Giannopoulos, and Vitali D. Milman,** Asymptotic Geometric Analysis, Part II, 2021

260 **Lindsay N. Childs, Cornelius Greither, Kevin P. Keating, Alan Koch, Timothy Kohl, Paul J. Truman, and Robert G. Underwood,** Hopf Algebras and Galois Module Theory, 2021

259 **William Heinzer, Christel Rotthaus, and Sylvia Wiegand,** Integral Domains Inside Noetherian Power Series Rings, 2021

258 **Pramod N. Achar,** Perverse Sheaves and Applications to Representation Theory, 2021

257 **Juha Kinnunen, Juha Lehrbäck, and Antti Vähäkangas,** Maximal Function Methods for Sobolev Spaces, 2021

256 **Michio Jimbo, Tetsuji Miwa, and Fedor Smirnov,** Local Operators in Integrable Models I, 2021

255 **Alexandre Boritchev and Sergei Kuksin,** One-Dimensional Turbulence and the Stochastic Burgers Equation, 2021

254 **Karim Belabas and Henri Cohen,** Numerical Algorithms for Number Theory, 2021

253 **Robert R. Bruner and John Rognes,** The Adams Spectral Sequence for Topological Modular Forms, 2021

252 **Julie Déserti,** The Cremona Group and Its Subgroups, 2021

251 **David Hoff,** Linear and Quasilinear Parabolic Systems, 2020

250 **Bachir Bekka and Pierre de la Harpe,** Unitary Representations of Groups, Duals, and Characters, 2020

249 **Nikolai M. Adrianov, Fedor Pakovich, and Alexander K. Zvonkin,** Davenport–Zannier Polynomials and Dessins d'Enfants, 2020

248 **Paul B. Larson and Jindrich Zapletal,** Geometric Set Theory, 2020

247 **István Heckenberger and Hans-Jürgen Schneider,** Hopf Algebras and Root Systems, 2020

246 **Matheus C. Bortolan, Alexandre N. Carvalho, and José A. Langa,** Attractors Under Autonomous and Non-autonomous Perturbations, 2020

245 **Aiping Wang and Anton Zettl,** Ordinary Differential Operators, 2019

244 **Nabile Boussaïd and Andrew Comech,** Nonlinear Dirac Equation, 2019

243 **José M. Isidro,** Jordan Triple Systems in Complex and Functional Analysis, 2019

242 **Bhargav Bhatt, Ana Caraiani, Kiran S. Kedlaya, Peter Scholze, and Jared Weinstein,** Perfectoid Spaces, 2019

241 **Dana P. Williams,** A Tool Kit for Groupoid C^*-Algebras, 2019

240 **Antonio Fernández López,** Jordan Structures in Lie Algebras, 2019